METABOLIC PROFILING: ITS ROLE IN BIOMARKER DISCOVERY AND GENE FUNCTION ANALYSIS

METABOLIC PROFILING: ITS ROLE IN BIOMARKER DISCOVERY AND GENE FUNCTION ANALYSIS

edited by

George G. Harrigan
Pharmacia Corporation
Chesterfield, MO 63198
U.S.A.

Royston Goodacre
Department of Chemistry, UMIST,
PO Box 88, Manchester, M60 1QD,
United Kingdom.

KLUWER ACADEMIC PUBLISHERS
Boston / Dordrecht / London

Distributors for North, Central and South America:
Kluwer Academic Publishers
101 Philip Drive
Assinippi Park
Norwell, Massachusetts 02061 USA
Telephone (781) 871-6600
Fax (781) 681-9045
E-Mail: kluwer@wkap.com

Distributors for all other countries:
Kluwer Academic Publishers Group
Post Office Box 322
3300 AH Dordrecht, THE NETHERLANDS
Telephone 31 786 576 000
Fax 31 786 576 474
E-Mail: services@wkap.nl

Electronic Services <http://www.wkap.nl>

Library of Congress Cataloging-in-Publication Data

A C.I.P. Catalogue record for this book is available
from the Library of Congress.

Printed on acid-free paper.

Printed in the United States of America.

The Publisher offers discounts on this book for course use and bulk purchases. For further information, send email to mimi.breed@wkap.com.

TO BETH, SEAN AND EVAN and
TO ELIZABETH, TAMARA AND PICKLES

Contents

Contributors

Aram S. Adourian. Beyond Genomics Inc., 40 Bear Hill Road, Waltham, MA 02451, USA

Jonas Almeida. Department of Biometry and Epidemiology and Department of Biochemistry and Molecular Biology, Medical University of South Carolina, Charleston, SC 29435, USA

László G. Boros. Harbor-UCLA Research and Education Institute, 1124 West Carson Street RB1, Torrance, CA 90502, USA

Alan P. Breau. Global Drug Metabolism, Pharmacia Corporation, 4901 Searle Parkway, Skokie, IL 60077, USA

Glenn H. Cantor. Investigative Toxicology, Pharmacia Corporation, 301 Henrietta St., Kalamazoo, MI 49007, USA

Marta Cascante. Department of Biochemistry and Molecular Biology, University of Barcelona, Marti/I Franques 1, Barcelona 08028, Catalonia, Spain

Jaideep Chaudhary. Atairgin Technologies, 101 Theory, Irvine, CA 92612, USA

Yeun-Li Chung. Cancer Research UK Biomedical Magnetic Resonance Research Group, Department of Biochemistry and Immunology, St. George's Hospital Medical School, London, SW17 ORE, UK

Sarah Clarke. Institute of Biological Sciences, University of Wales, Aberystwyth, SY23 3DD, UK

Tim Compton. Atairgin Technologies, 101 Theory, Irvine, CA 92612, USA

Eugene Davidov. Beyond Genomics Inc., 40 Bear Hill Road, Waltham, MA 02451, USA

David I. Ellis. Institute of Biological Sciences, University of Wales, Aberystwyth, SY23 3DD, UK

Oliver Fiehn. Max-Planck-Institute of Molecular Plant Physiology, 14424 Potsdam/Golm, Germany

Royston Goodacre. Department of Chemistry, University of Manchester Institute of Science and Technology, PO Box 88, Sackville St., Manchester, M60 1QD, UK

Dayan Goodenowe. Phenomenome Discoveries Inc., 204-407 Downey Road, Saskatoon SK S7N 4L8, Canada

Julian L. Griffin. Division of Biomedical Sciences, Imperial College of Science, Technology and Medicine, London, SW7 2AZ, UK

John R. Griffiths. Cancer Research UK Biomedical Magnetic Resonance Research Group, Department of Biochemistry and Immunology, St. George's Hospital Medical School, London, SW17 ORE, UK

George G. Harrigan. Global HTS, Pharmacia Corporation, 700 Chesterfield Parkway, Chesterfield, MO 63198, USA

Douglas B. Kell. Department of Chemistry, University of Manchester Institute of Science and Technology, PO Box 88, Sackville St., Manchester, M60 1QD, UK

Wai-Nang Paul Lee. Harbor-UCLA Research and Education Institute, 1124 West Carson Street RB1, Torrance, CA 90502, USA

Edward W. Marple. Beyond Genomics Inc., 40 Bear Hill Road, Waltham, MA 02451, USA

Stephen Naylor. Beyond Genomics Inc., 40 Bear Hill Road, Waltham, MA 02451, USA

Matej Oresic. Beyond Genomics Inc., 40 Bear Hill Road, Waltham, MA, 02451, USA

Jeff A. Parrott. Atairgin Technologies, 101 Theory, Irvine, CA 92612, USA

Elizabeth Sang. Department of Biochemistry, Oxford University, Oxford, OX1 3QV, UK

Joachim Spranger. German Institute of Human Nutrition (DifE), 14558 Bergholz-Rehbrücke, Germany

Marion Stubbs. Cancer Research UK Biomedical Magnetic Resonance Research Group, Department of Biochemistry and Immunology, St. George's Hospital Medical School, London, SW17 ORE, UK

Seetharaman Vaidyanathan. Institute of Biological Sciences, University of Wales, Aberystwyth, SY23 3DD, UK

Jan van der Greef. Beyond Genomics Inc., 40 Bear Hill Road, Waltham, MA 02451, USA, TNO Pharma, PO Box 2215 2031 CE, Zernikedreff 9, 2333 CK, Leiden, Netherlands and Leiden University/Amsterdam Center for Drug Research, Division of Analytical Sciences, NL-2300 RA Leiden, Netherlands

Rob van der Heijden. Leiden University/Amsterdam Center for Drug Research, Division of Analytical Sciences, NL-2300 RA Leiden, Netherlands

Elwin Verheij. TNO Pharma, PO Box 2215 2031 CE, Zernikedreff 9, 2333 CK, Leiden, Netherlands

Jack Vogels. TNO Pharma, PO Box 2215 2031 CE, Zernikedreff 9, 2333 CK, Leiden, Netherlands

Eberhard O. Voit. Department of Biometry and Epidemiology, Medical University of South Carolina, Charleston, SC 29435, USA

Preface

An impressive amount of data has emerged from genomics and proteomics ventures yet this has often served more to highlight the complexity of cellular regulation than to elucidate mechanism. It is also evident that biochemical control is not strictly hierarchical and that intermediary metabolism can contribute to control of regulatory pathways. Metabolic studies are therefore increasingly contributing to gene function analyses, and an increased interest in metabolites as biomarkers for disease progression or response to therapeutic intervention is also evident in the pharmaceutical industry. Front Line Strategic Consulting recently stated (September, 2002) that, worldwide, the metabolomics industry is anticipated to grow at a compound rate of over 40% over the next five years and that the market for metabolomic technology sales will exceed $255 million by 2007.

The purpose of this book is to offer guidelines as to the technology currently available for such pursuits and the bioinformatics and database strategies now being developed. We present evidence that metabolic profiling is a valuable addition to genomics and proteomics strategies devoted to drug discovery and development, and believe that metabolic profiling may be seen to offer numerous advantages, as will be highlighted in this volume.

There has been a number of recent literature reviews devoted to metabolic profiling and an increasing number of related conferences, including the First International Congress of Plant Metabolomics (Wageningen, 2002) and the Systeomics Conference (San Francisco, 2002). The genesis of this book can be traced to the First Annual Cambridge HealthTech Institute Conference on Metabolic Profiling (Chapel Hill, 2001) and a subsequent conference report published in *Drug Discovery Today*. There has, however, been no definitive volume that describes progress in this area, particularly in the health care arena, and this book aims to address this. It is not, however, intended as a complete review of all that is going on in pharmaceutically directed metabolic profiling. We have not addressed *in silico* modeling, for example, and of course there has already been several excellent volumes on metabolic control/regulation and the technological impact thereof.

We hope that the book will be of interest to researchers in the pharmaceutical industries (Big Pharma and small venture capital endeavors), academe-industry technology transfer organizations and academic researchers. The bias of the book is to healthcare and the pharmaceutical industry but metabolic profiling has been extensively applied in the agriculture and microbiological arenas. Some of the book contributors have developed their expertise in these areas and have much to offer our colleagues in biomedicine. We are of course greatly indebted to our contributors without whom this volume would not have been possible. They have succeeded in ensuring a volume of high scientific merit and quality.

Finally, we would like to acknowledge our colleagues who contributed valuable comments and support, and helped in proofreading. Roxanne LaPlante and Elizabeth Collantes diligently proofread many chapters and helped with manuscript preparation. They, along with Margann Wideman, Ying Ping Zhang, Nancy Wall and Gilles Goetz, have made valuable contributions to the development of metabolic profiling at Pharmacia. We also thank David Lester for valuable suggestions at the onset of this book. We would like to thank the University of Wales, Aberystwyth and the UK BBSRC and UK EPSRC for allowing us the academic freedom and financial assistance to investigate metabolic profiling, and look forward to continuing this research in the University of Manchester Institute of Science and Technology (UMIST).

George G. Harrigan

Royston Goodacre

Chapter 1

INTRODUCTION

Metabolic Profiling:Pathways in Drug Discovery

George G. Harrigan[1] and Royston Goodacre[2,3]
[1]Global HTS, Pharmacia Corporation, 700 Chesterfield Parkway, Chesterfield, MO 63198, USA [2]Institute of Biological Sciences, University of Wales, Aberystwyth, SY23 3DD, UK [3]Department of Chemistry, University of Manchester Institute of Science and Technology, PO Box 88, Sackville St., Manchester, M60 1QD,UK

1. INTRODUCTION

Since the completion of the first whole genome sequence, that of the microorganism *Haemophilus influenzae* (Fleischmann *et al.*, 1995), we have increasingly realized the paucity of our knowledge with respect to the function of novel genes. The completion of draft sequences for the human genome has accelerated demand for determining the biochemical function of orphan genes and for validating them as molecular targets for therapeutic intervention. The search for biomarkers that can serve as indicators of disease progression or response to therapeutic intervention has also increased. Functional analyses have emphasized gene expression studies (transcriptomics) and protein profiling (proteomics). Considerably less emphasis has been placed on profiling the end products of gene expression, the metabolome. To date, the use of metabolite analyses as a tool in describing biochemical networks has been aimed primarily at accurate quantitation of substrates and products in an individual enzymatic pathway. There has, however, been an increasing emphasis on producing spectral "fingerprints" of metabolic profiles that can be correlated to a phenotype of interest without identification of specific metabolite classes. Further, by parallel analytical testing of sample tissues or biofluids a number of chemically different metabolites can be readily identified and quantitated. Cataloguing such data can then yield valuable information on the responses of specific metabolic pathways to genetic mutations or environmental stimuli. Because such information can be integrated with transcriptomics and proteomics, metabolic analyses are now accepted as an integral part of functional studies in bacteria, yeast and plants.

Whilst the metabolome is complementary to transcriptomics and proteomics, it may be seen to offer special advantages. In particular, we know from the theory underlying metabolic control analysis (MCA) (Kell and Westerhoff, 1986; Fell, 1996; Mendes *et al.*, 1996), as well as from experiment (Fiehn *et al.*, 2000; Raamsdonk *et al.*, 2001), that while changes in the levels of individual enzymes may be expected to have little effect on metabolic fluxes, they can, and do, have significant effects on the concentrations of a variety of individual metabolites. In addition, as the 'downstream' result of gene expression, changes in the metabolome are amplified relative to changes in the transcriptome and the proteome.

The increasing interest in applying metabolite analyses to functional analyses in man is the primary, but not exclusive, focus of this volume. The following chapters discuss developments in the use of spectroscopic tools for the acquisition of metabolite data, pattern recognition and modeling studies based on such data, the integration of metabolome data with transcriptomics and genomics and, a vitally important topic in this bioinformatics age, current research strategies in database and visualization technology.

2. METABOLIC PROFILING

The further development of spectroscopic and spectrometric tools for high throughput analyses of selected biochemical pathways is crucial to the acquisition of metabolome data sets of sufficient quality for metabonomics and metabolomics. Whilst *metabolic fingerprinting* or *metabonomics* assumes it is not necessary to determine levels of all individual metabolites for classification or response readouts, *metabolic profiling* or *metabolomics* absolutely requires the identification and quantitation of as broad a class of metabolites as possible. The term metabolic profiling is also often used to encompass all form of metabolite analyses; and in the early stages of such work it is not surprising that terminologies are still evolving. It may also be worth pointing out at this stage that analytical strategies for metabonomics generally tend to be applied to biofluids and are devoted primarily to biomarker discovery whereas metabolomics tends to be applied to tissue and is devoted to both biomarker discovery and gene function analysis.

The technology available for analyzing metabolite concentrations and fluxes currently focuses primarily on mass spectrometry (MS) and nuclear magnetic resonance (NMR) spectroscopy. Both, however, lack the resolving power to distinguish all cell or tissue metabolites in a single spectral measurement. Capturing metabolic information by spectral monitoring of as wide a range of metabolic pathways as possible and correlating it with changes in the transcriptome and proteome thus still represents an intriguing challenge; especially when one also considers that the totality of the proteome can not yet be measured and that current DNA array experiments are not wholly reproducible. However, with current estimates of over 500 human diseases with direct defects in metabolism, and with more complex pathologies such as cancer and inflammation also known to involve pronounced metabolic changes, metabolic profiling clearly represents a worthwhile pursuit for the pharmaceutical industry. As illustrated throughout this volume considerable progress is being made.

Metabolic fingerprints, which reflect levels of non-targeted and generally incomplete set of metabolites in a biological matrix but without necessarily providing explicit quantitative information, are proving increasingly helpful particularly as high-throughput screening techniques are continue to develop. Much of the pioneering animal biofluid fingerprinting work has been through the Imperial College, London, UK group of Nicholson. Their work on NMR studies, related primarily to toxicology and which they refer to as 'metabonomics', has been extensively reviewed (*e.g.* Lindon *et al.*; 2000) and is increasingly seen as an integral component of toxicology screening in the pharmaceutical industry. This is illustrated by the establishment of COMET (Consortium on Metabonomics in Toxicology), a consortium of six pharmaceutical companies under the oversight of Imperial College. Breau and Cantor (Chapter 4) present an overview of the potential of NMR based metabonomics and give examples of how this technology can be applied in the pharmaceutical research and development process.

While the COMET program is devoted mainly to biofluid analysis, NMR spectroscopy can of course be expanded to metabolite analyses in tissue extracts, solid tissue itself, and even *in vivo* experiments. Griffin and Sang (Chapter 3) demonstrate how NMR derived metabolic profiles from tissue can contribute to an understanding of Duchenne muscular dystrophy and, by extrapolation, to proteomic analyses in other diseases. Chung *et al.* (Chapter 5) demonstrate the utility of *in vivo* NMR technologies in revealing profound differences in metabolic profiles of tumor xenografts differing in only a single transcription factor, the hypoxia-inducible factor 1β. These two latter chapters present critical developments in the ability to generate biological hypotheses from metabolomic data sets.

Other technologies are, of course, emerging and becoming available for the acquisition of metabolome data sets including vibrational spectroscopic techniques. Raman spectroscopy can be considered in its infancy as far as biological applications are concerned (Clarke and Goodacre, Chapter 6).

Continued technical developments suggest however that this technology has much promise in metabolic fingerprinting and in diagnostics and will be of great interest because of its non-invasiveness. Ellis *et al.* (Chapter 7) demonstrate the fingerprinting utility of Fourier-transform infrared spectroscopy (FT-IR). As they demonstrate, FT-IR is a rapid measurement technology with sufficient resolving power to distinguish between microbial cells even at the sub-species level. Its application outside of cancer diagnostics has been limited but recent literature examples include the analyses of body fluids from diabetes and arthritis patients, brain material infected with transmissible spongiform encephalopathies, and follicular fluid analysis for investigating oocyte development. Chemical imaging options are also available but FT-IR remains an unfortunately under-utilized and under-appreciated technology in this respect.

Mass spectrometry of course, represents the acquisition technology most commonly applied to metabolome data sets. It allows investigations of a greater range of metabolites than NMR and vibrational spectroscopy, including those metabolites present in low abundances, and the current 'gold standard' approach uses gas chromatography (GC) separation. Vaidyanathan and Goodacre (Chapter 2) demonstrate that MS is equally applicable to metabolome and proteome research. The focus of this chapter is on rapid profiling for purposes of microbial characterization. It is demonstrated that direct infusion electrospray ionization-MS is ideally suited for high-throughput classification or determining sample origin.

Fourier-transform ion cyclotron resonance MS is also finding application in metabolome research as presented by Goodenowe (Chapter 8) and more conventional forms of MS are finding increasing utility as a tool for measurements of stable-isotope labeled metabolomes allowing for metabolite flux analysis. The methodology behind this approach is highlighted by Boros *et al.* (Chapter 9) and its value to the pharmaceutical industry discussed.

Beyond Genomics, Inc. believe that a strategy allying metabolomic data with measurements of transcripts and especially proteins can generate much more detailed functional descriptions of processes involved in disease progression. They refer to this strategy as Systems Biology. Their metabolite analyses of mammalian biofluids and tissues are based on the use of parallel NMR, liquid chromatography-MS and GC-MS platforms targeting specific metabolite classes including, but not limited to, lipids, steroids, eicosanoids and bile acids. This approach was applied to a study on transgenic mice that overexpress the human ApoE3 gene. These mice serve as a model of atherosclerosis and coronary artery disease. Pattern recognition algorithms applied to spectroscopic analyses of plasma from wild type and transgenic mice allowed rapid identification of significant metabolic differences.

Specifically, certain triglycerides were elevated in the ApoE3 transgenic mice relative to wild type strains whereas lysophosphatidylcholine was decreased. As such, this implies a role for metabolic profiling in the discovery of disease biomarkers or biomarker fingerprints that may have value in clinical settings. Correlating the above metabolome data with proteomic and genomic studies has allowed comprehensive annotation of the metabolic pathways implicated in ApoE3-mediated atherosclerosis (Davidov *et al.,* 2002). Van der Greef and colleagues (Chapter 10) give a comprehensive assessment of how metabolomics can contribute to Systems Biology. Specific examples presented include studies in antifungal drug target analysis, toxicology, osteoarthritis and multiple sclerosis.

Fiehn and Spranger (Chapter 11) illustrate approaches to utilizing metabolic patterns in human disease, particularly type 2 diabetes mellitus. They point out that existing biomarkers for type 2 diabetes mellitus have limited value for assessments of individual risks an that biological variability imposes constraints on the validation of analytical strategies used.

The role of bioactive lipids in reproductive and other diseases is presented by Chaudhary and colleagues from Atairgin Technologies (Chapter 12). Here the authors point out the diagnostic and therapeutic significance of lipids as biomarkers. Advances in genomics also allow correlations of profiles of genes involved in lipid metabolism with results from global and focused profiling of bioactive lipids; a strategy Atairgin refer to as lipogenomics.

Many of the pattern recognition strategies currently pursued in metabolome analyses are based on "unsupervised" techniques such as principal components analysis and this approach is extensively utilized in metabolic profiling. Goodacre and Kell (Chapter 13) present the case for applying supervised machine learning approaches to metabolome analysis. More specifically, they advocate the use of evolutionary computation, and in particular, genetic programming. While the examples presented by Goodacre and Kell are primarily from the microbiological and agricultural arena, it is noteworthy that Griffin and Sang (Chapter 3) have also utilized genetic programming in their NMR analyses of tissue metabolites.

The rationale for integrating genomic and metabolomic data reflects, at least in part, the well recognized challenges in utilizing gene expression data to explain complex biochemical networks. A major disadvantage of gene expression clustering, for example, is that it can miss relationships between pathway genes if they are regulated to differing degrees. This was an issue recently addressed by Voit in an excellent demonstration of the contribution of the mathematically-based biochemical systems theory (BST), in understanding regulation of metabolic pathways (Voit and Radivoyetich, 2000). Voit pointed out that in an investigation of glycolysis in heat-shocked yeast the genes for glucose transport and phosphorylation are upregulated 5-20 fold yet phosphofructokinase expression is essentially unaltered. Clearly clustering genes by relative expression would fail, in this instance, to reveal pathway correlations between phosphofructokinase and hexokinase. BST was able, however, to demonstrate that the observed, and somewhat non-intuitive, gene expression profile satisfied the primary metabolic goals of increased ATP, trehalose and NADPH production in heat shocked yeast. It is anticipated that modeling metabolic profiling data by BST will aid interpretation of corresponding gene expression data. Voit and Almeida now present here (Chapter 14) results of an intriguing investigation on combining genetic algorithms with BST to establish pathway structures and aid gene function discovery. The partnership between modeling and metabolic profiling offers considerable potential to contribute to biomarker discovery and gene function analysis.

Cleary as metabolome data acquisition increases resources, both financial and intellectual, will have to be increasingly devoted to database infrastructure and technology and user requirements must be clarified. This is a point emphasized by Hardy and Fuell (Chapter 15). However although metabolic profiling, as pointed out by Mendes and colleagues (Chapter 16), has created new challenges for bioinformatics, there is now an opportunity to establish standardization at, what still is, an early stage in the metabolomic endeavor. Mendes suggest that the metabolomics community can achieve such standardization, particularly if it can take advantage of standardization based on programming languages such as Extensible Mark-up Language (XML) and Systems Biology Mark-up Language (SBML). It is also an opportune time to give thought to how we view metabolic pathways. Such pathways are frequently presented as linear constructs and to some, metabolic pathways are simply "the titles of chapters of biochemistry text books". A major concern is that pathways are often represented without adequate attention to side reactions. Mendes promotes the concept of metabolite neighborhoods defined as *consisting of a central metabolite, all the reactions that include it as a substrate or product, plus all metabolites that take part in those reactions* as an alternative and complementary view to looking at pathway and as a means of emphasizing the interconnected nature of intermediary metabolism. It is argued that metabolome database technologies should incorporate both pathway and neighborhood analyses

An intriguing complement to the databases discussed within this book will be the Dictionary of the Human Metabolome (Beecher, Chapter 17), a project sponsored by Paradigm Genetics. This will serve as a central repository for all publicly available information (and proprietary Paradigm Genetic information) on small molecule human metabolites. It is estimated that the number of entries will range from 2-10,000 metabolites.

3. CONCLUDING REMARKS

Metabolite analyses have demonstrated that the metabolome and its components can serve as biomarkers or biomarker fingerprints and metabonomics is becoming an increasing established contributor in the pharmaceutical arena. However metabonomic strategies tend to focus on biofluids and it is far from straightforward to relate those changes to alterations in gene expression in tissue. Even in simpler organisms or in areas where research has been most focused *i.e.* plants (there are an estimated 200,000 different metabolites in the plant kingdom (Fiehn, 2002)) there is no immediately obvious way to correlate changes in metabolome data sets with changes in gene expression profiles. Oliver and colleagues have proposed an analytical strategy termed FANCY (Functional Analysis by Co-responses in Yeast) for determining the roles of unknown genes based on analysis of metabolite profiles (Teusink *et al.*, Oliver *et al.*, 1998). Kose *et al.* (2001) have investigated role of pair wise correlations or metabolite cliques in plant metabolome analyses as a means to establish links between different biosynthetic pathways. This work highlights that the traditional biochemical 'two-dimensional thinking' of metabolism occurring by simple pathways has to be replaced by recognition that metabolism occurs in an integrated complex *n*-dimensional (with respect to each metabolite) network.

A major confounding factor in correlating gene and metabolite expression centers around the fact that the hierarchical paradigm of strict genetic control over metabolic fluxes is dubious as exemplified in a case study on the regulation of glycolysis (ter Kuile and Westerhoff, 2000). This is amplified by Voit's demonstration, referred to earlier, that an understanding of metabolism is often required to give meaning to gene expression profiles.

Investigation of the metabolome in biofluids and in tissues upon genetic or environmental perturbation may well then be viewed as a necessity for biomarker discovery and gene function analysis. Much of these analyses will be dependent upon improvements in current metabolomic data acquisition, storage, retrieval and interpretation. However, even with current technologies, there is still room for optimism that biological hypotheses are being generated from metabolomic datasets. Raamsdonk *et al.* (2001) have established that NMR based strategies have value in investigating mutations that yield silent phenotypes in yeast, while Johnson *et al.* (2000) have used FT-IR to investigate environmental stresses on tomato plants. Steps in integrating metabolome data with transcriptome and proteome continue apace as exemplified by the association networks alluded to by van der Greef *et al.* in Chapter 10. Database technology continues to develop and languages such as XML or SBML offer an opportunity for the standardization of metabolome database strategies, as discussed at the recent First International Congress of Plant Metabolomics (Wageningen, 2002). Eventually, the combination of metabolomic analysis with other profiling methods, especially proteomics, and integrative technologies such as MCA and BST could enable pathway discovery and evaluation of gene function.

This book tries to capture the excitement surrounding these developments and to present metabolic profiling as an expanding and essential discipline to biomarker discovery and gene function analysis. We hope you enjoy it!

REFERENCES

Davidov E, Clish CB, Meyes M *et al.*, Systems biology approach: parallel analysis of the ApoE3-Leiden transgenic mouse model. *Nature Biotechnol* submitted (2002).

Fell DA. *Understanding the Control of Metabolism.* Portland Press, London (1996).

Fiehn O. Metabolomics – the link between genotypes and phenotypes. *Plant Mol Biol* 48: 155–171 (2002).

Fiehn O, Kopka J, Dörmann P *et al.* Metabolite profiling for plant functional genomics. *Nature Biotechnol* 18: 1157-1161 (2000).

Fleischmann RD, Adams MD, White O *et al.*, Whole-genome random sequencing and assembly of *Haemophilus influenzae* Rd. *Science* 269: 496-512 (1995).

Johnson HE, Gilbert RJ, Winson MK *et al.* Explanatory analysis of the metabolome using genetic programming of simple, interpretable rules. *Genetic Program Evolv Mach* 1: 243-258 (2000).

Kell DB, Westerhoff HV. Towards a rational approach to the optimization of flux in microbial biotransformations. *Trends Biotechnol* 4: 137-142 (1986).

Kose F, Weckwerth W, Linke T, Fiehn O. Visualising plant metabolomic correlation networks using clique-metabolite matrices. *Bioinformatics* 17: 1198-1208 (2001).

Lindon JC, Nicholson JK, Holmes E, Everett JR. Metabonomics: metabolic processes studied by NMR spectroscopy of biofluids. *Concepts Magn Reson* 12: 289-320 (2000).

Mendes P, Kell DB, Westerhoff HV. Why and when channeling can decrease pool size at constant net flux in a simple dynamic channel. *Biochim Biophys Acta* 1289: 175-186 (1996).

Oliver SG, Winson MK, Kell DB, Baganz F. Systematic functional analysis of the yeast genome. *Trends Biotechnol* 16: 373-378 (1998).

Raamsdonk LM, Teusink B, Broadhurst D *et al.* A functional genomics strategy that uses metabolome data to reveal the phenotype of silent mutations. *Nature Biotechnol* 19: 45-50 (2001).

ter Kuile BH, Westerhoff HV. Transcriptome meets metabolome: hierarchical and metabolic regulation of the glycolytic pathway. *FEBS Lett* 500: 169-171 (2001).

Teusink B, Baganz F, Westerhoff HV, Oliver SG. Metabolic control analysis as a tool in the elucidation of the function of novel genes. *Meth Microbiol* 26: 297-336 (1998).

Voit EO, Radivoyetich T. Biochemical systems analysis of genome-wide expression data. *Bioinformatics* 16: 1023-1037 (2000).

Chapter 2

METABOLOME AND PROTEOME PROFILING FOR MICROBIAL CHARACTERIZATION

Mass Spectrometric Applications

Seetharaman Vaidyanathan[1] and Royston Goodacre[1,2]
[1]*Institute of Biological Sciences, University of Wales, Aberystwyth, SY23 3DD, UK*
[1,2]*Department of Chemistry, University of Manchester Institute of Science and Technology, PO Box 88, Sackville St., Manchester, M60 1QD, UK*

1. INTRODUCTION

In the post-genomic era, there is greater emphasis in evaluating functional aspects of genes and gene products for cellular characterization. There is a better appreciation of the extent of microbial diversity, of how uncultured microorganisms might be grown, and a better understanding of how the metabolic potential of microorganisms can be maximized. This can be attributed to a paradigm shift from traditional biology to bioinformatics (Bull *et al.*, 2000). Evaluation and assessment of cellular activities is evolving from the traditional approaches of biochemistry and molecular biology, where cellular processes are investigated individually and often independently of each other, to a more global approach in which cellular composition is analyzed in its entirety, in order to establish a more holistic picture. While the genetic make-up of a cell characterizes an organism to some extent, there are questions that will remain unanswered by analyzing at the genetic level alone. The physiological response of a cell to fluctuations in its environment is the most obvious case in point. In such instances, analysis of gene products, such as mRNAs, proteins and metabolites will be of greater relevance (Oliver, 2002) as we strive to understand, and even define, an organism's phenotype. Developments in analytical techniques can now enable simultaneous high-throughput measurements of several analytes, at the level of transcripts (transcriptomics), proteins (proteomics), and metabolites (metabolomics). Mass spectrometry (MS), in particular, offers considerable potential for such analyses, especially for proteomes and metabolomes. The significance of analyzing proteomes and metabolomes for microbial characterization, the general analytical strategies available, and mass spectrometric applications are discussed in the following pages.

2. PROTEOME AND METABOLOME ANALYSIS FOR MICROBIAL CHARACTERIZATION

Characterization of microbes may serve diverse purposes. Usually, the objective is to identify micro-organisms for taxonomic classification, medical diagnosis, ecological purposes, or for detecting biohazards in food and the environment. There is especially a heightened interest in microbial identification for detection of biological warfare agents (Dando, 1994). Microbial characterization is also useful in bioprocess monitoring, operation and development. At a fundamental level, it serves in understanding cellular biochemistry and physiology, in identifying biosynthetic bottlenecks and regulatory networks, in growth media optimization, and in screening for production of bioactive molecules. Characterization of microbes (or in general, a cell, tissue or organ) may involve more than mere structural description. Being a functional entity whose functions vary for a given structure, it is essential to extend biological characterization to include functional aspects, manifested as phenotypes, and to elaborate their biochemical or molecular basis. Depending on the analytical objective, microbial characterization may involve, a) identification of a specific biomarker or set of biomarkers, for instance, in screening samples to detect the presence of a known micro-organism in a sample matrix, or identify the condition or state of a micro-organism, b) isolation and enrichment of an uncharacterized micro-organism of interest from a population prior to more elaborate biochemical investigations or c) assessment of the biochemical make-up of known micro-organisms under different sets of conditions to screen for novel metabolites and pathways for, as an instance, analyzing genetic mutants to assign function to 'orphan' genes.

Since the advent of DNA sequencing technology in 1977, and the first genomic sequencing of a free-living organism in 1995, that of *Haemophilus influenzae*, a bacterium of relevance to human health and disease (Fleischmann *et al.*, 1995), over 60 microbial genomes have been sequenced (http://www.tigr.org/tdb/mdb/mdbcomplete.html). The availability of the basic genetic make-up of these microbes is now paving the way for comparative assessments between and within microbial species (comparative genomics). This can offer insights into evolutionary traits, identify potential disease causing genes in pathogens and contribute to elucidating their pathogenic basis (Fraser *et al.*, 2000). Microbial characterization at the genetic level through the use of DNA probes or gene amplification by polymerase chain reaction (PCR) is not entirely robust, and does not provide complete answers at the functional level. The next level of analysis, that of gene expression profiles, the subject of functional genomics, involves the use of information made available from genomics to quantify the spatial and temporal accumulation patterns of mRNAs, proteins and important metabolites using high-throughput technologies. Such analyses would undoubtedly add further information that could contribute significantly to microbial characterization. In particular, the information offered at the proteome and the metabolome levels may have greater relevance in assessing function, given their proximity to functional responses of the cell, and manifested as phenotypes (Fig. 1).

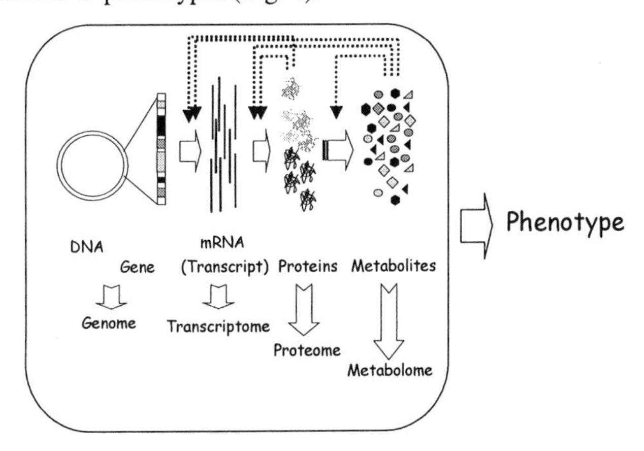

Figure 1. General schematic of "omic' organization. Dotted arrows indicate interactions regulating respective "omic' expression.

2.1 Protein Profiles and Proteomics

In translating genetic information to cellular function, proteins are at the "business end", and carry out cellular activities at the biochemical level. While the genome can indicate which proteins the cell has potential to produce, and mRNA profiles can give an approximation as to which proteins are being made, further information is required to understand what expressed proteins actually do in order to define cell function at the molecular level. The requirement for characterizing organisms beyond the genomic and transcriptomic levels arises from several considerations.

a) Genetic and transcript information alone is insufficient to annotate genomes precisely. Information at the protein level will be required to ascertain genomic annotations.
b) The existence of an open reading frame (ORF) does not necessarily imply the existence of a functional gene.
c) The relationship between genes and gene products (*i.e.* proteins) is not necessarily linear. A given gene can express more than one protein. Indeed, in eukaryotes, 6-8 proteins are expressed per gene.
d) Expression profiles at the protein level may throw more light on function than those at the transcript level as mRNA levels do not necessarily correlate with protein levels (Gygi *et al.*, 1999b).
e) Biochemical characterization is required to link genomic information with phenotypic variations.
f) Complex networks of interactions mediate cellular responses to signals, and the nature of the response is dependent on the cell type and states. These aspects cannot be accounted for through genomic or transcriptomic studies alone.
g) At a molecular level, cellular function is closely associated with activities of proteins. They are intermediary in translating the genetic information to phenotypic function and thus offer an ideal platform for characterizing cellular activities.

Proteins serve as structural and functional entities in cells, contributing as catalysts, secondary messengers and transporters, and are of considerable importance in cellular characterization. The 'proteome' has been defined as the entire protein complement of a genome, a cell, or a tissue type (Wilkins *et al.*, 1996). It can be dependent on growth conditions and therefore defined for a given set of conditions. Proteomic analysis involves assessment of total cellular protein content, in terms of protein type, nature and quantity. Assessment of subcellular localization and protein-protein interactions that help in defining cell function can also be included. Proteomics can thus translate to mapping cellular proteins in a spatial and temporal manner (Blackstock and Weir, 1999). Unlike the genome, where information content is conserved for a given organism, proteomic information is dynamic, depending on cell type, state and environmental conditions (Godovac-Zimmermann and Brown, 2001). The nature of a protein can also vary, depending on post-transcriptional (*e.g.* splice variants) and post-translational (*e.g.* phosphorylation, glycosylation, proteolysis) modifications. Furthermore, in many cases, multi-protein complexes act as functional units in carrying out cellular activities. There are thus different aspects of the proteome that can be analyzed for microbial and other cellular characterization. The fact that microbial cells are relatively easier to handle and manipulate, enables microbial proteomic analysis to serve as a test-bed for strategies potentially relevant to proteomic analysis of higher organisms (O'Connor *et al.*, 2000).

2.1.1 Strategies and Applications

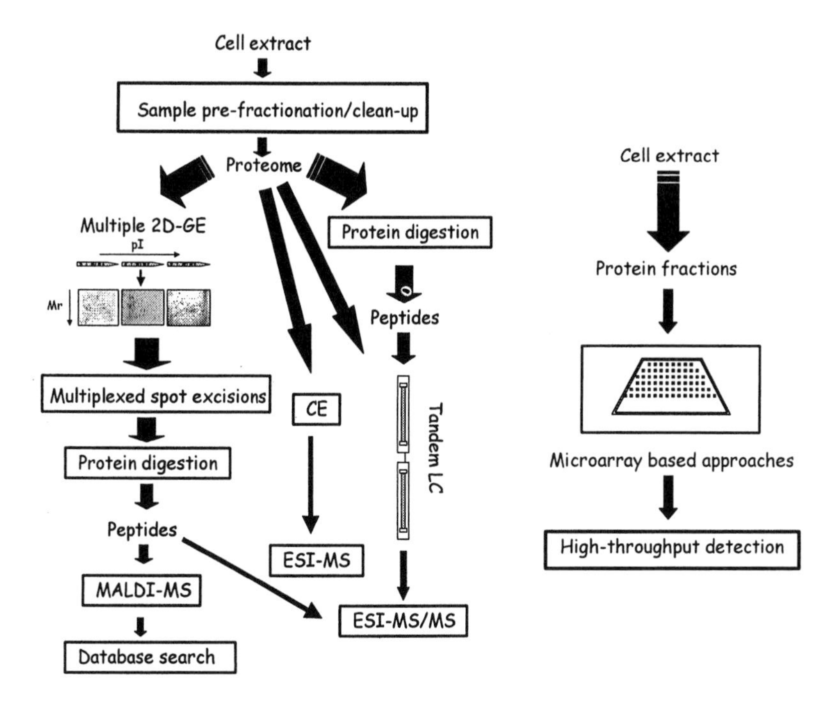

Figure 2. Strategies for proteome analysis (refer to text for abbreviations).

Three major analytical strategies are available for proteomic analysis; a) 2-D gel electrophoresis (2D-GE) based separation followed by mass spectrometric identification, b) liquid chromatography (LC) or capillary electrophoresis (CE) based separation of proteins/digested peptides, followed by MS, and c) immunoassay based microarray technology. Of these, the 2D-GE strategy is the most widely used tool for high-throughput proteomics. Here cellular extracts of proteins are separated based on charge in one dimension using isoelectric focusing, and then by size in the second dimension using sodium dodecyl polyacrylamide gel electrophoresis (SDS-PAGE). The gels can be used at an analytical scale to look for prominent signals or at a micro-preparative scale for identification and further characterization of separated proteins. The advent of immobilized pH gradient strips has improved resolution and reproducibility of analyses involving milligram quantities of proteins. In-gel sample loading and the development of sensitive protein stains (*e.g.* ammoniacal silver), radioactive stains, and fluorescent dye tags, have also contributed to recent improvements in the 2D-GE technique.

However, for comprehensive proteome analysis, it is essential that the analytical technique be capable of separating and resolving all different protein types, ideally with minimal sample preparation. The difficulties with 2D-GE in this regard arise from the fact that there are enormous variations in the levels of different proteins, often ranging from as much as 2,000,000 copies/cell to as little as 100 copies/cell, as observed in *Saccharomyces cerevisiae*. In other words variations in protein concentrations can be as much as 10,000 fold (Futcher *et al.*, 1999). Unlike genomics, where PCR can be used to amplify DNA, there are no ready-made amplification strategies in proteomics. This makes it difficult to detect low abundance proteins, although protein enrichment strategies such as the use of hydroxyapatite chromatography, prior to 2D-GE (Fountoulakis *et al.*, 1999) could contribute to minimizing this problem. Differential protein processing (resulting in more than one spot per protein) and co-migrating spots also present problems for quantitative protein expression and database matching (Gygi *et al.*, 2000) with 2D-GE approaches. Certain proteins constituting a proteome may also have a relatively homogenous distribution with respect to size (Mr) and charge (pI) necessitating analysis over narrower pI and Mr ranges. This can increase the number of gels that need to be run for a given sample. A further difficulty is in relating positions on different gels for comparative purposes. Despite these shortcomings, 2D-GE is still the most widely practiced option, and is useful in qualitatively, and perhaps semi-quantitatively, assessing proteomes, particularly with respect to the most abundant proteins.

2D-GE technology has been adopted in analyzing several microbial proteomes, as can be inferred from databases available at http://www.expasy.ch. The proteome of *H. influenza* provides an ideal start for assessing the utility of 2D-GE in proteomic analysis given its small size (about 1700 proteins). Using several strategies to capture proteins not amenable to conventional 2D-GE, Langen *et al.*, (2000) were able to detect about 1100 individual protein spots, of which more than 500, representing about 30% of the predicted number of ORFs, could be identified by peptide mass fingerprinting (*vide infra*). The analysis brought to light some relatively abundant proteins, several of which were of unknown function.

Comparative 2D-GE approaches can be useful in identifying proteins that are up- or down-regulated. Such differential display proteomics can be useful in bioprocess operations, for example. Acetate and formate are major products in *Escherichia coli* fermentations but an excess of either can retard growth. The steady-state expression levels of proteins, as viewed by 2D-GE, revealed a differential proteomic response when *E. coli* was cultivated under acetate or formate stress (Kirkpatrick *et al.*, 2001) suggesting differential regulation of proteins, for seemingly similar environmental stimuli. Differential 2D-GE proteomic display patterns were also observed between immobilized and free floating *E. coli* cells, providing insights into possible reasons for the high resistance of the former to stress conditions (Perrot *et al.*, 2000). Similarly, differential regulation of a multitude of proteins has been observed between early and developed biofilms of *Bacillus cereus* (Oosthuizen *et al.*, 2002).

Proteomic analysis is also significant in studying pathogenic bacteria when aiming to elucidate protein candidates for vaccine and diagnostic applications, to screen for novel drug targets or to study the effects of drugs on cellular physiology (Cordwell *et al.*, 2001). This is shown in applications of proteomic analysis to *Helicobactor pylori* (see McAtee *et al.*, 1998; Jungblut *et al.*, 2000; Haas *et al.*, 2002; Lock *et al.*, 2002), the causative agent of gastritis, ulcer and stomach carcinoma.

S. cerevisiae is an excellent eukaryotic model in which 40% of single-gene determinants of human diseases find homologs (Oliver, 2002). The recent availability of a library of deletant mutants corresponding to its 6000+ genes (Delneri *et al.*, 2001; Giaever *et al.*, 2002) will mean that proteomic methods pioneered with yeast will be helpful in analyzing higher organisms. *S. cerevisiae* could also be exploited as a surrogate for functional genomics of higher organisms.

Another approach to large-scale assessment of protein profiles is the use of immunoassays on microarrays (Blagoev and Pandey, 2001; Schweitzer and Kingsmore, 2002). This analysis involves a scale-up of enzyme-linked immunoassays (ELISA) that have long been in use for protein analysis. Antibodies immobilized in an array format on specifically treated surfaces act as 'baits' allowing detection of proteins that bind to the relevant antibodies using *e.g.* fluorescence detection. Protein microarrays can be also developed for assessing protein-protein, enzyme-substrate, and other protein-metabolite interactions. In one study (Zhu *et al.*, 2000) 119 yeast protein kinases were arrayed in microwells and examined for kinase activity with 17 different substrates illustrating the value of microarrays in multiplexed protein functional assessments.

2.1.2 Protein-Protein Interactions and Protein Complexes

In the post-genomic era, definition of protein function has to include the context of a proteins interactions with other intracellular (and extracellular) proteins (Eisenberg *et al.*, 2000). In addition to qualitatively or quantitatively determining total cellular protein content, information regarding protein-protein interactions and post-translational modifications are required and hence must form a significant part of proteomic analysis. *S. cerevisiae* has been the prototype for characterizing protein complexes and protein-protein interactions. A powerful and simple method for identification and analysis of protein-protein interactions is the yeast two-hybrid system which is a genetic method (Fields and Song, 1989). Using this strategy, over 1200 protein-protein interactions have been identified in *He. pylori,* connecting 46% of its proteome (Rain *et al.*, 2001).

Multi-protein complexes carry out many cellular activities. Characterization of these complexes can provide fundamental biological information. Epitope-tagging can be used to identify protein interactions and analyze protein complexes (Mann *et al.*, 2001). The use of two epitope tags in a tandem affinity purification (TAP) strategy (Rigaut *et al.*, 1999) provides a means to minimize binding to non-specific proteins and improve recovery of protein complexes. Using the TAP-tag strategy, 232 distinct multi-protein complexes were identified in *S. cerevisiae*, enabling functional description of a eukaryote proteome at a high level of organization (Gavin *et al.*, 2002). Other than protein-protein interactions, the subcellular localization of proteins should provide a valuable database for defining protein function. Using high-throughput immunolocalization of epitope-tagged gene products, the subcellular localization of over 2500 *S. cerevisiae* proteins has been determined (Kumar *et al.*, 2002).

2.2 Profiling the Metabolome

The metabolome, which is the metabolic complement of the cell, is further down the line from gene to function (Fig. 1) and more closely reflects the activities of the cell at the functional level. Like the proteome, but unlike the genome it is contextual and varies from condition to condition. Physical interactions governing cellular responses to perturbations when combined with proteomic and genomic data can be useful in modeling cellular pathways (Ideker *et al.*, 2001). Such integrative approaches would benefit from metabolomic databases (Fell, 2001). In fact, it has been shown, for example (ter Kuile and Westerhoff, 2001), that glycolytic flux is rarely regulated by gene expression alone, raising doubts as to whether transcriptomics and proteomics suffice to assess biological function, and emphasizing the need for investigations at the metabolomic level. For simple organisms, like prokaryotes it may even be possible to identify gene function and regulatory networks based on mRNA expression and metabolomics, without the need for proteomics (Phelps *et al.*, 2002). Associating cellular activities at the metabolite level provides a more interactive medium for assessing cell function and effecting better microbial characterization. This can be appreciated by noting developments in the field of metabolic engineering (Varma *et al.*, 1993; Nielsen, 2001; Stafford and Stephanopoulos, 2001).

Table 1. Relative Genomic, Proteomic and Metabolomic Content of Selected Microorganisms (from http://BioCyc.org/)

Microorganism	Genes	Proteins	Protein complexes	Metabolites (Compounds)
Escherichia coli K12	4392	4464	605	794
Agrobacterium tumefaciens C58	5469	5430	54	696
Bacillus subtilis 168	4221	4118	32	576
Chlamydia trachomatis D-UW-3-CX	939	913	8	323
Caulobacter crescentus	3818	3788	51	699
Helicobacter pylori 26695	1609	1580	19	450
Haemophilus influenzae Rd	1746	1632	135	460
Mycobacterium tuberculosis H37Rv	3966	3943	40	594
Mycoplasma pneumoniae M129	706	694	4	241
Pseudomonas aeroginosa PA01	5643	5586	48	630
Saccharomyces cerevisiae S288C	6526	6241	42	458
Treponema pallidum Nichols	1082	1036	3	304
Vibrio choloreae N16961	3950	3853	39	656

An indication of the relative diversity of information, in terms of 'omic' content for selected microorganisms, can be inferred from Table 1. The number of metabolites is typically an order of magnitude less than the number of genes or proteins suggesting that it may be easier to comprehensively cover metabolomes in large scale analyses.

Alterations in the cell in response to environmental or developmental stimuli, or to a genetic mutation, should result in changes in the quasi-steady-state levels of intermediate metabolites of pathways or in the final accumulation levels of terminal metabolites. Capturing these changes would require monitoring the metabolites and their levels, both spatially and temporally. Since the metabolic complement is even more dynamic than the proteome, analysis can be envisaged at different levels. While it may be ideal to have information on the status of the entire metabolic complement of a cell, there may be instances when it would suffice to derive information on only a portion of the total metabolome. For instance, it may suffice to monitor selectively only the relevant metabolites that contribute to a specific pathway that may be directly associated with function. In some instances, it may even be sufficient to monitor changes in the overall or partial metabolic pool structure and classify samples, without the need for determining the levels of individual metabolites. Fiehn (2001) classifies these approaches, as follows:

a) *Metabolite target analysis*: analysis restricted to metabolite(s) of, for instance, a particular enzyme system directly affected by a perturbation.
b) *Metabolite profiling*: analysis focused on a group of metabolites, say a class of compounds, such as carbohydrates, amino acids, or those associated with a specific pathway.
c) *Metabolomics*: a comprehensive analysis of the entire metabolome, under a given set of conditions.
d) *Metabolic fingerprinting*: classification of samples based on their biological relevance or origin.

Other terminologies exist such as *metabolic profiling* (often used synonymously with *metabolite profiling)* and *metabonomics* (Nicholson *et al.*, 1999). Both involve classification of samples based on their metabolic status. Terminologies are still evolving and there may be overlap in their definition. However, the above definitions highlight options available for monitoring the metabolome. Large-scale analysis of the metabolome, akin to genomic and proteomic analysis, is still in its formative years. Most approaches are being investigated and reported in plant cell systems (Fiehn *et al.*, 2000) and have not yet been widely pursued in microbes. However, investigations on manipulations on metabolic fluxes and pathway networks, have been more popular at the microbial level. Several of these metabolic engineering investigations could be classified as metabolite profiling (Cameron and Chaplen, 1997; Nielsen, 2001). Advances effected by metabolic engineering include improvements in pathways yielding aromatic metabolites and new products, improvements in cell physiology, developments in NMR-based methods for the monitoring of intracellular metabolites and metabolic flux, and the application of metabolic control analysis and metabolic flux analysis to a variety of systems.

2.2.1 Strategies and Applications

Metabolites are chemical entities and can be analyzed using standard chemical analyses (Fig. 3). For metabolite target analysis and metabolite profiling, the analysis can be focused on specific metabolites and involves selective analysis and well-developed calibrations. This can be achieved with conventional techniques involving separations by gas chromatography (GC) or high performance liquid-chromatography (HPLC) coupled to suitable detection systems. For metabolites containing chromophores, ultraviolet (UV) detection can be used, while non-UV-absorbing metabolites (*e.g.* sugars) can be detected by refractive index. MS based detection may also be applied (*vide infra*). Selectivity of analysis can be afforded by the choice of column (*e.g.* ion exchange, reversed-phase or affinity) that may be specific to the class of compounds to which the targeted metabolites belong. GC separations followed by flame ionization detection or MS detections can be employed. However, metabolites may require derivatization to generate volatile analytes prior to GC separation. Although efficient derivatization methods are available sample throughput can be adversely affected.

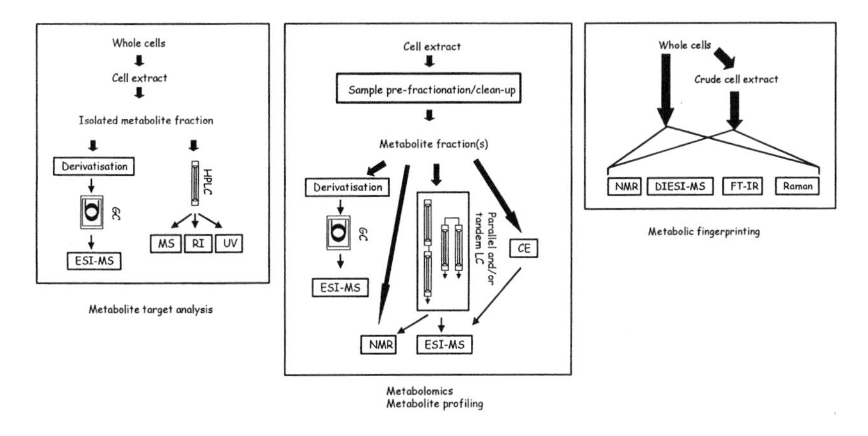

Figure 3. Strategies for metabolomic analysis. DIESI: direct-infusion ESI (refer to text for other abbreviations).

Another technique that can provide valuable information on metabolomes is nuclear magnetic resonance (NMR) spectroscopy. It is based on the fact that nuclei such as ^{1}H, ^{13}C and ^{31}P can exist at different energy levels in a strong magnetic field, as they possess nuclear spin. If such nuclei are subject to a magnetic field and pulsed with radio frequency energy, the absorption and re-emission of energy as they change energy levels can be measured as chemical shift, the NMR spectrum being a series of peaks representing the chemical environments within a molecule. It is thus valuable in obtaining structural information. ^{1}H-, ^{13}C- and ^{31}P NMR can also be used to trace metabolites along biosynthetic and other pathways. However, ^{1}H NMR spectra, particularly of mixtures, can be complex. ^{13}C or ^{31}P NMR are ideally suited for monitoring biochemical activities, but ^{13}C NMR requires growing cultures in media containing isotopically labeled substrates in order to improve sensitivity of detection. ^{31}P NMR can be used for *in vivo* studies, but requires high cell densities for sensitive measurements.

Due to high turnover of metabolites and rapid kinetics (Godovac-Zimmermann and Brown, 2001) it is often essential to minimize time delays in analysis. Kinetic analysis of the metabolome can be carried out non-invasively and in a quantitative manner using ^{13}C stable isotope labeling and NMR (Schmidt *et al.*, 1999; Dauner *et al.*, 2002). Differences between glucose and fructose fed fermentations in *Zymomonas mobilis* were found to relate to global alterations in intracellular levels of phosphorylated metabolites (De Graaf *et al.*, 1999) detected by ^{31}P NMR, an approach that can be classified as metabolite profiling.

For comprehensive analysis of the metabolome (metabolomics), it is essential to employ strategies that have a wider coverage in terms of the type and number of metabolites analyzed. Sample preparation may have to be elaborate incorporating sample clean up and pre-fractionation steps in order to capture even minor changes through comprehensive metabolomic analysis. Combinations of several analytical techniques may have to be employed for such analysis (van der Greef, Chapter 10).

Most extraction procedures reported in the literature are not comprehensive, resulting in only a portion of the metabolome being analyzed and modeled. However the information obtained even in such investigations points to the value of monitoring metabolomes comprehensively. In an early attempt at metabolomic analysis (Tweeddale *et al.*, 1998) trends were detected in the pool sizes of several of the 70 most abundant metabolites extracted from *E. coli* growing in glucose-limited chemostats at different growth rates. Cell extracts containing the metabolites were separated and analyzed by thin layer chromatography. A similar approach has also revealed metabolic shifts in response to superoxide stress (Tweeddale *et al.*, 1999). Analysis of the metabolome of *E. coli* cells grown under conditions to generate high or low density cultures showed dramatic changes in the metabolome profiles that indicates differentiation in a scale similar to that of exponential and stationary-phase bacteria (Liu *et al.*, 2000).

Metabolic fingerprinting is rapid and would be ideally suited for rapid characterizations if prominent changes in the metabolome can be captured in a reproducible manner. Techniques capable of handing a large number of samples with minimal sample preparation, but still capable of providing relevant chemical information would be ideally suited for rapidly generating fingerprints. In this regard, crude extracts or whole cells can be analyzed using direct infusion MS, NMR, Fourier-transform infrared (FT-IR), or Raman spectroscopy. A metabolic fingerprinting approach using NMR has been shown to have potential in detecting "silent genes", genes that do not overtly show a discernible phenotype, such as changes in growth rate (Raamsdonk *et al.*, 2001). The study also highlights that it may be possible to reveal functions when "metabolic snapshots" from strains deleted for unstudied genes are compared to those deleted for known genes.

Other than the techniques discussed, cell extraction followed by CE (Terabe *et al.*, 2001) can also be adopted for characterizing the metabolome.

3. MASS SPECTROMETRY OF PROTEOMES AND METABOLOMES IN MICROBIAL CHARACTERIZATION

MS involves the detection of gas phase ions, based on two of its properties, its mass and its charge. Since its discovery by Thomson in the early 20th century, there has been little change in the principle of analysis, in that all mass spectrometers measure the mass-to-charge (*m/z*) ratios of gaseous ions from a sample using electric and/or magnetic fields. However, vast improvements have been made in the analyzers employed, and the methods used to generate gas phase ions have evolved to enable detection of even non-volatile and labile macromolecules.

3.1 Mass Spectrometers

MS has been a major technique in chemical analysis over the last century and has been used to identify and characterize chemical compounds. Simplistically, a mass spectrometer consists of, a) an ionization source to generate the gas phase ions from the sample, b) a mass analyzer to separate the ions based on its *m/z* ratio and c) a detector to detect the separated ions (Fig. 4).

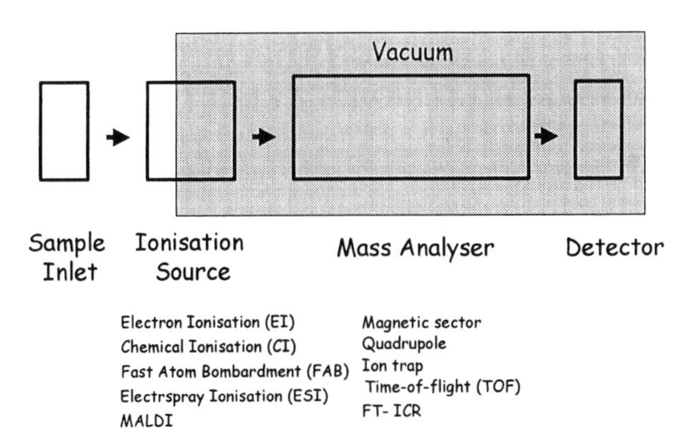

Figure 4. Schematic of a mass spectrometer (typical ionization methods and mass analyzers are listed).

Traditionally, ions are generated by directly passing a beam of electrons through a gas sample or a volatilized liquid sample in the gas-phase (electron impact – EI), or through ion-molecule reactions of a reagent gas, such as methane, isobutane, or ammonia, under pressure (chemical ionization – CI). These traditional methods of ion generation require the analyte to be volatile and thermally stable. Due to the high energy imparted to the analyte in these methods, fragmentation of the molecular ion frequently occurs, especially with EI, giving rise to more than one peak in the mass spectrum, for a given analyte. In an alternative even harsher approach to introducing non-volatile materials into a vacuum, pyrolysis has been used to thermally degrade materials prior to MS analysis (Meuzelaar *et al.*, 1982). Subsequently, ion generation methods, such as field-desorption (FD), fast-atom-bombardment (FAB), secondary-ionization (SI), and plasma desorption (PD) have appeared. These are "softer" in that they produce little or no fragmentation, and are based on desorption or desolvation techniques. This enables the analysis of non-volatile compounds and hence these methods are suited for routine analysis of biological samples. The advent of electrospray ionization (ESI) and matrix-assisted laser desorption/ionization (MALDI) techniques has enabled even more efficient soft-ionization, and has broadened the scope of biomolecular analysis. These latter techniques have generally superceded earlier soft-ionization methods and today occupy center stage in biomolecular analysis.

Sample desolvation and ionization in ESI takes place when a liquid sample is allowed to flow through a narrow capillary, the tip of which is connected to a voltage supply, to generate charged aerosols that can be desolvated by the assistance of temperature, and a coaxial flow of gas (Fig. 5A). In MALDI, samples are presented as dried spots, mixed with a matrix. Pulsed application of a laser at a wavelength at which the matrix absorbs enables desorption of the sample-matrix co-crystals from the surface into the analyzer as charged ions (Fig. 5B).

Ion separation is carried out in a mass analyzer, which is usually one of the four types, namely a) (magnetic) sector b) quadrupole filters (Q, Fig. 5C) and ion traps (Fig. 5D), c) time-of-flight (TOF, Fig. 5E) or d) Fourier-transform ion cyclotron resonance (FT-ICR).

One or more of these analyzers (of the same type or of different types) can be coupled to create tandem mass spectrometers (MS/MS). Each mass spectrometer can be used to scan, select one or all ions, and between two mass analyzers these ions can be subjected to collision with neutrals and gases or acceleration, deceleration, decomposition *etc.* This leads to several possible MS/MS experiments. For example, MS1 can be used to select a precursor ion which can be collided with inert gases (collision activated dissociation) in a "collision cell" resulting in fragment ions (product ions) that can be scanned using MS2 over the desired *m/z* range (product-ion scans). Alternatively, MS1 can be used to scan while MS2 is fixed at a chosen *m/z* value, so that the precursor ions of a selected product are detected (precursor-ion scan); there is a response at the detector only when a precursor of the selected ion passes through MS1. In another option, both

MS1 and MS2 can be used to scan with a fixed mass difference, enabling detection of constant neutral loss or gain. Two common tandem MS geometries are the triple quadrupole and the Q-TOF™. Tandem MS experiments are also possible with ion-traps and FT-ICR analyzers. Here, successive MS experiments are separated in time instead of space, enabling MS^n possibilities. Combinations with sector instruments are also possible, but these are generally costly.

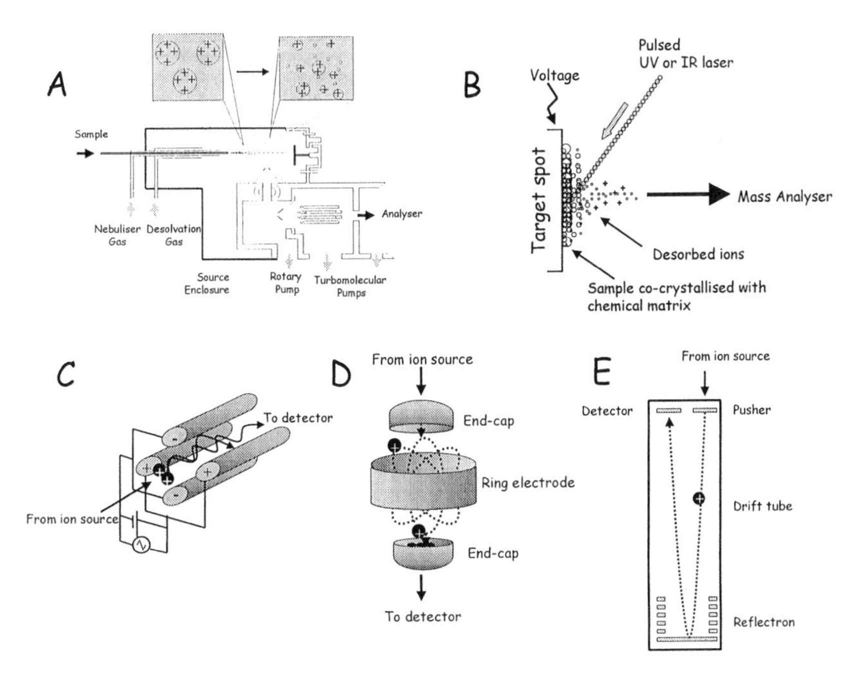

Figure 5. Schematic of (A) ESI in the Z-spray arrangement, (B) MALDI, (C) quadrupole ion filter, (D) ion-trap, and (E) TOF

3.2 Metabolome Profiling by Mass Spectrometry

The sensitivity and selectivity of MS make it an ideal candidate for detecting metabolites. However, comprehensive metabolomic analysis would require sample pre-fractionation prior to MS. As indicated earlier, LC or GC is usually employed for this purpose. For LC separations, reversed-phase, anion-exchange or affinity based methods can be employed depending on the type of metabolites profiled. Traditionally, LC or GC was followed by on-line analysis and detection using EI-MS. More recently, soft-ionization MS, especially ESI-MS, has been coupled to LC and GC separations, to minimize fragmentation. Using these methods it has been possible, for example, to profile and quantify intracellular concentrations of glycolytic intermediates (Buchholz *et al.*, 2001), amino acid isotopomers (Dauner and Sauer, 2000), neutral and acidic sugars from bacterial acid extracts (Wunschel *et al.*, 1997) and even to investigate intracellular metabolic fluxes (Christensen and Nielsen, 2000). As an alternative, direct infusion of crude extracts from the cell or fermentation broth into the mass spectrometer without prior analyte separation has been employed and shown to be informative for profiling secondary metabolites from actinomycetes (Higgs *et al.*, 2001) and *Penicillium* cultures (Smedsgaard and Frisvad, 1996). For a test set of 43 actinomycetes, direct infusion ESI-MS was found to show a generally good agreement (close to 80% correlation) with an HPLC based approach (Higgs *et al.*, 2001). The use of ESI-MS minimizes fragmentation and enables the possibility of identifying the metabolites using tandem MS (Smedsgaard and Frisvad, 1996). Even when the identity of metabolites is unknown, such approaches can be useful for metabolic fingerprinting and will provide valuable information in rapid high throughput screening programs.

MALDI-MS has been largely employed for characterizing relatively large molecular weight components, such as proteins. The occurrence of matrix adducts at low molecular weights has minimized the applicability of MALDI-MS for profiling low- molecular weight metabolites. However, MALDI-MS methods have been used to directly quantify isotope labeling patterns of low molecular weight metabolites, enabling elucidation of key fluxes in the central metabolism of lysine producing *Corynebacterium glutamicum* batch cultures (Wittmann and Heinzle, 2001). Secondary metabolites from whole cells have also been detected and profiled using MALDI-MS methods (Erhard *et al.*, 1997; Leenders *et al.*, 1999). There have been several investigations on the application of MS, usually after pre-fractionation steps, for profiling metabolites and cell fractions for specific characterizations. For instance, peptidoglycan fragments, containing glycopeptide mixtures, produced by enzymatic digestion have been characterized after LC fractionation, using MALDI-MS and nano ESI-MS, both in the positive and negative ion modes (Bacher *et al.*, 2001).

3.3 Protein Profiling by Mass Spectrometry

For 2D-GE resolved proteomes, the next stage of analysis is identification of the protein spots. MS is the method of choice for this purpose. MALDI and ESI (usually nano-ESI) are the common ionization methods employed. There are two main approaches, following excision of 2D-GE separated protein spots.

In 'peptide mass fingerprinting' (Henzel *et al.*, 1993) the protein spots are subjected to in-gel digestion by a sequence-specific protease, usually trypsin, after destaining, reduction, alkylation and washing steps. This is followed by analysis of the eluted peptides by MALDI-MS (Shevchenko *et al.*, 1996). The set of masses from the MS analysis is then compared to theoretically expected tryptic peptide masses in a database to identify the protein. The requirement for protein digestion prior to mass spectrometric analysis arise from the relative difficulty in eluting gel-separated proteins and the poor mass accuracy and resolution of high molecular weight proteins, compared to those of peptides. A match of at least five peptide masses and 15% protein coverage would indicate unambiguous identification, with mass accuracies of 10-50 ppm (Mann *et al.*, 2001).

The second approach involves peptide sequencing using tandem MS followed by analysis of fragmented peptides. This approach usually uses ESI-MS/MS. The peptides in liquid phase are electrosprayed into the mass spectrometer, a precursor ion corresponding to the ionized peptide mass is chosen using the first mass analyzer (usually a quadrupole) and this is fragmented by the application of a collision gas (such as argon) under pressure to give product ions that are separated by a second mass analyzer (usually a TOF), thus generating a fragmentation pattern that can be used to sequence the peptide and in turn identify the protein. Peptide chemistry dictates that there is a propensity for certain fragments to occur in preference to others, enabling the deduction of rules for sequence identity. Larger peptides often fragment efficiently providing long ion series, but due to multiple charge distribution of the precursor ion intensity, may have poor sensitivity (Mann *et al.*, 2001).

More than 500 proteins (corresponding to about 30% of the predicted number of ORFs) have been identified in the proteome of *H. influenzae*, using peptide mass fingerprinting (Langen *et al.*, 2000). Mass spectrometric identification by peptide mass fingerprinting and by tandem MS combined with database search has been used to identify proteins from brewing yeast strains that have homology to those of *S. cerevisiae* from a protein database (Joubert *et al.*, 2001).

MALDI-MS and ESI-MS are sensitive techniques. However, the nature of the ionization process, which is still not completely understood, necessitates that the sample analyzed is clean (*i.e.* presence of little or no unwanted chemical species other than the analyte of interest). In samples containing mixtures of analytes, the chances of observing and quantifying analytes of interest decreases as the number of mixture analytes increases due to ion suppression effects and competition for ionization. It is therefore common to employ a sample clean-up or an analyte separation stage prior to MS, when analyzing protein mixtures or protein digests. ESI-MS is especially suited to coupling with LC, and the application of LC-MS or LC-MS/MS (Pflieger *et al.*, 2002) for proteomic studies on peptide mixtures derived from 2D-GE is now becoming common. LC-MS (MS/MS) (Li *et al.*, 1999) and CE-MS can also be used directly on intact protein mixtures for proteomic analysis, obviating the need for 2D-GE separation. Measurement of several proteins in cell lysates of *E. coli* was possible by combining capillary isoelectric focusing with FT-ICR-MS (Jensen *et al.*, 2000). An even advanced approach is the combination of similar or different separation strategies with MS *e.g.* LC-LC-MS (MS/MS). Reversed-phase, ion-exchange (Davis *et al.*, 2001; Washburn *et al.*, 2001), affinity and hydrophobic interaction columns (Langen *et al.*, 2000) have been applied in isolation or in tandem prior to mass spectral analysis. The use of capillary LC systems provides greater efficiency of separation. Multiple capillary systems using serially connected dual-capillary columns can be more effective as it eliminates time delays for column regeneration. Over 100,000 peptides could be detected in yeast cytosolic tryptic digests using this strategy combined with FT-ICR-MS (Shen *et al.*, 2001). Chromatography can also be used as a pre-concentration step to increase the sensitivity of detection for peptides derived from protein digests of 2D-GE spots (Langen *et al.*, 2000; Timperman and Aebersold, 2000). MS can also be performed after the first dimension of 2D-GE, by performing surface analysis on immobilized pH gradient strips using MALDI-MS, to construct a virtual 2D-gel (Loo *et al.*, 2001). Most of these advanced strategies, however, are still in the proof-of-principle stage, and although shown to have significant potential further validation and improvements with respect to practical application will be needed before they come into routine operation.

MS can be useful in the identification of post-translational modifications in proteins and analyzing protein complexes. For example, phosphorylation which is a common post-translational modification, can be detected in protein and peptide mixtures, based on the fact that phosphopeptides and phosphorylated proteins are heavier than their unphosphorylated counterparts, and the mass spectra can be scanned for neutral losses of 98 (H_3PO_4) or 80 (HPO_3). Depending on the nature of the sample several strategies can be employed to detect and even quantify the phosphorylated proteins (Mann *et al.*, 2002). More than 1,000 phosphopeptides were detected using a MS-strategy applied to the analysis of whole cell lysates of *S. cerevisae* (Ficarro *et al.*, 2002). It has also been possible to identify post-translational modifications such as loss of initiating methionine, acetylation, methylation, and proteolytic maturation in yeast ribosomal proteins (Lee *et al.*, 2002). Systematic unambiguous identification of post-translational modifications in proteins from their mass spectral profiles may require the application of software tools (Wilkins *et al.*, 1999).

Protein complexes have also been characterized using MS (Link *et al.*, 1999; Jensen *et al.*, 2000; Gavin *et al.*, 2002; Lee *et al.*, 2002). In a high-throughput strategy tandem MS was used to systematically identify protein complexes in the yeast proteome (Ho *et al.*, 2002). Starting with 10% of predicted yeast proteins as bait over 3,000 proteins, corresponding to 25% of the yeast proteome, were detected.

Quantification of proteins expressed is an important aspect in proteomic analysis. This is a difficulty with 2D-GE, as the staining intensities of proteins on gels can be poorly resolved. Quantification is therefore error prone and not easy to assess in a reproducible manner. Alternatively, methods based on the incorporation of stable isotopes open the possibility of quantification using MS. The methods rely on the principle that stable isotope incorporation shifts the mass of the peptides by a predictable amount. The ratio of the analyte between the isotope incorporated and the non-incorporated state can than be determined accurately by the measured peak ratio between the underivatized and the derivatized sample. Using the above principle, combined with tandem MS of protein digests, and the targeted isolation of selected peptides from complex mixtures using isotope coded affinity tags, quantitative protein expression was compared between yeast cells grown with ethanol or glucose as carbon source (Gygi *et al.*, 1999a). Such selective protein labeling followed by multidimensional chromatography and tandem MS should enable quantification of global protein expression (Griffin and Aebersold, 2001), and used for microbial characterization.

3.4 Mass Spectrometry of Crude Cell Extracts and Whole Cells

For rapid and high-throughput microbial characterizations, required for such purposes as monitoring food and environmental safety, medical diagnosis, and detection of biological warfare agents, short and simple protocols that still provide the required information can be useful. In this regard, microbial fingerprinting using MS of crude cell extracts and whole cells offers potential. Such methods, usually involve little sample preparation steps, and offer information sufficient for discriminatory and identification purposes. Early investigations in this direction involved the application of pyrolysis MS (Magee, 1993; Tas and van der Greef, 1994; Goodfellow, 1995). However, the highest *m/z* value reproducibly attainable in the commonest pyrolysis MS instruments is very small (typically < *m/z* 200), moreover, due to the *in vacuo* thermal degradation step essentially all information on the structure or identity of the molecules producing the pyrolyzate is lost (Goodacre and Kell, 1996b), unless the metabolite is of very low molecular weight (Goodacre *et al.*, 2000). Finally, whilst mathematical transformation methods have been developed to compensate for inherent instrument drift (Goodacre and Kell, 1996a; Goodacre *et al.*, 1997) this method has fallen out of favor. The application of desorption techniques, such as plasma desorption (Cotter, 1988) and fast-atom bombardment (FAB) (Heller *et al.*, 1987; Drucker, 1994) allowed the analysis of non-volatile fractions, and extended the measurable mass range. The advent of soft-ionization techniques, such as MALDI and ESI has enabled the mass spectrometric analysis of both large molecular weight compounds such as proteins and nucleic acids (Roepstorff, 1997; Yates, 2000), and low molecular weight metabolites (Black and Fox, 1996), giving the potential for rapid characterization of microorganisms with greater accuracy, selectivity, and sensitivity for better discrimination. Consequently, the application of MALDI-MS and ESI-MS techniques for microbial classification and identification has been the subject of several investigations (Claydon *et al.*, 1996; Krishnamurthy *et al.*, 2000; Demirev *et al.*, 2001b; Vaidyanathan *et al.*, 2001, 2002a, 2002b; Wang *et al.*, 2002) and reviews (van Baar, 2000; Fenselau and Demirev, 2001; Lay, 2001). In particular, MALDI-MS of whole cells, is particularly attractive because of its ability to characterize the proteome, albeit only a fraction of it, directly without 2D-GE separation and match these proteins to sequence databases (Demirev *et al.*, 2001a; Ryzhov and Fenselau, 2001).

Whole cell MALDI-MS techniques involve mixing microbial cells, either after washing the media off the cells or directly from colonies grown on agar plates, with a matrix solution, usually cinnamic acid derivatives, such as α-cyano-4-hydroxycinnamic acid for low mass proteins, or 3,5-dimethoxy-4-hydroxycinnamic acid (sinapinic acid) for high mass proteins, depositing them on a target plate and analyzing by MALDI-MS. Different strategies of mixing and spotting may have to be followed in order to increase the information content at different mass ranges (Vaidyanathan *et al.*, 2002b), and such effects on MALDI-MS spectra need to be more fully investigated to understand the ionization process better. Proteins contribute predominantly to the mass spectral signals in whole cell MALDI-MS analyses. Low mass proteins (typically less than 30 kDa) are readily determined in whole cell MALDI-MS. However, it is also possible to observe high mass proteins in some cases (Madonna *et al.*, 2000; Vaidyanathan *et al.*, 2002b), albeit with relatively poor mass resolution and accuracy. Although such rapid methods are not comprehensive in terms of information content, a sizeable number of signals (typically 40-50 peaks for Gram-negative, and 10-20 for Gram-positive microorganisms) are observed, and have been shown to have discriminatory value (Hathout *et al.*, 1999; Lay, 2000). Cold-shock proteins (Holland *et al.*, 1999), cell-wall or membrane associated proteins (Dai *et al.*, 1999), as well as those from the cytosol (Ryzhov and Fenselau, 2001) have all been implicated as contributors to the spectral signals in whole cell MALDI-MS. Majority of the investigations have concentrated on methods development and on demonstrating the applicability of the technique to relatively well characterized microorganisms. Consequently, the number and diversity of microorganisms investigated for whole-cell MALDI-MS is not yet comprehensive. The availability of proteome databases enables comparison of the spectral data sets to proteins in the database for identification and matching purposes based on mass alone (Demirev *et al.*, 2001a; Ryzhov and Fenselau, 2001). However, given the incompleteness of proteome databases and variability in the size of the databases for different microorganisms, statistical models (Pineda *et al.*, 2000) will have to be developed and used to verify microorganism identification by matching to such databases. It should also be remembered that the signals observed in whole cell MALDI-MS cannot be taken to correlate to proteins in the proteome databases *per se* because many of the proteins in the public proteome databases are derived from genome sequence data and do not necessarily account for aspects like post-translational modifications. Besides, fragmentation and observation of multiple charge states (at least singly and doubly charged proteins in MALDI-MS) may contribute to the signals. In these regards, it may be more useful to construct and compare protein mass databases (Wang *et al.*, 2002).

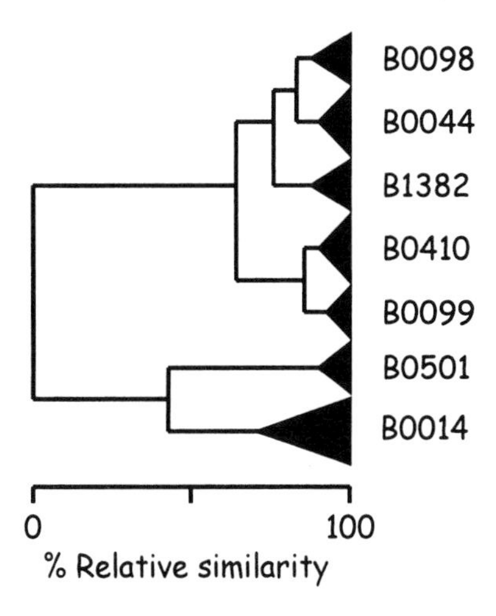

Figure 6. Dendrogram derived from positive-ion ESI mass spectra of whole cells showing sub-species level discrimination for different isolates of *B. subtilis*. (Adopted from Vaidyanathan *et al.*, 2001).

Alternatively in ESI-MS, the samples are in the liquid phase, and ionization is carried out at atmospheric pressure thus making it amenable to on-line automated integration with analyte separation techniques such as LC and CE. This widens the scope of its applicability considerably. Indeed, ESI-MS has already been shown to be useful for microbial characterization. However, the introduction of bacteria has largely been *via* specific cell fractions or lysates; *viz*, phospholipids (Smith *et al.*, 1995; Black *et al.*, 1997; Fang and Barcelona, 1998), lipopolysaccharides (Li *et al.*, 1998), lipooligosaccharides (Gibson *et al.*, 1994), muramic acids (Wunschel *et al.*, 1997), and proteins (Krishnamurthy *et al.*, 1999). Moreover, these have usually been presented to the ESI-MS after LC separation. Although such pre-fractionation approaches would be expected to yield better information content, ESI-MS of intact viruses (Siuzdak, 1998) and whole bacterial cells (Goodacre *et al.*, 1999) has been proven to produce information-rich spectra, Indeed, direct infusion ESI-MS has demonstrated sub-species level discrimination for different isolates of *Bacillus subtilis* (Fig. 6; Vaidyanathan *et al.*, 2001). One advantage of using ESI-MS on whole cells or crude cell extracts is the possibility of using tandem MS for global characterization (Xiang *et al.*, 2000) and biomarker identification (Demirev *et al.*, 2001b). Although at least three classes of macromolecules (proteins, phospholipids, and glycolipids) appear to contribute (Table 2), apart from smaller metabolites, phospholipids seem to dominate the spectral information of crude bacterial cell extracts (Fig. 7) (Vaidyanathan *et al.*, 2002a).

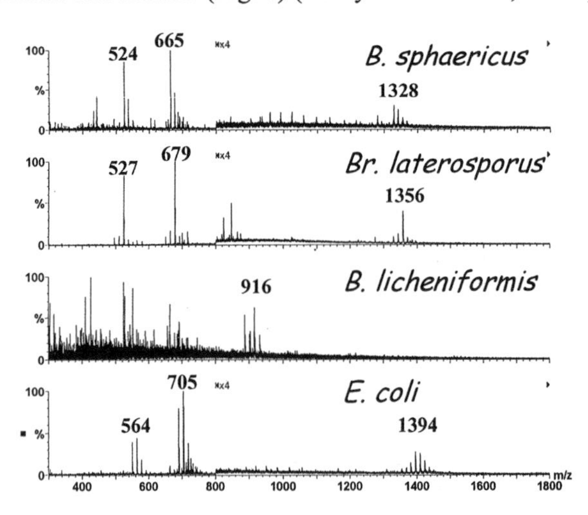

Figure 7. Positive-ion ESI mass spectra of crude cell extracts for four bacteria. (Adopted from Vaidyanathan *et al.*, 2002a)

Further research is still required to improve the spectral information so as to increase the number and range of metabolites and macromolecules with minimal sample preparation steps, in order to develop rapid high-throughput mass spectral fingerprinting strategies for microbial characterization. In such rapid analytical approaches it will be imperative to take into account the experimental and microbial growth conditions in assessing the spectral information for microbial characterization.

Table 2. Prominent Biomarkers Observed in Positive-ion ESI Mass Spectra of Crude Extracts for Representative Strains of Four Bacteria (Adopted from Vaidyanathan *et al*., 2002a)

Microorganism	Biomarkers	
	m/z	Likely origin(s)
E. coli UB5201	1422, 1408, 1394, 1380, 719, 705, 691, 665, 577, 564, 550	Phospholipids
	1096, 1056, 1047, 1018, 983, 964, 950, 928, 920, 891,864, 858, 839	Protein(s) (28, 10.4, 7.7 kDa)
B. sphaericus DSM 28	1328, 1342, 665, 524,	Phospholipids
	1137, 1024, 960, 905	Protein (30-32 kDa)
B. licheniformis NTCT 10341	932, 916, 904, 888,	Glycolipids
	689, 527, 705, 665, 524	Phospholipids
Brevibacillus. laterosporus NTCT 7579	1370, 1356, 1342, 716, 679, 665, 524	Phospholipids

4. CONCLUSION

Proteomes and metabolomes constitute complements of the cell that are closest to function and ultimately the phenotype. Hence they have the potential to reflect cellular activities better than transcriptomics as they provide information not directly obtainable from the genome *per se*. Profiling the microbial proteome and/or metabolome would enable a better understanding of physiology for optimal bioprocess operations, could discover novel metabolites and pathways, and could allow identification of specific traits in microorganisms that will be useful for more detailed characterization studies. Their dynamic nature (their profiles are time and growth condition dependent) is both useful and problematic. On one hand it offers several levels at which proteomes and metabolomes can provide information, while on the other, care must be taken in interpreting results in the context of growth conditions. Significant advances in database management and chemometrics are needed. In the post genomic era, the priorities for development of analytical strategies are shifting towards proteomic analysis. There are several aspects of the proteome, such as post-translational modifications and protein-protein interactions that are attracting closer study. This is being reflected in mass spectrometric applications, especially those of soft-ionization techniques that are particularly suited for proteomic analysis. However, even though the constituents of the metabolic complement, *i.e.* metabolites, can be accurately quantified using standard analytical techniques, relatively fewer strategies have been developed for more comprehensive holistic profiling. Although trends are changing a majority of reported metabolic investigations have concentrated on assessing only a portion of the metabolome. MS can play a useful role in a more comprehensive metabolome profiling, but will have to be coupled with pre-fractionation and clean-up steps that may have to be multiplexed for high-throughput analysis. Alternatively, rapid analytical approaches including direct-infusion MS on crude cell extracts or whole cells may be used to obtain metabolic fingerprints, which may suffice in some instances, providing partial information on the metabolic status of the cell.

ACKNOWLEDGEMENTS

The authors are indebted to the UK BBSRC (Engineering and Biological Systems Committee) for financial support.

REFERENCES

Bacher G, Korner R, Atrih A *et al.* Negative and positive ion matrix-assisted laser desorption/ionization time-of-flight mass spectrometry and positive ion nano-electrospray ionization quadrupole ion trap mass spectrometry of peptidoglycan fragments isolated from various *Bacillus* species. *J Mass Spectrom* 36: 124-139 (2001).

Black GE, Fox A. Liquid chromatography with electrospray ionization tandem mass spectrometry - profiling carbohydrates in whole bacterial cell hydrolysates. In *Biochemical and Biotechnological Applications of Electrospray Ionization Mass Spectrometry.* Vol. 619. Snyder PA (Ed) pp. 81-105, American Chemical Society, Washington, DC (1996).

Black GE, Snyder AP, Heroux KS. Chemotaxonomic differentiation between the *Bacillus cereus* group and *Bacillus subtilis* by phospholipid extracts analyzed with electrospray ionization tandem mass spectrometry. *J Microbiol Meth* 28: 187-199 (1997).

Blackstock WP, Weir MP. Proteomics: quantitative and physical mapping of cellular proteins. *Trends Biotechnol* 17: 121-127 (1999).

Blagoev B, Pandey A. Microarrays go live - new prospects for proteomics. *Trends Biochem Sci* 26: 639-641 (2001).

Buchholz A, Takors R, Wandrey C. Quantification of intracellular metabolites in *Escherichia coli* K12 using liquid chromatographic-electrospray ionization tandem mass spectrometric techniques. *Anal Biochem* 295: 129-137 (2001).

Bull AT, Ward AC, Goodfellow M. Search and discovery strategies for biotechnology: the paradigm shift. *Microbiol Mol Biol Rev* 64: 573-606 (2000).

Cameron DC, Chaplen FWR. Developments in metabolic engineering. *Curr Opin Biotechnol* 8: 175-180 (1997).
Christensen B, Nielsen J. Metabolic network analysis of *Penicillium chrysogenum* using C-13-labeled glucose. *Biotechnol Bioeng* 68: 652-659 (2000).
Claydon MA, Davey SN, Edwards JV, Gordon DB. The rapid identification of intact microorganisms using mass spectrometry. *Nature Biotechnol* 14: 1584-1586 (1996).
Cordwell SJ, Nouwens AS, Walsh BJ. Comparative proteomics of bacterial pathogens. *Proteomics* 1: 461-472 (2001).
Cotter RJ. Plasma desorption mass spectrometry - coming of age. *Anal Chem* 60: A781 (1988).
Dai Y, Li L, Roser DC, Long SR. Detection and identification of low-mass peptides and proteins from solvent suspensions of *Escherichia coli* by high performance liquid chromatography fractionation and matrix-assisted laser desorption/ionization mass spectrometry. *Rapid Comm Mass Spectrom* 13: 73-78 (1999).
Dando M. *Biological Warfare in the 21st Century*. Brassey's Ltd, London (1994).
Dauner M, Sauer U. GC-MS analysis of amino acids rapidly provides rich information for isotopomer balancing. *Biotechnol Prog* 16: 642-649 (2000).
Dauner M, Sonderegger M, Hochuli M *et al.* Intracellular carbon fluxes in riboflavin-producing *Bacillus subtilis* during growth on two-carbon substrate mixtures. *Appl Environ Microbiol* 68: 1760-1771 (2002).
Davis MT, Beierle J, Bures ET *et al.* Automated LC-LC-MS-MS platform using binary ion-exchange and gradient reversed-phase chromatography for improved proteomic analyses. *J Chromatogr* 752: 281-291 (2001).
de Graaf AA, Striegel K, Wittig RM *et al.* Metabolic state of *Zymomonas mobilis* in glucose-, fructose-, and xylose-fed continuous cultures as analysed by C-13- and P- 31-NMR spectroscopy. *Arch Microbiol* 171: 371-385 (1999).
Delneri D, Brancia FL, Oliver SG. Towards a truly integrative biology through the functional genomics of yeast. *Curr Opin Biotechnol* 12: 87-91 (2001).
Demirev PA, Lin JS, Pineda FJ, Fenselau C. Bioinformatics and mass spectrometry for microorganism identification: proteome-wide post-translational modifications and database search algorithms for characterization of intact *H-pylori. Anal Chem* 73: 4566-4573 (2001a).
Demirev PA, Ramirez J, Fenselau C. Tandem mass spectrometry of intact proteins for characterization of biomarkers from *Bacillus cereus* T spores. *Anal Chem* 73: 5725-5731 (2001b).
Drucker DB. Fast atom bombardment mass spectrometry of phospholipids for bacterial chemotaxonomy. In *Mass Spectrometry for the Characterization of Microorganisms*. Vol. 541. Fenselau C (Ed) pp. 18-35, American Chemical Society, Washington, DC (1994).
Eisenberg D, Marcotte EM, Xenarios I, Yeates TO. Protein function in the post-genomic era. *Nature* 405: 823-826 (2000).
Erhard M, von Dohren H, Jungblut P. Rapid typing and elucidation of new secondary metabolites of intact cyanobacteria using MALDI-TOF mass spectrometry. *Nature Biotechnol* 15: 906-909 (1997).
Fang J, Barcelona MJ. Structural determination and quantitative analysis of bacterial phospholipids using liquid chromatography electrospray ionization mass spectrometry. *J Microbiol Meth* 33: 23-35 (1998).
Fell DA. Beyond genomics. *Trends Genet* 17: 680-682 (2001).
Fenselau C, Demirev PA. Characterization of intact microorganisms by MALDI mass spectrometry. *Mass Spectrom Rev* 20: 157-171 (2001).

Ficarro SB, McCleland ML, Stukenberg PT *et al.* Phosphoproteome analysis by mass spectrometry and its application to *Saccharomyces cerevisiae. Nature Biotechnol* 20: 301-305 (2002).

Fiehn O, Kopka J, Dörmann P *et al.* Metabolite profiling for plant functional genomics. *Nature Biotechnol* 18: 1157-1161 (2000).

Fiehn O. Combining genomics, metabolome analysis, and biochemical modelling to understand metabolic networks. *Compar Funct Genom* 2: 155-168 (2001).

Fields S, Song OK. A novel genetic system to detect protein-protein interactions. *Nature* 340: 245-246 (1989).

Fleischmann RD, Adams, MD, White O *et al.* Whole-genome random sequencing and assembly of *Haemophilus- influenzae* Rd. *Science* 269: 496-512 (1995).

Fountoulakis M, Takacs MF, Berndt P *et al.* Enrichment of low abundance proteins of *Escherichia coli* by hydroxyapatite chromatography. *Electrophoresis* 20: 2181-2195 (1999).

Fraser CM, Eisen J, Fleischmann RD *et al.* Comparative genomics and understanding of microbial biology. *Emerg Infect Dis* 6: 505-512 (2000).

Futcher B, Latter GI, Monardo P *et al.* A sampling of the yeast proteome. *Mol Cell Biol* 19: 7357-7368 (1999).

Gavin AC, Bosche M, Krause R *et al.* Functional organization of the yeast proteome by systematic analysis of protein complexes. *Nature* 415: 141-147 (2002).

Giaever G, Chu AM, Ni L *et al.* Functional profiling of the *Saccharomyces cerevisiae* genome. *Nature* 418: 387-391 (2002).

Gibson BW, Phillips NJ, John CM, Melaugh W. Lipooligosaccharides in pathogenic *Haemophilus* and *Neisseria* species - mass spectrometric techniques for identification and characterization. *ACS Symposium Series* 541: 185-202 (1994).

Godovac-Zimmermann J, Brown LR. Perspectives for mass spectrometry and functional proteomics. *Mass Spectrom Rev* 20: 1-57 (2001).

Goodacre R, Heald JK, Kell DB. Characterisation of intact microorganisms using electrospray ionisation mass spectrometry. *FEMS Microbiol Lett* 176: 17-24 (1999).

Goodacre R, Kell DB. Correction of mass spectral drift using artificial neural networks. *Anal Chem* 68: 271-280 (1996a).

Goodacre R, Kell DB. Pyrolysis mass spectrometry and its applications in biotechnology. *Curr Opin Biotechnol* 7: 20-28 (1996b).

Goodacre R, Shann B, Gilbert RJ *et al.* Detection of the dipicolinic acid biomarker in Bacillus spores using Curie-point pyrolysis mass spectrometry and Fourier transform infrared spectroscopy. *Anal Chem* 72: 119-127 (2000).

Goodacre R, Timmins ÉM, Jones A *et al.* On mass spectrometer instrument standardization and interlaboratory calibration transfer using neural networks. *Anal Chim Acta* 348: 511-532 (1997).

Goodfellow M. Inter-strain comparison of pathogenic microorganisms by pyrolysis mass spectrometry. *Binary Comp Microbiol* 7: 54-60 (1995).

Griffin TJ, Aebersold R. Advances in proteome analysis by mass spectrometry. *J Biol Chem* 276: 45497-45500 (2001).

Gygi SP, Rist B, Gerber SA *et al.* Quantitative analysis of complex protein mixtures using isotope-coded affinity tags. *Nature Biotechnol* 17: 994-999 (1999a).

Gygi SP, Rochon Y, Franza BR, Aebersold R. Correlation between protein and mRNA abundance in yeast. *Mol Cell Biol* 19: 1720-1730 (1999b).

Gygi SP, Corthals GL, Zhang Y *et al.* Evaluation of two-dimensional gel electrophoresis-based proteome analysis technology. *Proc Natl Acad Sci USA* 97: 9390-9395 (2000).

Haas G, Karaali G, Ebermayer K *et al.* Immunoproteomics of *Helicobacter pylori* infection and relation to gastric disease. *Proteomics* 2: 313-324 (2002).
Hathout Y, Demirev PA, Ho YP *et al.* Identification of *Bacillus* spores by matrix-assisted laser desorption ionization-mass spectrometry. *Appl Environ Microbiol* 65: 4313-4319 (1999).
Heller DN, Cotter RJ, Fenselau C. Profiling of bacteria by fast atom bombardment mass spectrometry. *Anal Chem* 59: 2806-2809 (1987).
Henzel WJ, Billeci TM, Stults JT *et al.* Identifying proteins from 2-dimensional gels by molecular mass searching of peptide-fragments in protein-sequence databases. *Proc Natl Acad Sci USA* 90: 5011-5015 (1993).
Higgs RE, Zahn JA, Gygi JD, Hilton MD. Rapid method to estimate the presence of secondary metabolites in microbial extracts. *Appl Environ Microbiol* 67: 371-376 (2001).
Ho Y, Gruhler A, Heilbut A *et al.* Systematic identification of protein complexes in *Saccharomyces cerevisiae* by mass spectrometry. *Nature* 415: 180-183 (2002).
Holland RD, Duffy CR, Rafii F *et al.* Identification of bacterial proteins observed in MALDI TOF mass spectra from whole cells. *Anal Chem* 71: 3226-3230 (1999).
Ideker T, Thorsson V, Ranish JA *et al.* Integrated genomic and proteomic analyses of a systematically perturbed metabolic network. *Science* 292: 929-934 (2001).
Jensen PK, Pasa-Tolic L, Peden KK *et al.* Mass spectrometic detection for capillary isoelectric focusing separations of complex protein mixtures. *Electrophoresis* 21: 1372-1380 (2000).
Joubert R, Strub JM, Zugmeyer S *et al.* Identification by mass spectrometry of two-dimensional gel electrophoresis-separated proteins extracted from lager brewing yeast. *Electrophoresis* 22: 2969-2982 (2001).
Jungblut PR, Bumann D, Haas G *et al.* Comparative proteome analysis of *Helicobacter pylori. Mol Microbiol* 36: 710-725 (2000).
Kirkpatrick C, Maurer LM, Oyelakin NE *et al.* Acetate and formate stress: opposite responses in the proteome of *Escherichia coli. J Bacteriol* 183: 6466-6477 (2001).
Krishnamurthy T, Davis MT, Stahl DC, Lee TD. Liquid chromatography/microspray mass spectrometry for bacterial investigations. *Rapid Comm Mass Spectrom* 13: 39-49 (1999).
Krishnamurthy T, Rajamani U, Ross PL *et al.* Mass spectral investigations on microorganisms. *J Toxicol Toxin Rev* 19: 95-117 (2000).
Kumar A, Agarwal S, Heyman JA *et al.* Subcellular localization of the yeast proteome. *Genes Dev* 16: 707-719 (2002).
Langen H, Takacs B, Evers S *et al.* Two-dimensional map of the proteome of *Haemophilus influenzae. Electrophoresis* 21: 411-429 (2000).
Lay JO. MALDI-TOF mass spectrometry of bacteria. *Mass Spectrom Rev* 20: 172-194 (2000).
Lay JO. MALDI-TOF mass spectrometry and bacterial taxonomy. *Trends Anal Chem* 19: 507-516 (2000).
Lee SW, Berger SJ, Martinovic S *et al.* Direct mass spectrometric analysis of intact proteins of the yeast large ribosomal subunit using capillary LC/FTICR. *Proc Natl Acad Sci USA* 99: 5942-5947 (2002).
Leenders F, Stein TH, Kablitz B *et al.* Rapid typing of *Bacillus subtilis* strains by their secondary metabolites using matrix-assisted laser desorption ionization mass spectrometry of intact cells. *Rapid Comm Mass Spectrom* 13: 943-949 (1999).
Li J, Thibault P, Martin A *et al.* Development of an on-line preconcentration method for the analysis of pathogenic lipopolysaccharides using capillary electrophoresis-electrospray mass spectrometry - Application to small colony isolates. *J Chromatogr* 817: 325-336 (1998).

Li WQ, Hendrickson CL, Emmett MR, Marshall AG. Identification of intact proteins in mixtures by alternated capillary liquid chromatography electrospray ionization and LC ESI infrared multiphoton dissociation Fourier transform ion cyclotron resonance mass spectrometry. *Anal Chem* 71: 4397-4402 (1999).

Link AJ, Eng J, Schieltz DM *et al.* Direct analysis of protein complexes using mass spectrometry. *Nature Biotechnol* 17: 676-682 (1999).

Liu XQ, Ng C, Ferenci T. Global adaptations resulting from high population densities in *Escherichia coli* cultures. *J Bacteriol* 182: 4158-4164 (2000).

Lock RA, Coombs GW, McWilliams TM *et al.* Proteome analysis of highly immunoreactive proteins of *Helicobacter pylori. Helicobacter* 7: 175-182 (2002).

Loo RRO, Cavalcoli JD, Van Bogelen RA *et al.* Virtual 2-D gel electrophoresis: visualization and analysis of the *E-coli* proteome by mass spectrometry. *Anal Chem* 73: 4063-4070 (2001).

Madonna AJ, Basile F, Ferrer I *et al.* On-probe sample pretreatment for the detection of proteins above 15 KDa from whole cell bacteria by matrix-assisted laser desorption/ ionization time-of-flight mass spectrometry. *Rapid Comm Mass Spectrom* 14: 2220-2229 (2000).

Magee JT. Whole-organism fingerprinting. In *Handbook of New Bacterial Systematics.* Goodfellow M, O'Donnell AG (Ed) pp. 383-427, Academic Press, London (1993).

Mann M, Hendrickson RC, Pandey A. Analysis of proteins and proteomes by mass spectrometry. *Ann Rev Biochem* 70: 437-473 (2001).

Mann M, Ong SE, Gronborg M *et al.* Analysis of protein phosphorylation using mass spectrometry: deciphering the phosphoproteome. *Trends Biotechnol* 20: 261-268 (2002).

McAtee CP, Lim MY, Fung K *et al.* Characterization of a *Helicobacter pylori* vaccine candidate by proteome techniques. *J Chromatogr* 714: 325-333 (1998).

Meuzelaar HLC, Haverkamp J, Hileman FD. *Pyrolysis Mass Spectrometry of Recent and Fossil Biomaterials.* Elsevier, Amsterdam (1982).

Nicholson JK, Lindon JC, Holmes E. 'Metabonomics': understanding the metabolic responses of living systems to pathophysiological stimuli via multivariate statistical analysis of biological NMR spectroscopic data. *Xenobiotica* 29: 1181-1189 (1999).

Nielsen J. Metabolic engineering. *Appl Microbiol Biotechnol* 55: 263-283 (2001).

O'Connor CD, Adams P, Alefounder P *et al.* The analysis of microbial proteomes: strategies and data exploitation. *Electrophoresis* 21: 1178-1186 (2000).

Oliver SG. Functional genomics: lessons from yeast. *Philos Trans R Soc Lond Ser B* 357: 17-23 (2002).

Oosthuizen MC, Steyn B, Theron J *et al.* Proteomic analysis reveals differential protein expression by *Bacillus cereus* during biofilm formation. *Appl Environ Microbiol* 68: 2770-2780 (2002).

Perrot F, Hebraud M, Charlionet R *et al.* Protein patterns of gel-entrapped *Escherichia coli* cells differ from those of free-floating organisms. *Electrophoresis* 21: 645-653 (2000).

Pflieger D, Le Caer JP, Lemaire C *et al.* Systematic identification of mitochondrial proteins by LC-MS/MS. *Anal Chem* 74: 2400-2406 (2002).

Phelps TJ, Palumbo AV, Beliaev AS. Metabolomics and microarrays for improved understanding of phenotypic characteristics controlled by both genomics and environmental constraints. *Curr Opin Biotechnol* 13: 20-24 (2002).

Pineda FJ, Lin JS, Fenselau C, Demirev PA. Testing the significance of microorganism identification by mass spectrometry and proteome database search. *Anal Chem* 72: 3739-3744 (2000).

Raamsdonk LM, Teusink B, Broadhurst D *et al.* A functional genomics strategy that uses metabolome data to reveal the phenotype of silent mutations. *Nature Biotechnol* 19: 45-50 (2001).

Rain JC, Selig L, De Reuse H *et al.* The protein-protein interaction map of *Helicobacter pylori*. *Nature* 409: 211-215 (2001).

Rigaut G, Shevchenko A, Rutz B *et al.* A generic protein purification method for protein complex characterization and proteome exploration. *Nature Biotechnol* 17: 1030-1032 (1999).

Roepstorff P. Mass spectrometry in protein studies from genome to function. *Curr Opin Biotechnol* 8: 6-13 (1997).

Ryzhov V, Fenselau C. Characterization of the protein subset desorbed by MALDI from whole bacterial cells. *Anal Chem* 73: 746-750 (2001).

Schmidt K, Nielsen J, Villadsen J. Quantitative analysis of metabolic fluxes in *Escherichia coli*, using two-dimensional NMR spectroscopy and complete isotopomer models. *J Biotechnol* 71: 175-189 (1999).

Schweitzer B, Kingsmore SF. Measuring proteins on microarrays. *Curr Opin Biotechnol* 13: 14-19 (2002).

Shen YF, Tolic N, Zhao R *et al.* High-throughput proteomics using high efficiency multiple-capillary liquid chromatography with on-line high-performance ESI FTICR mass spectrometry. *Anal Chem* 73: 3011-3021 (2001).

Shevchenko A, Wilm M, Vorm O, Mann M. Mass spectrometric sequencing of proteins from silver stained polyacrylamide gels. *Anal Chem* 68: 850-858 (1996).

Siuzdak G. Probing viruses with mass spectrometry. *J Mass Spectrom* 33: 203-211 (1998).

Smedsgaard J, Frisvad JC. Using direct electrospray mass spectrometry in taxonomy and secondary metabolite profiling of crude fungal extracts. *J Microbiol Meth* 25: 5-17 (1996).

Smith PBW, Snyder P, Harden CS. Characterization of bacterial phospholipids by electrospray-ionization tandem mass spectrometry. *Anal Chem* 67: 1824-1830 (1995).

Stafford DE, Stephanopoulos G. Metabolic engineering as an integrating platform for strain development. *Curr Opin Microbiol* 4: 336-340 (2001).

Tas AC, van der Greef J. Mass spectrometric profiling and pattern recognition. *Mass Spectrom Rev* 13: 155-181 (1994).

ter Kuile BH, Westerhoff HV. Transcriptome meets metabolome: hierarchical and metabolic regulation of the glycolytic pathway. *FEBS Lett* 500: 169-171 (2001).

Terabe S, Markuszewski MJ, Inoue N *et al.* Capillary electrophoretic techniques toward the metabolome analysis. *Pure Appl Chem* 73: 1563-1572 (2001).

Timperman AT, Aebersold R. Peptide electroextraction for direct coupling of in-gel digests with capillary LC-MS/MS for protein identification and sequencing. *Anal Chem* 72: 4115-4121 (2000).

Tweeddale H, Notley-McRobb L, Ferenci T. Effect of slow growth on metabolism of *Escherichia coli*, as revealed by global metabolite pool ("metabolome") analysis. *J Bacteriol* 180: 5109-5116 (1998).

Tweeddale H, Notley-McRobb L, Ferenci T. Assessing the effect of reactive oxygen species on *Escherichia coli* using a metabolome approach. *Redox Rep* 4: 237-241 (1999).

Vaidyanathan S, Rowland JJ, Kell DB, Goodacre R. Discrimination of aerobic endospore-forming bacteria *via* electrospray-ionization mass spectrometry of whole cell suspensions. *Anal Chem* 73: 4134-4144 (2001).

Vaidyanathan S, Kell DB, Goodacre R. Flow-injection electrospray ionization mass spectrometry of crude cell extracts for high-throughput bacterial identification. *J Am Soc Mass Spectrom* 13: 118-128 (2002a).

Vaidyanathan S, Winder CL, Wade SC *et al.* Sample preparation in matrix-assisted laser desorption ionization mass spectrometry of whole bacterial cells and the detection of high mass (> 20 kDa) proteins. *Rapid Comm Mass Spectrom* 16: 1276-1286 (2002b).

van Baar BLM. Characterization of bacteria by matrix-assisted laser desorption/ionisation and electrospray mass spectrometry. *FEMS Microbiol Rev* 24: 193-219 (2000).

Varma A, Boesch BW, Palsson BO. Biochemical production capabilities of *Escherichia-coli. Biotechnol Bioeng* 42: 59-73 (1993).

Wang ZP, Dunlop K, Long SR, Li L. Mass spectrometric methods for generation of protein mass database used for bacterial identification. *Anal Chem* 74: 3174-3182 (2002).

Washburn MP, Wolters D, Yates JR. Large-scale analysis of the yeast proteome by multidimensional protein identification technology. *Nature Biotechnol* 19: 242-247 (2001).

Wilkins MR, Gasteiger E, Gooley AA *et al.* High-throughput mass spectrometric discovery of protein post- translational modifications. *J Mol Biol* 289: 645-657 (1999).

Wilkins MR, Pasquali C, Appel RD *et al.* From proteins to proteomes: large scale protein identification by two-dimensional electrophoresis and amino acid analysis. *Biotechnology* 14: 61-65 (1996).

Wittmann C, Heinzle E. Application of MALDI-TOF MS to lysine-producing *Corynebacterium glutamicum* - A novel approach for metabolic flux analysis. *Eur J Biochem* 268: 2441-2455 (2001).

Wunschel DS, Fox KF, Fox A *et al.* Quantitative analysis of neutral and acidic sugars in whole bacterial cell hydrolysates using high-performance anion-exchange liquid chromatography electrospray ionization tandem mass spectrometry. *J Chromatogr* 776: 205-219 (1997).

Xiang F, Anderson GA, Veenstra TD *et al.* Characterization of microorganisms and biomarker development from global ESI-MS/MS analysis of cell lysates. *Anal Chem* 72: 2475-2481 (2000).

Yates JR. Mass spectrometry - from genomics to proteomics. *Trends Genet* 16: 5-8 (2000).

Zhu H, Klemic JF, Chang S *et al.* Analysis of yeast protein kinases using protein chips. *Nature Genet* 26: 283-289 (2000).

Chapter 3

CAN ^{1}H NMR DERIVED METABOLIC PROFILES CONTRIBUTE TO PROTEOMIC ANALYSES?

A Study on Duchenne Muscular Dystrophy

Julian L. Griffin[1] and Elizabeth Sang[2]
[1]Division of Biomedical Sciences, Imperial College of Science, Technology and Medicine, London, SW7 2A, UK [2]Department of Biochemistry, Oxford University, Oxford, OX1 3QV, UK

1. INTRODUCTION

In the post genomic era, moving from the mapping of the genome of an organism to understanding its functional genomics is a massive undertaking. Many genes are not under translational control, suggesting that an approach based solely on transcriptomics is inadequate. Furthermore, analysis of the mammalian proteome remains an immense technical challenge. ^{1}H nuclear magnetic resonance (NMR) derived metabolic profiles, however can define the phenotype of a pathology or genetic modification (Oliver, 2001; Gavaghan *et al.,* 2001; Raamsdonk *et al.*, 2001; Nicholson *et al.*, 2002) providing a simple and new pathway into functional genomics.

In this chapter we describe one such investigation into a mouse model of Duchenne Muscular Dystrophy (DMD). In humans, this disease is an X-linked recessive disorder, affecting 1 in ~3500 male births, and characterized by progressive muscle wasting followed by death (Duchenne, 1868; Emery, 1989). DMD is caused by a failure to express dystrophin, a protein that cross links actin to the sarcolemma membrane (Hoffman *et al.*, 1987; Ahn and Kunkel, 1993). While the DMD gene and the dystrophin protein have been characterized, the purpose of the various isoforms of dystrophin found in non-muscle tissue as well as the truncated forms of the protein found throughout the body are not understood (Nudel *et al.*, 1989; Lederfein *et al.*, 1992), illustrating the difficulty with moving from the genome to an understanding of the proteome. In this chapter we demonstrate that in the mdx mouse, a model of DMD (Bullfield *et al.*, 1984), dystrophic tissue can readily be separated from control tissue using ^{1}H NMR derived metabolic profiles (Griffin *et al.,* 2001a, 2001b). Furthermore, we demonstrate how these metabolic profiles can be used to understand the interactions of dystrophin and utrophin, another key protein associated with the disease. Finally we describe how this information may be used to treat this devastating disease.

1.1 Muscular Dystrophies and DMD

The muscular dystrophies are a group of inherited disorders characterized by variable degrees of progressive weakness due to muscle wasting. The histological features of dystrophic muscle include variation in fiber size, focal areas of necrosis, regenerating fibers with central nuclei, invasion of macrophages and replacement of muscle tissue by fat and connective tissue. DMD is the most common form of dystrophy and the associated muscle wasting is progressive and severe. Patients are often wheelchair-bound by their early teens and die in their twenties, usually from respiratory or cardiac failure (Emery, 1989).

In 1982, the gene for DMD was isolated and localized to Xp21. The gene was subsequently cloned and sequenced (Murray *et al.*, 1982; Monaco *et al.*, 1986; Burghes *et al.*, 1987). It is the largest known gene, consisting of 2.4 million base pairs. It is composed of at least 85 exons (which have been conserved throughout evolution) and introns make up 98% of the gene. The gene encodes many different tissue-specific dystrophin isoforms generated by use of alternative promoters and differential splicing (Feener *et al.*, 1989, Ahn and Kunkel, 1993). In 1987, the product of the gene was identified as dystrophin (Hoffman *et al.,* 1987).

Dystrophin is a member of the β-spectrin/α-actinin protein family, a family characterized by an NH_2-terminal actin-binding domain followed by a variable number of spectrin-like repeats. It is a large 427 kDa cytoskeletal protein localized to the cytoplasmic face of the sarcolemma in muscle cells. Dystrophin is organized into distinct structural domains including an NH_2-terminal actin-binding domain, a central rod-like domain composed of 24 spectrin-like repeats interrupted by hinge regions, a cysteine-rich domain, and a COOH-terminal domain (Hoffman *et al.*, 1987). In skeletal muscle, dystrophin associates with a large multisubunit complex of membrane glycoproteins, collectively called the dystrophin-glycoprotein complex (DGC). The DGC spans the sarcolemma of skeletal and cardiac muscle providing a link between the actin cytoskeleton and the extracellular matrix (ECM) (Ervasti *et al.*, 1990; Campbell *et al.*, 1995). As well as the DGC, there are two further groups of dystrophin associated proteins, the intracellular syntrophins and dystrobrevins, and the transmembrane sarcoglycan-sarcospan complex. Lack of dystrophin has been shown to lead to a reduction in the levels of all the other DGC components (Ervasti *et al.*, 1990; Ohlendieck and Campbell, 1991; Campbell *et al.*, 1995). With loss of the DGC, and therefore the cytoskeleton-ECM link, the muscle-cell membrane is compromised (Petrof *et al.*, 1993).

Most DMD mutations introduce premature stop codons which disrupt the reading frame resulting in very little or complete loss of dystrophin whereas patients with Becker Muscular Dystrophy, a less severe form muscular dystrophy, have mutations which result in reduced expression of dystrophin or generate only a truncated partially functional protein (Monaco *et al.*, 1988).

The mdx mouse (Bulfield *et al.*, 1984; Sicinnski *et al.*, 1989) has been extensively used in research as a model for DMD. Mdx mice show many features of DMD but at later times in their relative life span when compared to DMD patients.

1.2 Dystrophin Function

Despite increased knowledge of the pathogenesis of muscular dystrophies, controversies still remain about the function of dystrophin. As dystrophin is part of the complex of proteins providing a membrane cytoskeleton-ECM link, it is hypothesized to play a role in stabilizing the sarcolemma and maintaining muscle fiber integrity during muscle contraction (Petrof *et al.*, 1993; Weller *et al.*, 1990; Pasternak *et al.*, 1995; Menke *et al.*, 1995). Mdx mouse skeletal muscle has a disrupted membrane skeleton with regions of the sarcolemma devoid of costameres (Williams *et al.*, 1999] and cytoskeletal γ-actin has been found not to be stably associated with the sarcolemma (Rybakova *et al.*, 2000). The structure of dystrophin-deficient muscle is therefore compromised, which may leave the sarcolemma susceptible to damage. Dystrophic muscle is also vulnerable to increased transient micro-disruptions (Petrof *et al.*, 1993) which may lead to elevations in cytoplasmic free calcium.

Dystrophin may also participate in the regulation of intracellular calcium levels though there is much conflicting data in this area (reviewed in Blake *et al.*, 2002). There is some evidence that increased activity of the calcium leak channel in dystrophic muscle (Fong *et al.*, 1990) may lead to elevated levels of proteases in dystrophic tissue.

Although dystrophin is important in the maintenance of the membrane cytoskeleton-ECM link, this is not its sole function. Additional roles include modulating force and signal transduction processes, and aggregation of neurotransmitter receptors. Mutations in α-dystrobrevin have no effects on sarcolemmal structure and instead seem to cause muscular dystrophy by altering signaling (Grady *et al.*, 1999). Another signaling molecule of increasing interest to those studying muscular dystrophy is the vasodilator nitric oxide. In normal muscle, neuronal nitric oxide synthase (nNOS) associates with syntrophin in the DGC. However, nNOS has been shown to be absent from the skeletal-muscle sarcolemma of DMD patients and mdx mice and instead has a cytoplasmic localization (Brenman *et al.*, 1995). In mdx mice (Thomas *et al.*, 1998) and DMD patients (Sander *et al.*, 2000) the absence of nNOS at the sarcolemma may contribute to abnormal blood vessel constriction during muscle contraction thereby contributing to disease progression.

2. METHODS FOR ^{1}H NMR SPECTROSCOPIC DERIVED METABOLIC PROFILES

2.1 NMR Spectroscopic Techniques

Systemic metabolism can effectively be investigated using biofluid NMR spectroscopy (Nicholson *et al.*, 1995; Holmes *et al.*, 1998; Beckwith-Hall *et al.*, 1998), a topic discussed elsewhere in this book (Breau and Cantor, Chapter 4). However, in this study the primary focus has been the impact of dystrophin on tissue metabolic profiles in specific organs. The most convenient manner to produce high resolution ^{1}H NMR spectroscopic data from tissues is to prepare extracts, and a number of processes are commonly employed, including perchloric acid, acetonitrile/water and water/methanol/ chloroform extractions where in each case metabolites are released from tissue by mechanical agitation and chemical action (Belle *et al.*, 2002). Typically, ~100 mg wet weight tissue is extracted into 1 mL of extraction media, and following lyophilization, reconstituted in 0.5 mL of D_2O. These extracts can then be investigated using conventional NMR pulse sequences in the presence of solvent suppression. Provided the extracts produced by one extraction procedure are examined using the same pulse sequence with identical parameters, a spectral profile can be generated for that tissue, with relative intensities being proportional to the absolute concentration of a given metabolite within the sample. With recent advances in flow probe technology these extracts can be quickly examined under fully automated conditions, allowing the rapid phenotyping of a large number of tissues.

However, these extracts are biased by the very nature of the chosen extraction procedure. Metabolite concentrations in a given extraction medium often represent their relative solubilities within the solvent rather than true tissue content. Furthermore, examination of both lipophilic and aqueous metabolites requires multiple extractions. Even with these multiple extraction procedures a pellet remains, which can be shown to contain both aqueous and lipid metabolites (Millis *et al.*, 1997; Griffin *et al.*, 2001c). So, in neural tissue can all the glutamate present be extracted by an aqueous extraction procedure when a significant amount of the neurotransmitter is contained within lipid vesicles?

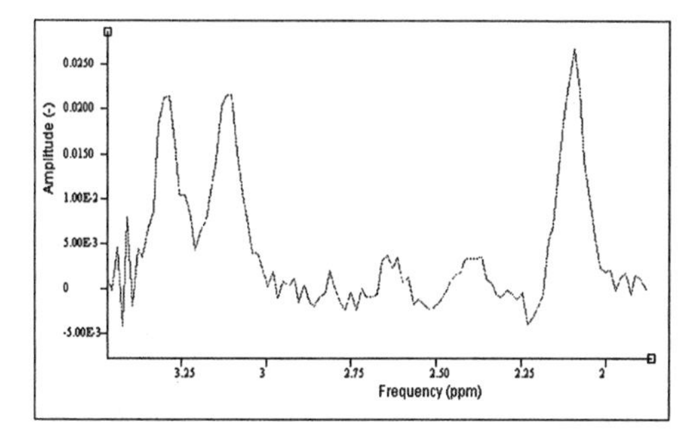

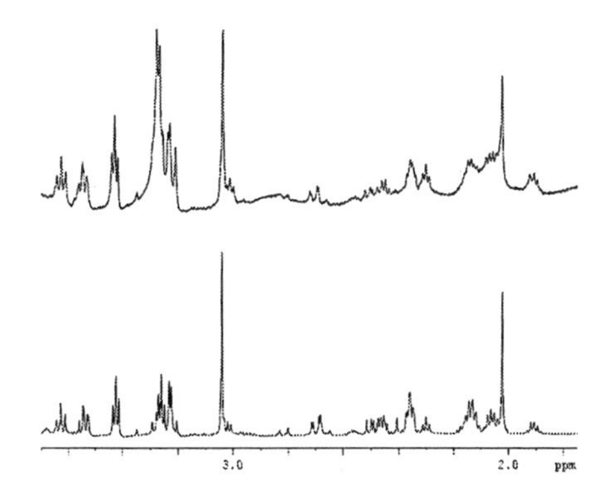

Figure 1. The improvement in spectral resolution obtainable using HRMAS ^{1}H NMR spectroscopy. Three spectra are shown (top) spectrum obtained *in vivo* from a region of human brain (courtesy of Dr. Caroline Rae, Department of Biochemistry, Sydney University, Australia), (middle) a HRMAS ^{1}H NMR spectrum of 10 mg of mouse brain cortex recorded at 600 MHz and (bottom) a spectrum of tissue extract from 100 mg of mouse brain cortex recorded at 600 MHz.

There is thus a need to examine tissue directly. However, intact tissue spectroscopy is impaired by a number of physical processes which broaden resonance line widths in ^{1}H NMR spectra. These include bulk magnetic susceptibility effects, dipole-dipole couplings and chemical shift anisotropy, with the last two effects being scaled by a $(3\cos^2\theta-1)$ term. Their effects on spectral resonances can be reduced dramatically by spinning the sample at the "magic angle" (54.7°) (Andrew *et al.*, 1959). The spectra obtainable from intact tissue using such magic angle spinning (MAS) NMR techniques are comparable in resolution to those obtainable using tissue extracts, especially for soft tissues such as those of the brain. (Fig. 1; Cheng *et al.*, 1998; Garrod *et al.*, 1999), testes (Griffin *et al.*, 2000), prostate (Tomlins *et al.*, 1998) and from intact cells (Weybright *et al.*, 1998; Griffin *et al.*, 2001d).

The preparation of samples for high resolution (HR) MAS ^{1}H NMR spectroscopy is relatively simple. When used in conjunction with 4 mm diameter zirconium oxide rotors and Teflon spacers tissue samples in the region of 5-15 mg wet weight are needed (Waters *et al.*, 2001). Samples are soaked in D_2O, acting as a field frequency lock, containing an optional trace of 3-trimethylsilyl[2,2,3,3-$^{2}H_4$] propionate (TSP) to act as a chemical shift reference and a defined peak resonance for shimming purposes, prior to insertion into the rotor alongside a further 5 μL of D_2O, spiked with TSP. The Teflon spacer allows sample packing within the sensitive region of the rotor, further improving spectral quality (Fig. 2). As a rule of thumb samples are spun at spin speeds equal to the spectral width of the sample. This ensures that the spinning side bands associated with the spectra are outside the region of interest. However, with increasing spin rate the samples may suffer degradation, especially for softer tissues or cells, as centripetal force increases with the square of spin speed. This has led to a number of research groups developing pulse sequences, such as TOSS and PASS, for use with HRMAS ^{1}H NMR spectroscopy to minimize tissue degradation (Hu *et al.*, 2002). Furthermore, to minimize enzymatic degradation samples, can be chilled to within a few degrees of freezing during the acquisition of spectra.

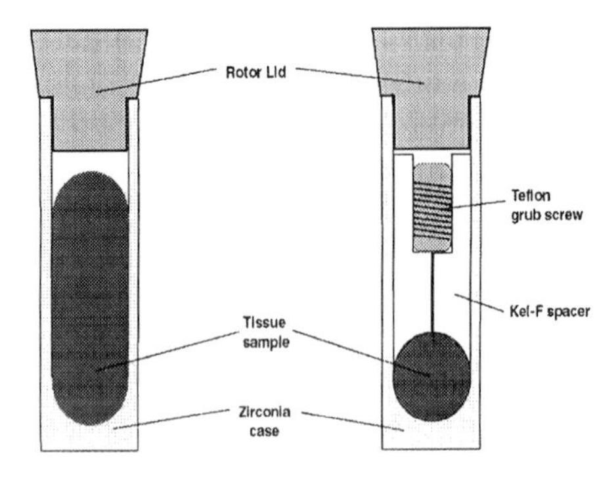

Figure 2. A schematic diagram showing sample position in a HRMAS NMR spectroscopy zirconium oxide rotor with and without Teflon spacer (Kel-F). Figure provided by Dr. N. Waters (PhD Thesis, Imperial College of Science Technology and Medicine, 2001).

Fats within restricted environments such as the cell membrane are subjected to dipolar couplings far greater than those removed by the modest spin speeds used in HRMAS ^{1}H NMR (Siminovitch *et al.*, 1988), and thus, the lipids observed using HRMAS ^{1}H NMR are relatively mobile (rotationally, if not translationally). These lipids still consist of a large number of different chemical moieties, and ^{1}H NMR spectroscopy detects the different chemical groups present within lipid compounds rather than returning a single resonance for each lipid as a whole. Thus, broad resonances associated with these lipids obscure large regions of spectra from lipid rich tissues, such as skeletal muscle and adipose and cardiac tissue (Millis *et al.*, 1997; 1999). However, to detect low molecular weight metabolites, edited pulse sequences can be used to tailor sensitivity to a particular compound class or to further separate spectral resonances in another dimension of the NMR experiment (Griffin *et al.*, 2001b).

Many of the 2-D pulse sequences developed to analyze complex mixtures or identify unknown metabolites can be used to separate resonances using effects arising from the NMR experiment directly. ^{1}H-^{1}H Correlation Spectroscopy (COSY), Total Correlation Spectroscopy (TOCSY) and J Resolved Spectroscopy (JRES) separate resonances in two dimensions courtesy of the couplings that exists between neighboring ^{1}H nuclei (Aue *et al.*, 1976; Nagayama *et al.*, 1980). Heteronuclear pulse sequences can also be employed, making use of coupling between ^{1}H nuclei and neighboring ^{13}C nuclei, the ^{13}C nucleus having a far greater chemical shift range than ^{1}H.

Alternatively, the cellular environment may be used as a contrast mechanism. Metabolites within restricted environments have short longitudinal (T_2) relaxation times, resulting in broad resonances in the spectra. However, if a delay is placed in the pulse sequence so that the signals from these restricted metabolites have decayed to an insignificant level, the signals from the unrestricted metabolites can be more readily detected. One pulse sequence that utilizes this procedure is the Carr-Purcell-Meiboom-Gill (CPMG) sequence, where magnetization is trapped in the XY plane by a train of 180° pulses while the signal from the broad resonances decays at a relatively faster rate than those from more mobile metabolites. Alternatively, the mobile metabolites may be selectively removed from the spectra by applying two equal but opposite signed field gradients along the sample (Wu *et al.*, 1995). Slow moving metabolites will experience a gradient of G after the first gradient and then negative G after the second, resulting in no net magnetization. However, the field gradients applied will not cancel out if the metabolite moves rapidly. The attenuation of a resonance intensity is given by

$$A_g = A_0.\exp[-g^2\delta^2\gamma^2D(\Delta-\delta/3-\tau/2]$$

where A is the signal intensity at gradient strength g, δ is the time the gradient is applied for, γ is the gyromagnetic ratio of the nucleus and Δ is the total diffusion time (Fig. 3).

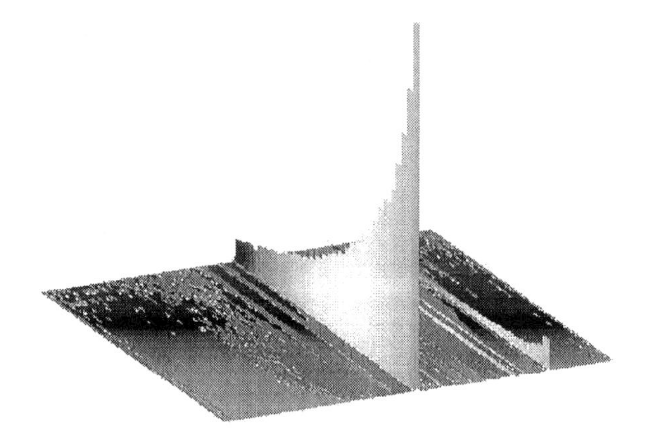

Figure 3. An oblique projection of the relative attenuation of water in a tissue sample during a Diffusion Ordered Spectroscopy (DOSY) experiment. As the field gradient strength is increased, the water resonance is attenuated to a greater extent. Displaying a series of these experiments in order of field gradient strength produces a distinctive exponential. From this decay the apparent diffusion coefficient can be calculated. Alternatively spectra can be acquired at one field gradient strength optimized to select for a particular class of compounds.

Thus, data sets can be produced that are dependent on the cellular environment of the metabolites. In this respect NMR spectroscopy has a huge intrinsic advantage over mass spectrometry for deriving metabolic profiles. The cellular environment can often dictate the role a metabolite has within a tissue; for example glutamate within the brain can be a neurotransmitter if found within vesicles, involved in energy generation by controlling flux across the mitochondrial membrane, or be excitotoxic when extracellular.

2.2 Data Reduction, Normalization, and Scaling

The pre-processing of NMR data sets is an integral part to the approach of deriving NMR based metabolic profiles (Beckwith-Hall *et al.*, 1998). These routines can be separated into those used for convenience or to simplify the problem examined, and those routines used to improve the diagnostic capabilities of the pattern recognition. Examples of the former processes are data reduction and normalization. High resolution spectra typically contain between 8k and 64k data points and in order to handle these data sets rapidly during computer based pattern recognition it is useful to reduce the data set in size. This can be carried out by separating the spectra into multiple integral regions (hixels or buckets) and carrying out automated integration. While some spectral information is lost, in particular coupling patterns, these simplified spectral representations are less susceptible to changes associated with pH or ionic strength which affect either biofluids or intact tissue, especially for resonances such as citrate, malate and oxaloacetate. In the following analyses all spectral profiles were produced using a bucket width of 0.04 ppm. It is also often convenient to exclude areas of the spectra where there are large degrees of variation associated with physical processes involved in the acquisition of the spectra. One such example is the water saturation region which can be variable and holds no metabolic information.

To remove concentration effects the data set can also be "normalized" so that each bucket region is represented as a ratio to the total integral region, an internal standard or a particular metabolite (including water in a non solvent suppressed spectrum). This can be particularly useful for solid state spectra where it is difficult to estimate the total tissue present within the sensitive region of the probe. Metabolic profiles derived by NMR are also subjected to a dynamic range problem (as well as an intrinsic dynamic range associated with the spectrometer). A high concentration metabolite will have a large numerical variance compared with a low concentration metabolite, although the fold changes may be greater for the lower concentration metabolite. Thus, the spectral regions must be scaled in order to maximize the information obtainable from both high and low concentration metabolites. In the subsequent analyses presented here three scalings have been used. All the data has been "mean centered", where the average of all the spectra are subtracted from each individual spectrum. Variance is expressed with respect to the mean spectrum. Using this scaling alone, high concentration metabolites contribute most significantly to the description. To increase the representation of low concentration metabolites the variance of each integral region can be divided by the variance of the integral region, standardizing the variance for each integral region (referred to as unit variant preprocessing in this chapter). Alternatively, intermediate scalings can be used to interrogate the data set further; for example Pareto scaling where the variance is divided by the square root of variance (Eriksson *et al.*, 1998). In this manner both high concentration and trace metabolites can be identified by subsequent pattern recognition.

2.3 Pattern Recognition Techniques for Deriving Metabolic Profiles

To investigate the innate variation in a data set principal component analysis (PCA) has been applied. However, where a specific question is being posed, the supervised technique Prediction To Latent Structures through partial least squares (PLS) has been applied (Martens and Næs , 1989; Eriksson *et al.*, 1998). This technique can also be used as a means of data filtering, referred to as Orthogonal Signal Correction (OSC). Variation that is orthogonal to the trend of interest is removed using PLS. For the PLS models presented, the goodness of fit algorithm was used to determine whether a correlation was significant (Q^2>0.097). This algorithm was also used to determine how many PLS components were used in the model. Metabolic perturbations caused by a failure to express a protein were determined from loadings plots indicating chemical shift regions. Spectral regions having a modulus loadings score of greater than 50% of the maximum loading value for that model were identified as having the most significant changes.

3. APPLICATION OF NMR SPECTROSCOPY TO A MULTI-TISSUE DESCRIPTION OF DMD

3.1 PCA in Descriptions of DMD

The genetic basis of DMD is well documented, but there are still questions remaining as to the role of the various proteins produced by the gene, demonstrating the problem with moving from a genomic to a proteomic description of a disease. There are three full length isoforms of dystrophin: M-dystrophin found in muscle and, to a small extent, glial cells, P-dystrophin found in the Purkinje cells of the cerebellum and C-dystrophin found in the cerebral cortex (Nudel *et al.*, 1989). Furthermore, there are a number of truncated forms of dystrophin (Lederfein *et al.*, 1992). While previous studies have noted metabolic perturbations in both dystrophic muscle and cerebral tissue (Kemp *et al.*, 1993; Even *et al.*, 1994; Tracey *et al.*, 1995; Mokhatarian *et al.*, 1996; McIntosh *et al.*, 1998a) these had largely been confined to one tissue type or organ and focused only on a handful of metabolites that were predicted to be affected by the disease.

Using conventional solvent suppressed ^{1}H NMR spectra (Fig. 4) metabolic profiles can be quickly established in extracts of various dystrophic tissues using PCA (Griffin *et al.*, 2001a). Using a mean centered data set, dystrophic muscle tissue from cardiac tissue, the diaphragm and the soleus appear to be similarly perturbed metabolically by a failure to express dystrophin, characterized by an increase in taurine and lactate alongside a decrease in creatine in tissue extracts from dystrophic tissue (Table 1). This is in keeping with the single protein isoform known to be involved in all these tissue types. These metabolic perturbations were augmented by others when the data set was autoscaled prior to PCA, demonstrating that even a simple protein loss produced a multifactorial response in the metabolome (Table 1).

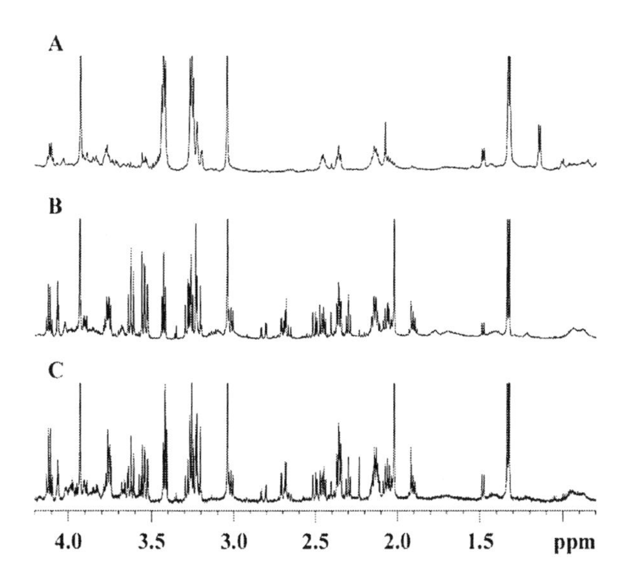

Figure 4. Typical high resolution ^{1}H NMR spectra of cardiac (A), cerebellar (B) and cerebral cortex (C) tissue extracts. Spectra were acquired at 600 MHz using a flow probe system coupled to an automated sample robot. Spectra from muscle tissue could visually be separated from cerebral tissues by eye but PCA was required to separate dystrophic and control tissue for the individual regions. After conversion of these spectra to a numerical format representing resonance areas and chemical shifts, the spectra formed the basis of the data matrices used in subsequent analyses.

Perhaps the most significant of all the metabolic changes is the increase in taurine detected in all muscle tissues. Previously, McIntosh and co-workers (1998a, 1998b) correlated muscle taurine content to regeneration in the mdx, MyoD and mdx:MyoD cross mice. Using *in situ* hybridization and autoradiography the transcription of key genes (myogenin and myf5) involved in muscle cell proliferation was monitored and correlated to the tissue content of taurine, with mdx tissue having the highest concentration of muscle taurine as well as the most number of muscle proliferative cells for the three mouse types. Although the increased taurine content appears beneficial to the muscle tissue, the metabolite has a number of diverse metabolic roles including osmolyte, cell membrane stabilizer, Ca^{2+} homeostatin and neurotransmitter (for a review see Huxtable, 1992), and it remains to be seen in dystrophic muscle cells what role the increased taurine plays.

Table 1. Metabolites Identified by PCA as Perturbed in Key Tissue Types. Cumulative total scores represent the combined percentage variance between the spectra from dystrophic and control tissue. + indicates an increase, - indicates a relative decrease.

Tissue	Processing	PC Separation	Score	Key Metabolites
Cardiac	Mean centered	PC2	19.4	lactate (+) β-hydroxybutyrate (+) taurine (+) creatine (-) glutamate (-)
	Univariant	PC2 PC5	24.2	leucine (+) isoleucine (+) valine (+) β-hydroxybutyrate (+) glycine (+) taurine (+) alanine (-) glutamate (-)
Diaphragm	Mean centered	PC2 PC4	32.4	taurine (+) lactate (+) creatine (-)
	Univariant	PC1 PC3	62.6	lactate (+) glutamate (+) taurine (+) hydroxyl *n*-valerate (+) glucose (-) valine (-)
Soleus	Mean centered	PC3 PC4	8.9	creatine (-) taurine (+)
	Univariant	PC2 PC3	40	creatine (-) taurine (+) lactate (-)
Cerebral Cortex	Mean centered	PC2 PC5	18.2	lactate (+) phosphocholine (+) taurine (-) myo-inositol (-)
	Univariant	PC1 PC2	63.6	lactate (+) glutamate (+) isoleucine (+) creatine (-) myo-inositol (-)
Cerebellum	Mean centered	PC3	7.3	creatine (-) glycine (-) lactate (-) *N*-acetyl aspartate (+) glutamate (+)
	Univariant	PC2 PC3	17.2	glutamate (+) leucine (+) isoleucine (+) valine (+) creatine (-) glycine (-) myo-inositol (-)

Metabolic abnormalities in cerebral tissue have previously been characterized in both sufferers of DMD and the mdx mouse using either direct chemical assay or ^{1}H, ^{31}P and ^{13}C NMR spectroscopy (Tracey *et al.*, 1996; Kato *et al.*, 1997; Rae *et al.*, 1998), and differences between the mouse types were readily apparent in the extracts of cerebellar tissue in our studies. However, separation was only detectable in cortical tissue for relatively low PCs, especially when using mean centered preprocessing, representing a small amount of the total metabolic variation in the tissue (Fig. 5).

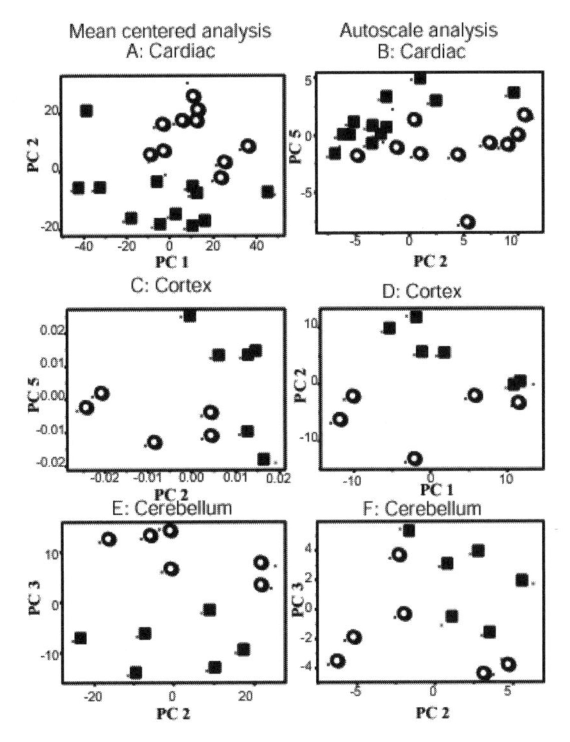

Figure 5. PCA of ^{1}H NMR spectra of extracts from cardiac, cerebral cortex and cerebellum tissue using either mean centered or univariant (autoscaled) preprocessing. Each symbol represents a NMR spectrum reduced to PC space. The number of PCs used was determined by a goodness of fit routine designed to prevent overfitting of the model. The scale indicates the loadings of each spectrum along that particular PC. **Key:** ■ dystrophic tissue ❍ control tissue.

While the cerebellum appears markedly affected by a lack of dystrophin, there is still much debate as to the role of C-dystrophin in cortical tissue (reviewed in Anderson *et al.*, 2002). The different metabolic profiles produced by a failure to express dystrophin in cerebral and cerebellar tissue are most likely caused by the proposed role of the isoforms of dystrophin as an anchorage motif for neurotransmitter receptors (Anderson *et al.*, 2002). With different regions of the brain having different proportions of neurotransmitter receptors this would be expected to vary greatly across the brain, and indeed within any one region of the brain. Investigations failed to find a distinct metabolic fingerprint for dystrophic cerebellar tissue using HRMAS ^{1}H NMR spectroscopy of the intact tissue, in part caused by the profound metabolic variation of cerebellar tissue across the structure when sampling at a 10 mg level rather than using the whole region as required for a tissue extract (Nicholls, Griffin and Mortishire-Smith, unpublished data]. Given the increasing miniaturization of NMR spectroscopy it remains to be seen whether a regio-metabolic profile of the cerebellum may deduce the role of P-dystrophin within the cerebellum.

3.2 Metabolic Profiles from Intact Tissue

One of the principal benefits that ^{1}H NMR spectroscopy has over mass spectrometry is that it is a potentially non-invasive technique, avoiding any sample destruction. With the advent of HRMAS ^{1}H NMR spectroscopy it is now possible to derive metabolic profiles from high resolution spectra without any sample preparation, using a biopsy sized piece of material. However, for muscle tissue the spectra are often dominated by broad lipid resonances, caused by both restricted rotational movement and the number of different lipids represented by the resonances. NMR spectroscopy detects lipids *via* the chemical groups contained within them and hence the resonance associated with saturated hydrocarbon groups will contain a number of metabolites. Thus, to reap all the information obtainable using HRMAS ^{1}H NMR spectroscopy it is necessary to use a range of pulse sequences to both increase sensitivity to non lipid metabolites and also probe the molecular environment differences between dystrophic and control tissue.

Using a conventional water presaturation pulse sequence, spectra are produced that are dominated by -CH_2**CH_2**CH_2- and **CH_3**CH_2- lipid moieties (Fig. 6). These same lipid moieties readily classified dystrophic and control cardiac tissue using PCA, in keeping with the known progression of dystrophic muscle degeneration (Gillet *et al.*, 1993). Waves of muscle necrosis are accompanied by extracellular lipid infiltration, and it is this increased lipid content that is detected by simple solvent suppression HRMAS ^{1}H NMR pulse sequences. To further investigate the lipid resonances, PCA was applied to spectra acquired with either a diffusion edited pulse sequence with large magnetic field gradients, to attenuate mobile metabolites, or a T_2 edited pulse sequence with a 40 ms delay, attenuating resonances relatively more from metabolites within restricted environments (Fig. 6). While lipid moieties still separated the two tissue types, intriguingly the chemical groups responsible for this separation were different. Dystrophic tissue contained more rotationally free lipids with CH_2**CH_2**CH_2, **CH_3**CH_2, CH=CH**CH_2**- and $COCH_2$**CH_2**- chemical groups and slow moving lipids with saturated fats containing relatively less $COCH_2$**CH_2** and **CH_3**CH_2 chemical groups.

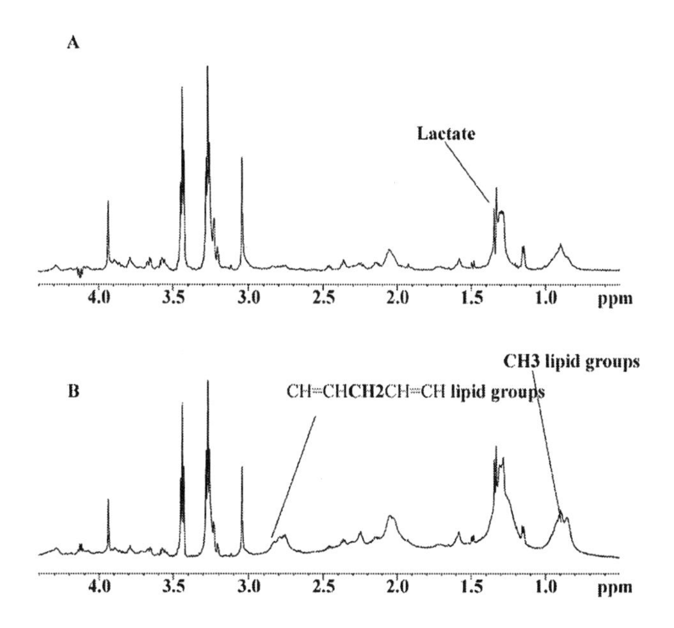

Figure 6. HRMAS ^{1}H NMR spectra of cardiac tissue from the mdx mouse. The spectra were acquired from the same piece of cardiac tissue using either a CPMG pulse sequence with a total T_2 spin echo delay of 40 ms (**A**) or a conventional solvent suppressed pulse sequence (based on the start of the 2 dimensional NOESY sequence) (**B**). The spin echo delay attenuates the signal from all resonances, but affects motionally constrained metabolites more. In the cardiac spectra the spin echo delay reduces the resonances associated with a number of broad lipid components, and in particular those associated with $\mathbf{CH_3}CH_2$ and $CH{=}CH\mathbf{CH_2}CH{=}CH$ lipid groups. However, the resonances associated with the methyl group of lactate are more pronounced in these spectra.

However, from the analysis of aqueous extracts it was clear that the concentrations of a number of low molecular metabolites were also perturbed. To examine this class of compounds *in situ* there are two different techniques to address this problem; using higher dimensional pulse sequences to separate resonances that are co-resonant in one dimensional pulse sequences or using the T_2 edited CPMG pulse sequences with longer T_2 delays. Using the former approach, PCA applied to COSY spectra also readily separated control and dystrophic tissue, even using large integral regions of 0.4 ppm by 0.4 ppm to limit the number of variables inspected by pattern recognition (Griffin *et al.*, 2001b). With ever increasing improvements in terms of both probe sensitivity and magnetic field strength, the time requirements of such two dimensional pulse sequences will decrease, allowing the detection of more low molecular weight metabolites directly from intact tissue. Both these effects will warrant a decrease in integral region sizes and if COSY spectra are interrogated at the same resolution as the one dimensional pulse sequences, this will require the use of data vectors of 64,000 variables.

3.3 PLS in Metabolic Profile Driven Proteomics

While DMD is not a simple metabolic disorder, and many of the metabolic perturbations detected using PCA describe the metabolic consequences of a failure to express dystrophin rather than the primary deficit itself, there is potentially a wealth of information concerning the protein interactions of dystrophin, the dystrophin related proteins and the dystrophin-glycoprotein complex (DGC) complex contained within the metabolic profiles. Furthermore, the mouse models of the muscular dystrophies are amongst the best characterized mammalian group of disorders (Tinsley *et al.*, 1996, 1998; Deconinck *et al.*, 1997). Thus, functional proteomics of the dystrophin family of proteins can be investigated by judicious selection of mouse models. For this purpose a pattern recognition technique that does more than simply classify data is required so as to correlate metabolic changes with protein expression. PLS (Eriksson *et al.*, 1998) allows such a series of correlations for extra variables to a data set, allowing the correlation of protein expression with metabolic profiles.

Utrophin is functionally related to dystrophin, and both proteins possess binding sites for actin and the DGC, albeit through different motifs (Rybakova *et al.*, 2002), effectively bridging the sarcolemmal membrane. During fetal development utrophin is found over the entire surface of muscle fibers, but is replaced by dystrophin during development, when utrophin becomes localized to the neuromuscular junction in adult skeletal muscle (Hoffman *et al.*, 1989; Khurana *et al.*, 1990; Khurana *et al.*, 1991). The dystrophic phenotype normally observed in mdx mice is absent when muscles overexpress utrophin (Tinsley *et al.*, 1996, 1998; Deconinck *et al.*, 1997). Furthermore, utrophin has a particularly favorable characteristic in dystrophic tissue; while the body may treat dystrophin as a foreign protein mounting an immune response, increasing the innate expression of utrophin does not have this pitfall (Burton and Davies, 2002). For this reason upregulation of utrophin expression in muscle is a favored gene therapy approach for treating DMD.

To investigate whether metabolic profiles could be used to predict the proteomics of a disease or therapy regime, the ^{1}H NMR derived metabolic profiles of cardiac and diaphragm tissue from four different mouse models were examined; (i) the mdx mouse which does not express dystrophin (ii) the $Tg_{full\ length}/Dmd^{mdx}$ transgenic mouse, termed Fiona, which expresses full-length utrophin in skeletal muscle but not heart, crossed with the mdx mouse to produce a mouse lacking dystrophin but having utrophin localized at the sarcolemma (Tinsley et al., 1998) (iii) the $Tg/Dmd^{mdx};utrn^{-/-}$ transgenic mouse, termed Gavin, which expresses a truncated utrophin transgene crossed onto a $Dmd^{mdx}/utrn^{-/-}$ double mutant background, resulting in a mouse with no dystrophin in skeletal muscle but with a truncated utrophin transgene and heart with no dystrophin or utrophin (Rafael *et al.*, 1998) (iv) the control mouse to (iii), $Tg_{truncated}/Dmd^{mdx}$, termed Nigel, which has no dystrophin but expresses utrophin (Table 2).

To avoid any age related changes being superimposed on the model, animals were studied at six months of age. This limited the number of mice available, especially Gavin mice, with only two becoming available during the study. However, a major benefit of PLS is that it models X-variable changes in terms of variations in Y, and so long as the two Gavin mice did not express "protein scores" vastly different from the other mouse models they could be included within the data set. In practical terms this allowed their inclusion in a PLS model of the effects of dystrophin on cardiac metabolism. Each NMR integral region represented an X-variable in the PLS model, while protein expression was represented as a Y vector. For both tissue types presence and absence of dystrophin was represented as 1 or 0 (dystrophin expression or dystrophin non-expression). This is the one component model (section 3.4). For diaphragm tissue utrophin expression was represented by a second Y variable as 1 (normal expression level) or 2 (promoted expression). This is the two component model (section 3.5).

Table 2. Mouse Models with Differing Dystrophin and Utrophin Expression

Name		Heart	Heart	Skeletal Muscle	Skeletal Muscle
		Dystrophin	Utrophin	Dystrophin	Utrophin
Control	Control	+	+	+	+
Mdx	*Mdx*	-	+	-	+
Fiona	$Tg_{full\ length}/Dmd^{mdx}$	-	+	-	++
Gavin	Tg/DMD_{mdx}	-	-	-	++
Nigel	$Tg_{truncated}/DMD^{mdx}$	-	+	-	++

Key – no protein expression, + normal protein expression, ++ promoted protein expression.

3.4 Building a One Component Model in the Heart

As with the previous PCA of spectral profiles for cardiac tissue, three different metabolic environments were investigated using different NMR techniques and pulse sequences; HRMAS ^1H NMR spectroscopy was carried out with and without a T_2 filter to investigate tissue lipid content and low molecular weight metabolites in intact tissue, and conventional NMR spectroscopy of aqueous extracts with its greater sensitivity for low molecular weight metabolites. For all three pulse sequences PLS models were readily built that separated dystrophic and control cardiac tissue (Fig. 7).

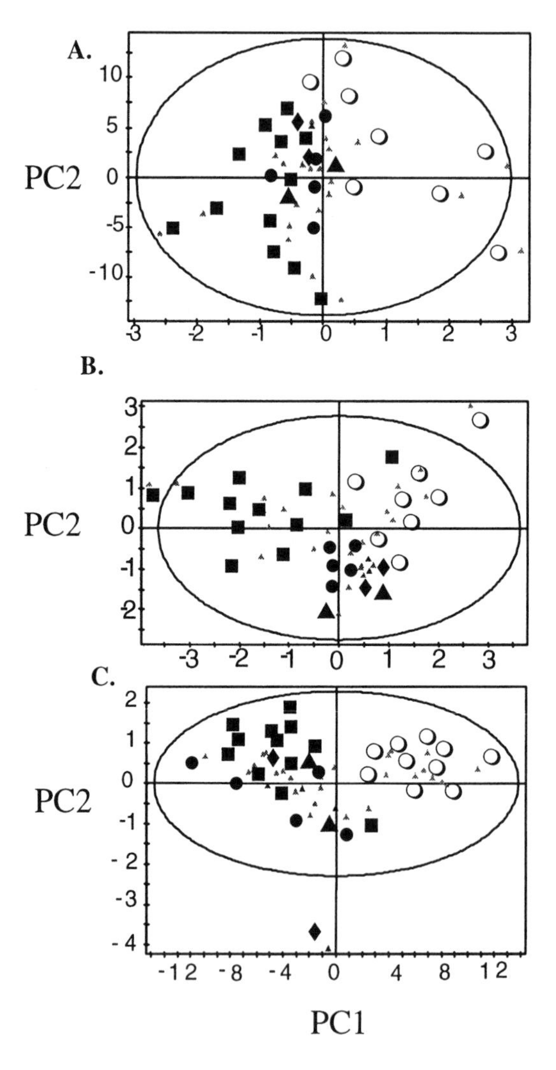

Figure 7. Three PLS plots of the effect of dystrophin expression on metabolic profile. Three data matrices were formed from high resolution spectra taken from intact tissue using either a standard pulse and acquire sequence (**A**) or a T_2 edited pulse sequence with a total relaxation time of 40 ms (**B**), or from acetonitrile/water extracts of the tissue (**C**). PLS regression models were built with Pareto prescaling, modeling expression of changes in metabolic profile against dystrophin expression. **Key**: ○ control; ■ mdx; ● Fiona; ◆ Gavin; ▲ Nigel

As with PCA, loadings contribution plots indicated that an increase in tissue taurine concentration was associated with dystrophic tissue, together with an increase in saturated lipid moieties and decreases in creatine and glucose within the tissue. Using the 50% maximum loading for a PLS component as the cut-off limit, 31 metabolites were identified as perturbed using data derived from all three preprocessing routines described (Table 3).

Table 3. PLS Models Built from Solvent Suppressed and CPMG HRMAS ^1H NMR Spectra of Intact Cardiac Tissue and from Spectra from Aqueous Tissue Extracts

Pulse sequence	PLS Model	Q^2	Metabolites Increased*	Metabolites Decreased*
HRMAS solvent suppress	autoscaled	0.26	lipid (CH_3CH_2, $CH_2C{=}C$, $C{=}CCH_2C{=}C$) lactate glutamate lysine taurine	glucose choline malonate, cholesterol β -hydroxybutyrate
	Pareto	0.34	lipid ($CH_2C{=}C$, CH_3CH_2) lactate taurine	choline malonate, cholesterol β -hydroxybutyrate
	mean centered	NO FIT		
HRMAS CPMG	autoscaled	0.30	lipid ($CH_2C{=}C$, CH_2CH_2CO, $CH_2CH_2CH_2$, CH_3CH_2) lactate	glucose acetate α-ketoglutarate lipid ($CH_2CH_2C{=}C$) β -hydroxybutyrate
	Pareto	0.43	taurine glutamate lactate	β -hydroxybutyrate
	mean centered	0.10	taurine lactate	
Aqueous Extract	Autoscaled	0.18	ATP inosine glycerol β-hydroxybutyrate *n*-butyrate	dimethylglycine valine aspartate glutamine acetate lysine leucine isoleucine
	Pareto	0.23	taurine lactate β -hydroxybutyrate	glutamine glutamate leucine isoleucine
	Mean centered	0.51	taurine lactate β -hydroxybutyrate	creatine glutamate

* in dystrophic tissue compared to the control group

In keeping with the mathematical basis of the three preprocessing techniques, Pareto scaling provided a half way point between the two extremes of no scaling (mean centering alone) and univariant scaling. Intriguingly, one metabolite, β-hydroxybutyrate, was consistently decreased in dystrophic intact cardiac tissue but increased in the aqueous extract, suggesting a "NMR invisible pool", possibly within mitochondria. Such effects have previously been observed for lactate and creatine within cerebral and cardiac tissues, respectively (Jouvensal *et al.*, 1999)

The most robust models were produced using the aqueous extract data set as the X variables for a PLS model as demonstrated by the goodness of fit algorithm (Q^2). During the validation of a PLS model, data sets are left out of the model and then used as a prediction set. This is repeated continually until the best model is converged upon. For the Pareto scaled data the Q^2 increased as low molecular weight metabolites contributed more to the spectral profiles (for solvent suppressed HRMAS spectra, $Q^2 = 0.119$; for CPMG HRMAS spectra $Q^2 = 0.169$; for aqueous extracts $Q^2 = 0.230$ for one component models).

As mentioned earlier PLS can also be applied as a means of data filtration *via* OSC which filters out data that is not correlated to a variable such as dystrophin expression. This is particularly useful for NMR spectral data acquired on different spectrometers or even at different magnetic field strengths. Applying this filter to the data set derived from aqueous extracts of the cardiac tissue with Pareto pre-scaling, two components of data not correlated to dystrophin expression were removed, producing a data set representing 33% of the remaining variation. However, this data set could be readily separated into dystrophic and control tissue in the first two PCs using PCA, representing 93% of the remaining spectral variation with a goodness of fit of $Q^2 = 0.682$. Applying PLS, a model was built correlated to protein expression, with a goodness of fit algorithm of $Q^2 = 0.875$, explaining 47% of the remaining spectral variation and 97% of the variation produced in mapping protein expression to Y = 0 or 1. In addition to the metabolites identified in the pre OSC filtered data set, two further regions were identified ($\delta = 3.06, 3.94$) corresponding to creatine.

Given the predictive capabilities of PLS, it is also possible to test the robustness of the models derived using data that has not been used to actually build the model. By separating the aqueous extracts of the cardiac tissue data set into a training set and a test set of ~75% and ~25% of the total data, the models produced were 90% successful at predicting class membership into dystrophic or control tissue for all mouse models. The technique will also predict the protein expression in new mouse models for DMD. PLS models were produced excluding either Fiona mice or a combination of Gavin and Nigel mice separately from different PLS models built using the remainder of the data, with the prediction success being 80% and 75 % for the two models, respectively.

Thus, using PLS, a supervised pattern recognition technique, to perform regressions between metabolic profile and protein expression, a common metabolic profile can be identified for dystrophic cardiac tissue despite the different genotypes of the mouse models used. Furthermore, the combined loading scores showing which metabolites contributed to this separation identified a number of metabolites not previously identified by PCA, including *n*-butyrate and glycerol, indicative of a failure to process ketone bodies and fatty acid derivatives, as well as a number of amino acids including glutamate and glutamine. Unlike the classification tool PCA, this approach could also be used to predict gene expression. Using the data set derived from cardiac tissue extracts we were able to demonstrate that models could be built capable of classifying test data, even for new mouse models of the disease that had not been included in the building of a PLS model. Furthermore, the approach could be used to follow treatments of DMD that cause a temporary change in protein expression (Krag *et al.*, 2001), such as gutted adenoviral vectors and plasmid vectors transfecting the missing dystrophin cDNA into a tissue (Cordier *et al.*, 2000; Wang *et al.*, 2000), aminoglycosides that suppress stop mutations (Barton-Davis *et al.*, 1999) and the introduction of stem cells (Gussoni *et al.*, 1999).

3.5 Building a Two Component Model for the Diaphragm

In theory PLS can model a whole series of proteins in terms of metabolite profiles by representing the proteomics data as the Y matrix of a PLS model, and using this to interrogate spectral profiles representing the X matrix. To investigate the reliability of the model we examined whether the techniques could be used for a two protein system, using extract spectra from diaphragm tissue extracts from control, mdx and Fiona mice. As the exact protein expression levels for dystrophin and utrophin were not available dystrophin expression was modeled as 1 (present) or 0 (absence) and utrophin expression as 1 (normal expression) or 2 (promoted expression), as a proof of concept experiment. Two component PLS models were built separating the data group into the three mouse groups with the Pareto pre-scaled data set providing the most robust model (goodness of fit $Q^2 = 0.412$; Fig. 8). For this model, the mean second PLS component loadings (± standard error), corresponding most closely to distinguishing dystrophic tissue, for each mouse type were significantly different for mdx and control mice according to an ANOVA test of variance followed by a Tukey-Kramer post test, but placed Fiona mice as intermediate (mdx = -8.7 ± 1.2, control = 7.5 ± 4.2, Fiona = 1.5 ± 2.8; $p < 0.01$ for differences between control and mdx). Again, as in cardiac tissue, taurine appeared to be a significant biomarker for a failure to express dystrophin. However, the first PLS component, corresponding most closely to the effects of utrophin expression, demonstrated decreased concentrations of glycerol, creatine and glucose and increased concentrations of short chain fatty acids and β-hydroxybutyrate in diaphragm tissue from Fiona mice.

The intermediary metabolic profile of Fiona mice in the PLS component correlated most with dystrophin expression suggesting that promoting utrophin expression, as occurs in skeletal muscle and diaphragm but not cardiac muscle of Fiona mice, can negate the metabolic effects of a failure to express dystrophin. This is also supported physiologically in the mdx mouse. The first wave of muscle necrosis only occurs in dystrophic tissue when utrophin expression falls below that found in neonatal muscle. Tinsley and co-workers (1998) have demonstrated that increased utrophin expression reverses the phenotype characteristic of mdx mouse muscle tissue. The utrophin gene has two known promotor regions and one enhancer region (Dennis *et al.*, 1996; Burton *et al.*, 1999), and a number of factors including heregulin and GA Binding Protein a/b can increase utrophin expression (Gramolini *et al.*, 1999). PLS derived metabolic profiles are ideal for screening other potential molecular activating factors. However, perhaps the most intriguing concept is that this may be a general approach for conducting proteomics using the high throughput capabilities of ^{1}H NMR derived metabolic profiles.

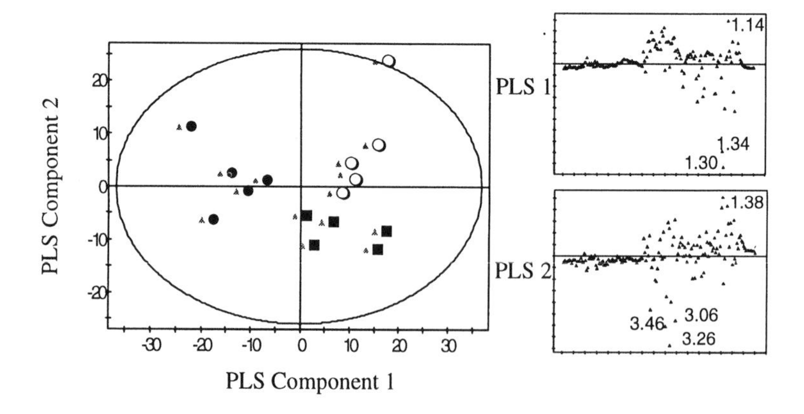

Figure 8. A two component PLS model was built to examine the metabolic profile of diaphragm tissue in terms of dystrophin and utrophin expression (Pareto scaled data shown). Right: loadings plots were derived for utrophin (top) and dystrophin (bottom) showing which metabolites contributed most to the model. Keys: ● Fiona; ■mdx, O control.

3.6 The Time Progression of DMD

One aspect of the disease that the metabolic profiles derived so far have largely ignored is the time progression of the pathology (in fact all the data presented are from ~ 6 month old adult mice). PLS, by virtue of its regression capabilities, can be used to correlate time progressions across

NMR derived metabolic profiles. Evens and co-workers (2002) have investigated the metabolic changes that occur in control and dystrophic muscle using NMR spectral profiles from both intact tissue and aqueous and lipid extracts. Using PLS models with the age of the animals modeled as the Y variable, time progression changes could be generated for both dystrophic and control tissues, showing a clear temporal progression across the tissues (Fig. 9).

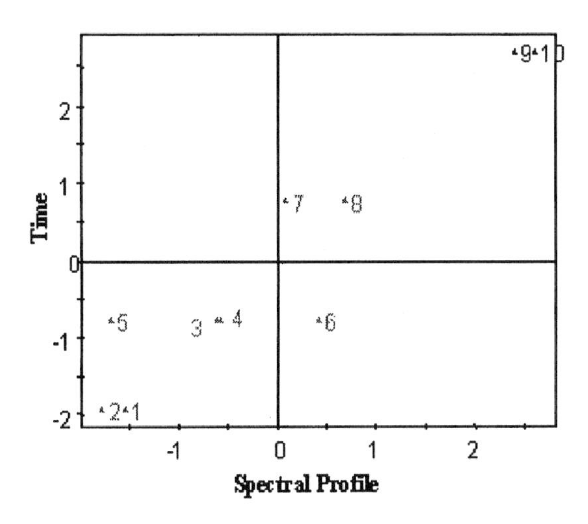

Figure 9. PLS model of spectral profiles correlated with age for extracts of mdx mouse cardiac tissue. Tissue was taken from 3 month (1-2), 6 month (3-6), 10 month (7-8) and 12 month (9-10) animals.

However, the metabolites that contributed to this separation were different for control and dystrophic tissues for all muscle tissues investigated. Furthermore all the dystrophic muscle displayed a decrease in tissue taurine concentration with age, while control tissue demonstrated an increase in short chain fatty acids. Similar time progressions were also noted for spectra from intact tissue and from the lipid extracts. In future studies the time progression of the disease will be a crucial aspect of any metabolic profile investigations, further expanding the data matrices that must be inspected by pattern recognition.

3.7 Other Pattern Recognition Techniques

Having derived a series of data matrices that categorize the metabolic changes that occur in dystrophic tissues, these can now be used to investigate alternative pattern recognition techniques and pre-processing scaling and normalization routines. Professor Douglas Kell, University of Manchester, investigating the ^{1}H NMR spectral profiles derived from aqueous extracts from cardiac tissue, was able to categorize tissue as being dystrophic or normal using only two variables from the data following applications of a genetic program to the data set (see Goodacre and Kell, Chapter 13 for discussion on applications of genetic programming to metabolome analysis). In a similar manner neural networks and decision trees can be built that also separate the data using a fraction of the variables modeled by PCA and PLS analyses. Thus, the results described in this chapter could have been derived using a range of different mathematical techniques, and the choice of which technique should be chosen in any new metabolomic/metabonomic study requires some thought. To identify a unique biomarker of a disease the genetic program approaches are probably the favored techniques. However, given the current sensitivity limitations of ^{1}H NMR spectroscopy, it is unlikely that very low concentration metabolites, unique to a particular disorder, can be detected. In such cases a description of how a pathology affects the overall metabolic profile of a tissue or biofluid seems preferable, and in which case one should consider techniques such as PCA or PLS.

4. FUTURE WORK

The techniques described in these studies are ideal for following a range of protein changes as measured by proteomics. For example, there are a number of dystrophin related proteins other than utrophin, and the failure to express dystrophin also affects expression of proteins in the dystrophin-glycoprotein complex. With further animal models of related disorders the action of dystrophin related proteins and dystrophin-glycoprotein complex proteins could be modeled in terms of metabolic phenotypes, providing a complete "metabolome" of the disease process. In particular the inclusion of a model of Beckers Muscular Dystrophy would be useful to model the effects of impaired protein function, as opposed to a complete lack of dystrophin as occurs in DMD. Similarly mouse models of X-linked cardiomyopathy, caused by mutations in the muscle-specific M-promoter of the DMD gene that abolish the cardiac gene expression of dystrophin while expression in skeletal muscle is preserved, could be examined. Patients have ventricular wall dysfunction, dilated cardiomyopathy and cardiac failure (Muntoni *et al.*, 1995). Widening the application of these data sets to other cardiac disorders may also provide useful information about this and other disease processes. Although cardiac hypertrophy is mild at 6 months in mdx mice (Nakamura *et al.*, 2001), some of the metabolic changes detected at this time point may be indicative of the early stages of this disease. The inclusion of models of cardiac hypertrophy would separate the metabolic changes that are associated with cardiac hypertrophy and those that are unique to the mdx mouse.

Perhaps the most intriguing application of these techniques may come in comparative functional genomics. These techniques are equally applicable to non-mammalian systems, and metabolic profiles have been used to probe "silent phenotypes" in yeast cells (Raamsdonk *et al.*, 2001). A nematode worm (*Caenorhabditis elegan*) model of DMD exists, where the dystrophin analogue dys-1 has been knocked out (Bessou *et al.*, 1998). These animals could be used to screen for small molecules which promote upregulation of dystrophin or its related proteins in this animal, providing a relatively cheap and high throughput technique for treating this terminal disease.

In conclusion, these initial studies have shown that even for a genetic disease where the primary lesion is not metabolic, a metabolic profile can still be derived and provide useful information; results shown here demonstrate that the question asked in the title can be answered in the affirmative. One intriguing question is why does the approach work at all. Metabolic pathways are highly interconnective, with a perturbation in one pathway potentially being propagated to all via these connections. To understand these connections will require a movement from the steady state measurements of metabolite pool sizes to the dynamic measurements of fluxes and rate constants.

ACKNOWLEDGEMENTS

This work was supported by the British Heart Foundation and the Royal Society, UK. The authors would also like to thank Drs. Richard Barton, Melanie Gulston and Ned Mason for useful discussions and suggestions.

REFERENCES

Ahn, AH, Kunkel, LM. The structural and functional diversity of dystrophin. *Nature Genet* 4: 283-291 (1993).

Andrew, ER, Bradbury A, Eades, RG. Removal of dipolar broadening of NMR spectra of solids by specimen rotation. *Nature* 183: 1802-1803 (1959).

Anderson JL, Head SI, Rae C, Morley JW. Brain function in Duchenne muscular dystrophy. *Brain* 125: 4-13 (2002).

Aue WP, Bartholdi E, Ernst RR. Two dimensional spectroscopy. Application to nuclear magnetic resonance. *J Chem Phys* 64: 2229-22 (1976).

Barton-Davis ER, Cordier L, Shoturma DI *et al.* Aminoglycoside antibiotics restore dystrophin function to skeletal muscles of mdx mice. *J Clin Invest* 104: 375-381 (1999).

Beckwith-Hall BM, Nicholson JK, Nicholls A *et al.* Nuclear magnetic resonance spectroscopic and principal component analysis investigations into biochemical effects of three model hepatotoxins. *Chem Res Toxicol* 11: 260-272 (1998).

Belle JE, Harris NG, Williams SR, Bhakoo KK. A comparison of cell and tissue extraction techniques using high-resolution ^{1}H-NMR spectroscopy. *NMR Biomed* 15: 37-44 (2002).

Bessou C, Giugia JB, Franks CJ *et al.* Mutations in the *Caenorhabditis elegans* dystrophin-like gene dys-1 lead to hyperactivity and suggest a link with cholinergic transmission. *Neurogenetics* 2: 61-72 (1998).

Blake DJ, Weir A, Newey SE, Davies KE. Function and genetics of dystrophin and dystrophin-related proteins in muscle. *Physiol Rev* 82: 291-329 (2002).

Brenman JE, Chao DS, Xia H *et al.* Nitric oxide synthase complexed with dystrophin and absent from skeletal muscle sarcolemma in Duchenne muscular dystrophy. *Cell* 82: 743-752 (1995).

Bullfield G, Siller WG, Wight PA, Moore KJ. X chromosome-linked muscular dystrophy (mdx) in the mouse. *Proc Natl Acad Sci USA* 81: 1189-1192 (1984).

Burghes AH, Logan C, Hu X *et al.* A cDNA clone from the Duchenne/Becker muscular dystrophy gene. *Nature* 328: 434-437 (1987).

Burton EA, Tinsley JM, Holzfeind PJ *et al.* A second promoter provides an alternative target for therapeutic up-regulation of utrophin in Duchenne muscular dystrophy. *Proc Natl Acad Sci USA* 96: 14025-14030 (1999).

Burton EA, Davies KE. Muscular dystrophy-reason for optimism? *Cell* 108: 5-8 (2002).

Campbell KP. Three muscular dystrophies: loss of cytoskeleton-extracellular matrix linkage. *Cell* 80: 675-679 (1995).

Cheng LL, Ma MJ, Becerra L *et al.* Quantitative neuropathology by high resolution magic angle spinning proton magnetic resonance spectroscopy. *Proc Natl Acad Sci USA* 94: 6408-6413 (1997).

Cordier L, Hack AA, Scott MO *et al.* Rescue of skeletal muscles of gamma-sarcoglycan-deficient mice with adeno-associated virus-mediated gene transfer. *Mol Ther* 1: 119-129 (2000).

Deconinck N, Tinsley J, De Backer F *et al.* Expression of truncated utrophin leads to major functional improvements in dystrophin-deficient muscles of mice. *Nature Med* 3: 1216-1221 (1997).

Dennis CL, Tinsley JM, Deconinck AE, Davies KE. Molecular and functional analysis of the utrophin promoter. *Nucleic Acids Res* 24: 1646-1652 (1996).

Duchenne GBA. *De l'Electrisation Localisee et son Application a la Therapeutique*. 2nd Edn. Balliere et Fils, Paris (1886).

Emery AE. Clinical and molecular studies in Duchenne muscular dystrophy. *Prog Clin Biol Res* 306: 15-28 (1989).

Eriksson L, Johansson E, Kettaneh-Wold N, Wold S. *Introduction to Multi- and Megavariate Data Analysis using Projection Methods (PCA & PLS).* Umetrics, Umea, Sweden (1999).

Ervasti JM, Ohlendieck K, Kahl SD *et al.* Deficiency of a glycoprotein component of the dystrophin complex in dystrophic muscle *Nature* 345: 315-319 (1990).

Even PC, Decrouy A, Chinet A. Defective regulation of energy metabolism in mdx-mouse skeletal muscles. *Biochem J* 304: 649-654 (1994).

Evens T. A bioinformatic approach to the progression of pathology in Duchenne muscular dystrophy. BSc Thesis, Imperial College of Science Technology and Medicine, London, UK (2002).

Feener CA, Koenig M, Kunkel LM. Alternative splicing of human dystrophin mRNA generates isoforms at the carboxy terminus. *Nature* 338: 509-511 (1989).

Fong P, Turner PR, Denetclaw WF, Steinhardt RA. Increased activity of calcium leak channels in myotubes of Duchenne human and mdx mouse origin. *Science* 250: 673-676 (1990).

Garrod S, Humpfer E, Spraul M *et al.* High resolution magic angle spinning ^{1}H NMR spectroscopic studies on intact rat renal cortex and medulla. *Magn Reson Med* 41: 1108-1118 (1999).

Gavaghan CL, Holmes E, Lenz E *et al.* An NMR-based metabonomic approach to investigate the biochemical consequences of genetic strain differences: application to the C57BL10J and Alpk:ApfCD mouse. *FEBS Lett* 484: 169-174 (2000).

Gillet B, Doan BT, Verre-Serrie C *et al. In vivo* 2D ^{1}H NMR of mdx mouse muscle and myoblast cells during fusion: evidence for a characteristic signal of long chain fatty acids. *Neuromusc Disord* 3: 433-438 (1993).

Grady RM, Grange RW, Lau KS *et al.* Role for alpha-dystrobrevin in the pathogenesis of dystrophin-dependent muscular dystrophies. *Nature Cell Biol* 1: 215-220 (1999).

Gramolini AO, Angus LM, Schaeffer L *et al.* Induction of utrophin gene expression by heregulin in skeletal muscle cells: role of the N-box motif and GA binding protein. *Proc Natl Acad Sci USA* 96: 3223-3227 (1999).

Griffin JL, Troke J, Walker LA *et al.* The biochemical profile of rat testicular tissue as measured by magic angle spinning ^{1}H NMR spectroscopy. *FEBS Lett* 486: 225-229 (2000).

Griffin JL, Williams HJ, Sang E *et al.* Metabolic profiling of genetic disorders: a multitissue (1)H nuclear magnetic resonance spectroscopic and pattern recognition study into dystrophic tissue. *Anal Biochem* 293: 16-21 (2001a).

Griffin JL, Williams HJ, Sang E, Nicholson JK. Abnormal lipid profile of dystrophic cardiac tissue as demonstrated by one- and two-dimensional magic-angle spinning (1)H NMR spectroscopy. *Magn Reson Med* 46: 249-255 (2001b).

Griffin JL, Walker L, Shore RF, Nicholson JK. High-resolution magic angle spinning ^{1}H-NMR spectroscopy studies on the renal biochemistry in the bank vole (*Clethrionomys glareolus*) and the effects of arsenic (As^{3+}) toxicity. *Xenobiotica* 31: 377-385 (2001c).

Griffin JL, Mann CJ, Scott J *et al.* Choline containing metabolites during cell transfection: an insight into magnetic resonance spectroscopy detectable changes. *FEBS Lett* 509: 263-266 (2001d).

Gussoni E, Soneoka Y, Strickland CD *et al.* Dystrophin expression in the mdx mouse restored by stem cell transplantation. *Nature* 401: 390-394 (1999).

Hoffman EP, Brown RH, Kunkel LM. Dystrophin: the protein product of the Duchenne muscular dystrophy locus. *Cell* 51: 919-928 (1987).

Hoffman EP, Beggs AH, Koenig M *et al.* Cross-reactive protein in Duchenne muscle. *Lancet* 2: 1211-1212 (1989).

Holmes E, Nicholson JK, Nicholls AW *et al.* The identification of novel biomarkers of renal toxicity using automated data reduction techniques and PCA of proton NMR spectra of urine. *Chemom Intel Lab Sys* 44: 245-255 (1998).

Hu JZ, Rommereim DN, Wind RA. High resolution ^{1}H NMR spectroscopy in rat liver using magic angle turning at a 1 Hz spinning rate. *Magn Reson Med* 47: 829-836 (2002).

Huxtable RJ. Physiological actions of taurine. *Physiol Rev* 72: 101-163 (1992).

Jouvensal L, Carlier PG, Bloch G. Evidence for bi-exponential transverse relaxation of lactate in excised rat muscle. *Magn Reson Med* 41: 624-626 (1991).

Kato T, Nishina M, Matsushita K *et al.* Increased cerebral choline-compounds in Duchenne muscular dystrophy. *Neuro Report* 8: 1435-1437 (1997).

Kemp GJ, Taylor DJ, Dunn JF *et al.* Cellular energetics of dystrophic muscle. *J Neurol Sci* 116: 201-206 (1993).

Khurana TS, Hoffman EP, Kunkel LM. Identification of a chromosome 6-encoded dystrophin-related protein. *J Biol Chem* 265: 16717-16720 (1990)

Khurana TS, Watkins SC, Chafey P *et al.* Immunolocalization and developmental expression of dystrophin related protein in skeletal muscle. *Neuromuscul Disord* 1: 185-194 (1991).

Krag TO, Gyrd-Hansen M, Khurana TS. Harnessing the potential of dystrophin-related proteins for ameliorating Duchenne's muscular dystrophy. *Acta Physiol Scand* 171: 349-58 (2001).

Lederfein D, Levy Z, Augier N *et al.* A 71-kilodalton protein is a major product of Duchenne muscular dystrophy gene in brain and other nonmuscle tissue. *Proc Natl Acad Sci USA* 89: 5346-5350 (1992).

Martens H, Næs T. *Multivariate Calibration*. John Wiley and Sons, Chichester (1989).

McIntosh LM, Garrett KL, Megeney L *et al.* Regeneration and myogenic cell proliferation correlate with taurine levels in dystrophin- and MyoD-deficient muscles. *Anat Rec* 252: 311-324 (1998a).

McIntosh LM, Baker RE, Anderson JE. Magnetic resonance imaging of regenerating and dystrophic mouse muscle. *Biochem Cell Biol* 76: 532-541 (1998b).

Menke A, Jockush H. Extent of shock-induced membrane leakage in human and mouse myotubes depends on dystrophin. *J Cell Sci* 108: 727-733 (1995).

Millis K, Maas E, Cory DG, Singer S. Gradient high resolution magic angle spinning nuclear magnetic resonance spectroscopy of human adipocyte tissue. *Magn Reson Med* 38: 399-403 (1997).

Millis K, Weybright P, Cambell N *et al.* Classification of human liposarcoma and lipoma using *ex vivo* proton NMR spectroscopy. *Magn Reson Med* 41: 257-267 (1999).

Mokhatarian A, Decrouy A, Chinet A, Even PC. Components of energy expenditure in the mdx mouse model of Duchenne muscular dystrophy. *Pfugers Arch* 431: 527-532 (1996).

Monaco AP, Neve RL, Colletti-Feener C *et al.* Isolation of candidate cDNAs for portions of the Duchenne muscular dystrophy gene. *Nature* 323: 646-650 (1986).

Monaco AP, Bertelson CJ, Liechti G *et al.* An explanation for the phenotypic differences between patients bearing partial deletions of the DMD locus. *Genomics* 2: 90-95 (1988).

Muntoni F, Wilson L, Marrosu G *et al.* A mutation in the dystrophin gene selectively affecting dystrophin expression in the heart. *J Cardio Invest* 96: 693-699 (1995).

Murray JM, Davies KE, Harper PS *et al.* Linkage relationship of a cloned DNA sequence on the short arm of the X-chromosome to Duchenne muscular dystrophy. *Nature* 300: 69-71 (1982).

Nagayama K, Kumar K, Wuthrich K, Ernst RR. Experimental techniques of two-dimensional correlated spectroscopy. *J Magn Reson* 40: 321-334 (1980).

Nakamura A, Harrod GV, Davies E. Activation of calcineurin and stress activated protein kinase/p38-mitogen activated protein kinase in hearts of utrophin-dystrophin knockout mice. *Neuromuscul Disord* 11: 251-259 (2001).

Nicholson JK, Foxall PJ, Spraul M *et al.* 750 MHz ^{1}H and ^{1}H-^{13}C NMR spectroscopy of human blood plasma. *Anal Chem* 67: 793-811 (1995).

Nicholson JK, Connelly J, Lindon JC, Holmes E. Metabonomics: a platform for studying drug toxicity and gene function. *Nature Rev Drug Discov* 1: 153-161 (2002).

Nudel U, Zuk D, Einat P *et al.* Duchenne muscular dystrophy gene product is not identical in muscle and brain. *Nature* 337: 76-78 (1989).

Oliver S. Guilt by association goes global. *Nature* 403: 601-603 (2001).

Pasternak C, Wong S, Elson EL. Mechanical function of dystrophin in muscle cells *J Cell Biol* 128: 355-361 (1995).

Petrof BJ, Shrager JB, Stedman HH *et al.* Dystrophin protects the sarcolemma from stresses developed during muscle contraction. *Proc Natl Acad Sci USA* 90: 3710-3714 (1993).

Raamsdonk LM, Teusink B, Broadhurst D *et al.* A functional genomics strategy that uses metabolome data to reveal the phenotype of silent mutations. *Nature Biotechnol* 19: 45-50 (2001).

Rae C, Scott RB, Thompson CH *et al.* Brain biochemistry in Duchenne muscular dystrophy: a ^{1}H magnetic resonance and neuropsychological study. *J Neurol Sci* 160: 148-157 (1998).

Rafael JA, Tinsley JM, Deconinck AE, Davies KE. Skeletal muscle-specific expression of a utrophin transgene rescues utrophin-dystrophin deficient mice. *Nature Genet* 19: 79-82 (1998).

Rybakova IN, Patel JR, Ervasti JM. The dystrophin complex forms a mechanically strong link between the sarcolemma and costameric actin. *J Cell Biol* 150: 1209-1214 (2000).

Rybakova IN, Patel JR, Davies KE *et al.* Utrophin binds laterally along actin filaments and can couple costameric actin with sarcolemma when overexpressed in dystrophin-deficient muscle. *Mol Biol Cell* 13: 1512-1521 (2002).

Sander M, Chavoshan B, Harris SA *et al.* Functional muscle ischemia in neuronal nitric oxide synthase-deficient skeletal muscle of children with Duchenne muscular dystrophy *Proc Natl Acad Sci USA* 97: 13818-13823 (2000).

Sicinnski P, Geneg Y, Ryder-Cook AS *et al.* The molecular basis of muscular dystrophy in the mdx mouse: a point mutation. *Science* 244: 1578-1580 (1989).

Siminovitch DJ, Ruocco MJ, Olejiniczak ET *et al.* Anisotropic 2H-nuclear magnetic resonance spin-lattice relaxation in cerebroside- and phospholipid-cholesterol bilayer membranes. *Biophys J* 54: 373-381 (1988).

Thomas GD, Sander M, Lau KS *et al.* Impaired metabolic modulation of alpha-adrenergic vasoconstriction in dystrophin-deficient skeletal muscle. *Proc Natl Acad Sci USA* 95: 15090-15095 (1998).

Tinsley JM, Potter AC, Phelps SR *et al.* Amelioration of the dystrophic phenotype of mdx mice using a truncated utrophin transgene. *Nature* 384: 349-353 (1996).

Tinsley J, Deconinck N, Fisher R *et al.* Expression of full-length utrophin prevents muscular dystrophy in mdx mice. *Nature Med* 4: 1441-1444 (1998).

Tomlins A, Foxall PJD, Lindon JC *et al.* High resolution magic angle spinning ^{1}H nuclear magnetic resonance analysis of intact prostatic hyperplastic and tumour tissues. *Anal Comm* 35: 113-115 (1998).

Tracey I. Scott RB, Thompson CH *et al.* Brain abnormalities in Duchenne muscular dystrophy: a ^{31}P magnetic resonance spectroscopy and neuropsychological study. *Lancet* 345: 1260-1264 (1995).

Tracey I, Dunn JF, Radda GK. Brain metabolism is abnormal in the mdx mouse model of Duchenne muscular dystrophy. *Brain* 119: 1039-1044 (1996).

Wang B, Li J, Xiao X. Adeno-associated virus vector carrying human minidystrophin genes effectively ameliorates muscular dystrophy in mdx mouse model. *Proc Natl Acad Sci USA* 97: 13714-13719 (2000).

Waters NJ, Garrod S, Farrant RD *et al.* High-resolution magic angle spinning (1)H NMR spectroscopy of intact liver and kidney: optimization of sample preparation procedures and biochemical stability of tissue during spectral acquisition. *Anal Biochem* 282: 16-23 (2000).

Waters NJ. High resolution magic angle spinning NMR spectroscopy and pattern recognition studies on drug-induced tissue damage. PhD thesis, Imperial College of Science, Technology and Medicine, London, UK (2001).

Weller B, Karparti G, Carpenter S. Dystrophin-deficient mdx muscle fibres are preferentially vulnerable to necrosis induced by experimental lengthening contractions *J Neurol Sci* 100: 9-13 (1990).

Weybright P, Millis K, Campbell N *et al.* Gradient, high-resolution, magic angle spinning ^{1}H nuclear magnetic resonance spectroscopy of intact cells. *Magn Reson Med* 39: 337-345 (1998).

Williams MW, Bloch RJ. Extensive but coordinated reorganisation of the membrane skeleton in myofibres of dystrophic (mdx) mice. *J Cell Biol* 144: 1259-1270 (1999).

Wu D, Chen A, Johnson CS. An improved diffusion-ordered spectroscopy experiment incorporating bipolar-gradient pulses. *J Magn Reson* 115: 260-264 (1995).

Chapter 4

APPLICATION OF METABONOMICS IN THE PHARMACEUTICAL INDUSTRY

Alan P. Breau[1] and Glenn H. Cantor[2]
[1]Global Drug Metabolism, Pharmacia Corporation, 4901 Searle Parkway, Skokie, IL 60077, USA [2]Investigative Toxicology, Pharmacia Corporation, 301 Henrietta St, Kalamazoo, MI 49007, USA

1. INTRODUCTION

One of the major precepts in the practice of modern medicine is to quantify endogenous constituents in biological fluids as an aid to diagnosis of disease and to monitor therapy. Measurement of endogenous biochemicals comprises clinical chemistry. Clinical chemistry tests are extremely valuable to the clinician; in fact major industries have developed to support them. In the development of many new drugs, the drug development process has also taken advantage of standard clinical chemistry tests to provide an early assessment about the potentially toxic and/or efficacious potential of drug candidates. The availability of these tests can streamline the drug discovery and development process by making decisive high quality data available at a time point well before a definitive morphologic response is evident. However, in the development of a novel drug, the clinical chemistry tests to address a target efficacy or toxicity are not always available, and this lack of an early indicator greatly increases the risk and expense of going forward with drug development.

In the 1970s, the metabolic profiling of biofluids by gas chromatography-mass spectrometry (GC-MS) and other chromatographic techniques was investigated for their potential utility to provide non-invasive signals that would describe a disease process. However, for various reasons, this approach was never widely used by the medical community (Holland *et al.*,

1986). In the 1980s, investigators began to study biofluids of untreated control and toxin treated animals using high-resolution nuclear magnetic resonance (NMR) technology (Nicholson *et al.*, 1983, 1989). The advantage of NMR over the GC-MS and chromatographic approaches included reproducibility of the analyses as well as the minimal sample preparation that was required. NMR provides simultaneous qualitative and quantitative determinations of a wide range of chemical classes in a reproducible and non-destructive way. The quantum mechanical basis of an NMR experiment makes it an ideal technique for the comparison of relative amounts of endogenous metabolites within a sample since the response of each compound in a sample in the NMR depends on the unique intrinsic properties of that compound and not, for example, its reaction in an ion source, as in mass spectrometry. This allows for the relative quantitation of peaks without the need for internal standards in the sample to correct for instrument/sample interactions. The biggest drawback to NMR is its lack of sensitivity relative to other technologies that limits the experiment to biofluid components that are in high physiological concentrations. The reproducibility of the NMR spectra made the comparison of spectra possible; however, the complexity of an NMR spectrum made the rapid analysis of multiple samples a time consuming and low throughput process. To overcome the problem associated with the analysis of these complex data, early investigators introduced pattern recognition algorithms to analyze and compare NMR spectra (Gartland *et al.*, 1991). These pattern recognition techniques have now evolved into models for specific organ toxicity (Holmes *et al.*, 1998) and often include the use of neural networks (Holmes *et al.*, 1994).

Since NMR based metabonomics relies on patterns of proton signals detected in a biofluid by an NMR instrument rather than the observation of a very specific marker, assigning a biological phenotype associated with these patterns must be done with caution. The goal of metabonomics is to analyze changes in endogenous metabolites produced by the body. A potential problem is that the administered compound, its potential metabolites and the dosing vehicle are all organic compounds that will also produce proton NMR signals. If the presence of these drug related components (DRC) in the biofluid is of sufficient abundance, they can limit the utility of the metabonomics experiment by producing signals in the proton spectrum that coincide with endogenous constituents of interest. An understanding of the elimination time course of the compound, its metabolites and vehicle is needed to ascertain if a particular biofluid collection time point will be susceptible to interferences from DRC. Limiting the dose administered, choosing NMR friendly dosing vehicles or disregarding early collection time points in the data analysis are some of the ways to deal with the DRC issue

(Beckwith-Hall *et al.*, 2002). Some investigators remove the portion of the spectra with DRC from all of the samples, but this reduces the number of variables in the data analyses and can be a particular problem when comparing effects of multiple drugs. Others have attempted to correct the problem by substituting a reference signal for the peaks that are apparently caused by DRC. Another consideration is the interpretation of a pattern that does not match anything in the known database. An "abnormal" pattern may be a unique toxicity or it may be a signal of the intended pharmacological activity of the administered compound, an efficacy biomarker.

With the aforementioned limitations in mind, metabonomics is unlike previous experiences with clinical chemistry or specific biological markers (biomarkers), due to its ability to simultaneously identify and quantify the relative abundance of endogenous biological constituents that produces patterns, rather than individual component measurements, to provide early predictors of eventual toxic or efficacious phenotypes. Acceptance of the concept that a pattern of non-obvious endogenous biochemical constituents can be used to predict a toxicity or possible efficacy is the main hurdle to convince regulators and management of the value of metabonomics. For example, it is easy to explain that the elevation of a liver enzyme as measured by standard clinical chemistry indicates liver toxicity, whereas it is more challenging to assert that the simultaneous increase in acetate, bile acids, creatine and taurine with the concomitant decrease in citrate, 2-ketoglutarate and succinate indicates liver toxicity (Beckwith-Hall *et al.*, 1998). The intent of metabonomics is not to replace established clinical chemistry, but rather, to quickly generate patterns to predict toxicity and efficacy for which no currently acceptable or convenient predictive clinical chemistry test or biomarker is available. A thorough chemometric analysis of metabonomics patterns can ultimately provide specific biomarker identification.

The formation of a consortium (Consortium on Metabonomics in Toxicology; COMET) consisting of 6 major pharmaceutical companies (Bristol Myers Squibb, Eli Lilly, Novo Nordisk, Roche, Pharmacia and Pfizer) and Imperial College, London, demonstrates an increasing acceptance of metabonomics. The Society of Toxicology 2002 Conference featured a symposium dedicated to metabonomics, and many conferences on metabonomics and systems biology are now available.

2. ADVANTAGES OF METABONOMICS

There is tremendous incentive for all pharmaceutical companies to increase their research productivity without a concomitant increase in resources. Most of these companies are trying to meet this challenge with investments in technologies that will provide earlier indications of toxicity or efficacy, the major reasons why compounds in pharmaceutical development fail to become approved drugs. Companies are investing in genomics, proteomics and metabonomics technologies in order to meet this challenge.

In genomic analyses, when an altered state arises, genomics would detect changes in mRNA transcription by measuring the difference in mRNA profiles/levels between controls and test subjects. While differences in mRNA profiles/levels may be important, research in yeast has shown that there is only a 0.4 correlation between global mRNA and protein expression, which is not good enough to help select the lead compound with a high degree of reliability (Gygi *et al.*, 1999). In pharmaceutical drug development it is imperative to select a quantitative marker or marker pattern that indicates the effect of the drug candidates consistently and in a time-predictable manner. Another critical factor in the application of genomics for toxicity prediction is that the timing of the sampling for genomics analyses must correlate with the turning on or off of the relevant gene transcription events. If sampling timing were not correct, genomic changes may not be detected or be detected below the statistical variation of the test group. In essence, the pharmacodynamics/pharmacokinetics relationships of gene expression that results from a toxic event need to be known prior to that event in order to establish when to optimally collect samples for genomics.

Proteomics compares the appearance or disappearance of specific proteins following the perturbation of an organism, for example during toxicity or a disease state. Proteomics' workhorse technology is a two-dimensional-gel electrophoresis (2D-GE). 2-D GE separates cellular proteins based on charge and mass, yielding a sheet of dark spots (stained proteins) suspended in a thin layer of polyacrylamide. For protein quantification, image analyzers with the associated software are often used to determine which proteins increase or decrease in abundance relative to control cell lysate, whereas identification of the protein spots is often done by a combination of protein sequencing and MS. Another problem of 2-D systems is the large variation in protein abundances, which can be five orders of magnitude or more in a given cell (Annan, 2000). Since there is no PCR analogy for proteins, trace level proteins cannot be amplified, and neither 2-D gels nor any other existing system can detect anything but a small fraction of the total protein content of a cell lysate in a single snapshot. Scarce enzymes such as protein kinases or telomerase are often critical control elements of cellular function or cell signaling pathways but are often not detected in 2-D maps. In addition to the difficulties of identifying low-level proteins with 2-D gel electrophoresis, analyzing membrane proteins is also difficult due to protein insolubility in aqueous media without high levels of detergent. These proteins are often involved in intra- and extra-cellular signalling, and make ideal drug targets. Therefore, their detection is critical. With recent advances in ICAT (Isotope Coded Affinity Tag) (Gygi *et al.*, 1999) and GIST (Global Internal Standard Tag) (Wang *et al.*, 2001) technologies the relative quantities of proteins can be measured directly by MS, but these techniques to date have not been used extensively outside of a few research groups. One new separation technology is protein arrays. Certain manufacturers have developed chip-based systems that immobilize antibodies or other proteins of interest. While the affinity chip based approach often does not give a representative view of the relative protein levels *in vivo* due to matrix and ionization differences, this approach has recently yielded a proteomic pattern biomarker from human serum that provides an early and accurate diagnosis of previously undiagnosed ovarian cancer (Petricoin *et al.*, 2002). Most proteomics analyses use tissues, not serum. Thus like genomics, this technology requires an invasive/destructive sampling technique since representative mRNA or proteins are not present in readily accessible biofluids.

Metabonomics relies on the changes in endogenous metabolism that occur as an organism tries to maintain homeostasis following the alteration of its normal state by a disease or drug/toxin insult. This technique can be used to measure the simultaneous variation in the levels of endogenous components in a biofluid from an animal or human following the administration of a drug candidate. Using principal component analysis (PCA), the variation patterns can clearly indicate a change in the constituent profile from the normal state and when these patterns are compared with existing metabonomics pattern databases can signify specific organ toxicity.

One of the major advantages of metabonomics is that it uses readily available biofluids that are collected non-invasively. This feature of metabonomics eliminates the need for a thorough understanding of pharmacokinetics/ pharmacodynamics or toxicokinetics/ toxicodynamics relationships and the optimal time to collect samples since the temporal progression of an alteration can be efficiently monitored through the use of sequential biofluid collections and analyses that span the post-treatment period. This ability to capture samples at the optimal time, even if that time varies among individual animals or is not known, reduces the numbers of animals that must be treated, thus reducing the amount of experimental compound required-a critical factor at the discovery and early development stages of the pharmaceutical research and development process. Also, the non-invasive sampling feature enables the toxicologist to not only identify that there is a lesion or abnormality but also to conveniently and efficiently monitor the time of onset and recovery from the lesion. This provides additional information that can be critical in assessing the risk/benefit aspects of a potential drug product. With modern instrumentation and expert pattern recognition systems, metabonomics can become a relatively high throughput, preliminary toxicity-screening tool that can supplement and enhance traditional histopathological examinations to determine compound induced toxicity.

3. IMPORTANCE OF METABONOMICS TO THE PHARMACEUTICAL INDUSTRY

Modern drug discovery and development is driven by the theory that the aberrant expression of a biochemical or physiological process reflects the underlying mechanism responsible for a diseased phenotype. Therefore, a compound that can restore the process to a normal state will cure the disease or alleviate symptoms. In the early discovery phase it is critically important to verify the biochemical mechanism that is the basis of a drug discovery project. Target validation is usually accomplished by the administration of a compound known to bind with the target from *in vitro* high throughput screen data, into an animal model of the disease. A key question that arises from these approaches is when the animals show signs of toxicity. If the etiology of the toxicity is due to the pharmacological properties of the compound then a pharmacological intervention of the chosen biochemical mechanism will probably produce this toxicity and limit the utility of this mechanism as a target for drug discovery. However, if the etiology of the toxicity is due to the intrinsic chemical properties of the compound and not its pharmacological properties, then it may be possible to identify an alternative compound or a different chemical template that will provide the desired pharmacological intervention without the general chemical class toxicity. Since metabonomics is able to provide information on the mechanisms of toxicity, it could differentiate between the pharmacological mechanism and an alternative mechanism of the toxicity. Thus metabonomics would be able to ascertain if the toxicity was an inevitable by-product of the pharmacological intervention or if the mechanism of toxicity was unrelated to the pharmacology expressed by the compound. A metabonomics biomarker that provides fast and accurate information on whether to validate or invalidate a particular drug target at an early stage of drug discovery would be extremely valuable to a pharmaceutical company. The correct decision to proceed forward or terminate the exploration of a drug target at an early point in the drug invention process has significant financial implications for a company since the expense of the subsequent testing phases of the discovery and development process are orders of magnitude higher than the early testing/target validation phase.

Once a drug target is validated, the discovery process proceeds to identify a variety of templates for small chemical compounds that may become the eventual clinical agent. For the eventual agent to be a successful drug, it must have acceptable ADME (absorption, distribution, metabolism and excretion), toxicity and physical chemistry properties, as well as appropriate activity against the target to modify the disease phenotype. Most major pharmaceutical companies have high throughput and computational screens in place to characterize the ADME and physical chemistry properties of their compounds in order to minimize the risk of compound failure due to solubility, bioavailability, potential adverse drug-drug interactions, pharmacokinetics or unfavorable protein binding. Except for the anti-infective therapeutic area, the major reasons for the eventual lack of success of compounds in clinical trials are usually unacceptable toxicity or lack of efficacy (Kennedy, 1997). Unfortunately information about clinical toxicity or lack of efficacy often occurs late in the drug development process after considerable sum resources have been expended. Metabonomics has the potential to provide this information at a much earlier stage in drug development through the identification of pre-clinical safety biomarkers, clinical safety biomarkers and clinical efficacy biomarkers. Thus, technologies such as metabonomics that greatly impact the development of safety and efficacy biomarkers will have the greatest impact on the economics of drug discovery and development.

4. APPLICATION OF METABONOMICS IN THE PHARMACEUTICAL INDUSTRY

4.1 Discovery Applications

4.1.1 Biomarkers to Establish Pharmacokinetics/ Pharmacodynamics Relationships

In drug discovery, the physiological endpoint from *in vivo* disease models is costly and resource intensive to obtain. The development of a metabonomics marker to provide a convenient surrogate market for toxicity or efficacy would play a key role in lead optimization. Since metabonomics is non-invasive, considerable savings in time, animals and valuable discovery compound would be realized. In an ideal situation, early *in vivo* readouts could be accomplished with less additional compound syntheses than is required by conventional testing. Vertex has shown that metabonomics analyses can be performed on samples from normal pharmacokinetics studies using catheterized animals with the stipulation that the database for target organ predictions is built using catheterized animals (Wang *et al.*, 2002). Pfizer is currently building a mouse metabonomics database in order to screen compounds at the early discovery stage without the need for scale up synthesis (Stevens *et al.*, 2002). They have shown dose and time dependent changes in the urinary profiles that are in agreement with observed alteration in clinical chemistry. They have also demonstrated the ability to follow the onset and recovery of a lesion in mice using metabonomics. These initial studies indicate that the mouse would be a viable system for predictive toxicological screening using metabonomics.

4.2 Early Development

Once a compound has passed all of the requirements in discovery, it becomes a candidate for early development. At this stage the primary objective is to ensure that the compound is safe for eventual administration into humans. It is also of considerable interest to have an available biomarker to obtain a definitive signal in early proof of concept clinical studies. The early proof of concept clinical trial is ideally designed to provide the decision on whether to terminate a compound/project or to commit full development efforts. The use of metabonomics for toxicity screening and efficacy biomarkers can make a significant impact on this phase of drug discovery/development.

4.2.1 Pre-screening Animals Prior to Entry into Toxicology Studies

Very often at the end of a toxicology study, lesions are reported in the placebo group or the incidence of lesions does not track with the increasing dose levels, (*i.e.* there are more lesions at a lower exposure than at a higher exposure). Metabonomics offers a tool to quickly and conveniently pre-screen the animals prior to entry into a study. At the beginning of the study and at various intervals during the study, urine can be collected for possible analysis afterwards. Then, animals with findings in the placebo or that are non-dose proportional can be examined with metabonomics to assess if these animals were predisposed to produce the observed lesion versus the animals in the study that did not produce the lesion. With such an analysis, it may be possible to rationalize positive findings in control animals, as well as the lack of dose proportionality. Data to support a predisposition to toxicity could greatly affect the decisions made with the toxicology findings.

4.2.2 Moderate Throughput Screening for Toxicity

Several groups (Robertson *et al.*, 2000; Holmes *et al.*, 2000) have shown that metabonomics can detect injuries in animals that are exposed to prototypical toxins. In fact, these researchers have shown that not only can metabonomics detect an abnormal animal following exposure, but that it is also possible to classify the target organ toxicity based on patterns by matching the patterns of an unknown(s) to those in a database. COMET are currently collaborating to systematically build a database to predict specific organ toxicity in rats and mice (Lindon *et al.*, 2002). As part of this effort each member company performed the COMET metabonomics protocol using hydrazine to demonstrate that these techniques are robust and reproducible. The in-life and the analytical results from this study indicated that all phases of the metabonomics experiment, conducted under the same protocol, were reproducible between these different laboratories.

Urine from hydrazine treated rats collected at each of the six companies was analyzed at Imperial College to show that the *in vivo* experimental techniques were reproducible. Split urine samples from one hydrazine study were analyzed at Imperial College and Roche to show that the NMR analysis was reproducible (Lindon *et al.*, 2002). Similarly, we recently analyzed urine samples from the hydrazine study conducted at Pharmacia to show that our NMR techniques were also reproducible. Urine samples were split between Pharmacia and Imperial College, analyzed at both institutions by ^{1}H NMR, and results were found to be highly similar (R. Bible, data not shown).

At the lead optimization stage in drug development, it may be possible to get an early assessment of the potential toxicity of a template through the judicious use of metabonomics. In most toxicology studies, rangefinder assessments are performed in order to set doses for the eventual formal toxicology study. At this stage, metabonomics would offer a way to screen compounds at high doses before the actual formal toxicology study takes place. With metabonomics, the rangefinder study can use fewer animals, smaller amounts of compounds and a greater number of dose increments to allow the company to quickly set dose levels in toxicology studies that will ensure optimal safety margins.

4.2.3 Reversibility of a Toxic Response

The lack of permanence of an injury greatly affects the impact of the lesion on the decision to proceed or abandon a drug candidate. If a short-term exposure to a compound provides therapeutic benefit with a reversible side effect, then a risk/benefit analysis may indicate that the compound is still a viable drug candidate. However, if short-term exposure to a compound provides therapeutic benefit with an irreversible side effect, then a risk/benefit analysis to indicate that the compound is a viable drug candidate is much more difficult to establish. Hence, it is very important in many circumstances to establish the reversibility of an injury. Metabonomics offers a non-invasive way to monitor the lesions progression and ultimate regression through the analysis of serial urine samples. The timing of the reversal can be easily established with a minimum of compound and in-life animal resources as compared to traditional methods.

4.2.4 Toxicity Mechanism Elucidation

A key question when a toxicology finding is noted is whether the toxicity results from the pharmacological activity of administered drug or whether toxicity is related to its chemical structure. An assessment of the injury-causing mechanism is key to this evaluation. The ramifications are profound: it can influence the decision as to whether to invalidate a target as a potential intervention point for therapy, or to continue with that target but with a different molecule or template. Metabonomics has already been shown to be able to quickly identify certain mechanisms of liver injury such as phospholipidosis (Nicholls *et al.*, 2000; Espina *et al.*, 2001). COMET is presently examining whether other metabonomics responses can predict other mechanisms of injury.

4.3 Clinical Applications

The metabonomics studies that have been reported in pre-clinical animals that are administered prototypical toxins are performed under controlled laboratory conditions. Differences in strain, diet, gender, hormonal cycle and diurnal rhythms can be detected in a metabonomics experiment (Bell *et al.*, 1991; Phipps *et al.*, 1998; Gavaghan *et al.*, 2001; Bollard *et al.*, 2001). However, a significant toxic event produces such profound biomarker changes that the subtle changes in biomarkers due to environmental variables in a controlled laboratory setting do not cause enough biomarker variation to preclude the use of metabonomics toxicity models to predict target organ toxicity by comparing metabonomics patterns from toxin treated and control animal databases (Holmes 2002).

It is not practical to implement the rigid control achievable in the laboratory in clinical settings. Given this assumption, it is reasonable to question whether clinical metabonomics will be feasible with the larger range of normal biomarker variation expected in people. It has been demonstrated that measuring the unique metabonomics pattern differences at pre-dose and post-dose within each individual rather than comparing each individual at pre-dose versus the "normal" population, may be a more reliable method to detect changes in the clinic (Antti *et al.*, 2002).

In one interesting study, van der Greef and co-workers ('t Hart *et al.*, 2002) have demonstrated that a metabonomics analysis of human urine utilizing NMR and PCA can distinguish between control male, control female, multiple sclerosis afflicted males and multiple sclerosis afflicted females. The subsequent metabonomics analysis of the plasma from these patients yielded several biomarkers for multiple sclerosis.

Human renal impairment can be diagnosed and monitored using NMR based techniques. Foxall *et al.*, (1989) reported changes in human urine following an accidental exposure to phenol. The urine in this study was analyzed 3, 4, 14, 25 and 42 days post exposure. As with animal metabonomics, changes in urinary metabolites occurred in a temporal fashion as patients recovered. The changes in the metabolites correlated with the recovery of renal function changes as measured by classical clinical chemistry.

Of clinical importance, the performance of, and complications in, a transplanted kidney can be non-invasively monitored using NMR (Foxall *et al.*, 1993). NMR spectra of urine from patients following renal transplant could distinguish between a functioning kidney, urinary tract infection, renal tubular ischemia and a non-functioning graft. An expert system for renal transplant patients has been developed to quickly and non-invasively distinguish, using only a urine sample, between a normal renal transplant, kidney transplant rejection due to graft *vs* host disease, and cyclosporine immunosuppressant toxicity. The ramifications of this diagnostic test are profound. The clinical intervention in graft *vs* host disease would be to increase dosage of immunosuppressant therapy whereas the intervention in diagnosis of cyclosporine toxicity would be to lower the dose.

These examples of NMR based metabonomics indicate that this technology can be a valuable tool to aid in clinical disease monitoring and early signs of toxicity. Since it is the objective of all pharmaceutical companies to expedite their research efforts, this tool will be greatly expanded in the near future to support initial proof of concept studies. The NMR spectra of human plasma (Nicholson *et al.*, 1995), and human urine (Bales *et al.*, 1984) have already been well characterized which will aid in the development of NMR based clinical chemistry techniques.

5. CONCERNS

While the potential of metabonomics to greatly assist in the discovery and development of drugs is exciting, the ramifications of the use of this new tool or any new tool need to be considered in an industry that operates in a regulated environment. The basic principle of metabonomics is that a specific pattern that consists of the up or down regulation of many endogenous constituents can serve as a biomarker for toxicity or efficacy. These patterns are initially derived from the empirical observation of changes following the administration of prototypical compounds. Unlike proteomics or genomics where there is usually a relatively straightforward molecular biology understanding of what is exactly up or down regulated and a clear linkage of that up/down regulation to biochemical function, the changes observed in a metabonomics experiment are not always straightforward to explain as the biochemical cause for the observed morphologic changes. In genomics the change in an mRNA will usually affect the protein that is coded by that specific message and in proteomics the change in the content of a specific protein will affect the metabolism or signalling function of that specific protein. While proteomics samples are obtained by biopsy where the origin of the sample is known, metabonomics from biofluids does not identify the exact tissue or organ responsible for a change of an endogenous constituent since plasma or urine collection is an average of the metabolism of the whole organism. Thus, in comparison to genomics or proteomics, metabonomics suffers from its inability in some cases to deliver biomarkers that can be well understood by the current state of knowledge of metabolism as to the mechanism of how these markers reflect the toxicity or efficacy that they predict. In these cases it will be a challenge to convince regulators that a pattern of seemingly non-related endogenous constituents serves as a marker to predict toxicity or efficacy when the conclusion is based on empirical observations and correlations.

In the future, will metabonomics patterns ever replace classical clinical chemistry or histopathology to identify the point where injury has occurred? If so then the "no observed effects" levels and safety margins would be based on the time of appearance of a predictive toxicology biomarker in an animal species rather than the appearance of a histopathological lesion or significant clinical chemistry perturbation. This application would bring Good Laboratory Practices and validation requirements to metabonomics.

In the future, as metabonomics information becomes part of regulatory dossiers to support safety or efficacy the issue of the validation of metabonomics patterns as biomarkers will arise. As alluded to earlier, in the conduct of a metabonomics analysis there are many decisions made during the analysis that can affect the outcome of the experiment. For example the presence of drug related components, drug or dosing vehicle, could compromise the analysis of a specific urine or plasma sample. How should a researcher decide to deal with this; omit the sample, replace that part of the spectrum with control spectrum or use an algorithm to attempt to electronically extract out the interfering substances from the NMR spectrum. In a regulated environment, these scientific decisions would be scrutinized for possible bias, yet there is no established method to deal with these issues, as metabonomics is a relatively new technology.

ACKNOWLEDGEMENTS

The authors would like to thank Dr. Charles Jeuell and Dr. Matthew McLean for their helpful reviews of this manuscript.

REFERENCES

Annan R. *Proteomics and Biological Mass Spectrometry*. Land O'Lakes Bioanalytical Conference, Devil's Head, WI (2000).

Antti H, Brindle J, Shockcor J *et al.* Multivariate statistical pre-treatment of NMR-based metabonomics data to remove inherent physiological variation confounding toxicity patterns. 41st Meeting of the Society of Toxicology, Nashville (2002).

Bales JR, Higham DP, Howe I *et al.* Use of high-resolution proton nuclear magnetic resonance spectroscopy for rapid multi-component analysis of urine. *Clin Chem* 30: 426-432 (1984).

Beckwith-Hall BM, Nicholson JK, Nicholls AW *et al.* Nuclear magnetic resonance spectroscopic and principal components analysis investigations into biochemical effects of three model hepatotoxins. *Chem Res Toxicol* 11: 260-272 (1998).

Beckwith-Hall BM, Holmes E, Lindon JC *et al.* NMR-based metabonomic studies on the biochemical effects of commonly used drug carrier vehicles in the rat. *Chem Res Toxicol* in press (2002).

Bell JD, Sadler PJ, Morris VC, Levander OA. Effect of aging and diet on proton NMR spectra of rat urine. *Magn Reson Med* 17: 414-422 (1991).

Bollard ME, Holmes E, Lindon JC *et al.* Investigations into biochemical changes due to diurnal variation and estrus cycle in female rats using high-resolution ^{1}H NMR spectroscopy of urine and pattern recognition. *Anal Biochem* 295: 194-202 (2001).

Espina JR, Shockcor JP, Herron WJ *et al.* Detection of *in vivo* biomarkers of phospholipodosis using NMR-based metabonomics approaches. *Magn Reson Chem* 28: 559-565 (2001).

Foxall PJD, Bending M, Gartland KPR, Nicholson JK. Acute renal failure following accidental cutaenous exposure to phenol: application of PMR urinalysis to monitor the disease process. *Human Toxicol* 9: 441-449 (1989).

Foxall PJD, Mellotte GJ, Bending MR *et al.* NMR spectroscopy as a novel approach to the monitoring of renal transplant function. *Kidney Intl* 43: 234-245 (1993).

Gartland KPR, Lindon JC, Beddell J, Nicholson JK. Application of pattern recognition methods to the analysis of toxicological data generated by NMR spectroscopy. *Mol Pharmacol* 39: 629-642 (1991).

Gavaghan CL, Nicholson JK, Connor SC *et al.* HPLC-NMR spectroscopic and chemometric studies on metabolic variation in Sprague-Dawley rats. *Anal Biochem* 291: 245-252 (2001).

Gygi SP, Rochon Y, Franza BR *et al.* Correlation between protein and mRNA abundance in yeast. *Mol Cell Biol* 19: 1720-1730 (1999).
Gygi SP, Rist B, Gerber SA *et al.* Quantitative analysis of complex protein mixtures using isotope-coded affinity tags. *Nature Biotechnol* 17: 994-999 (1999).
Holland JF, Leary JJ, Sweeley CC. Advanced instrumentation and strategies for metabolic profiling *J Chromator Biomed Appl* 379: 3-26 (1986).
Holmes E, Foxall PJD, Neild GH *et al.* Automatic data reduction and pattern recognition methods for analysis of ^{1}H nuclear magnetic resonance spectra of human urine from normal and pathological states. *Anal Biochem* 220: 284-296 (1994).
Holmes E, Nicholls AW, Lindon JC *et al.* Development of a model for classification of toxin-induced lesions using ^{1}H NMR spectroscopy of urine combined with pattern recognition. *NMR Biomed* 11: 235-44 (1998).
Holmes E, Nicholls AW, Lindon JC *et al.* Chemometric models for toxicity classification based on NMR spectra of biofluids. *Chem Res Tox* 13: 471-478 (2000).
Holmes E. Application of Metabonomics Expert Systems to the Study of Physiological and Toxicological Variation. 41st Meeting of the Society of Toxicology, Nashville (2002).
Kennedy T. Managing the drug discovery /development interface. *Drug Discov Today* 2: 436-444 (1997).
Lindon JC, Nicholoson JK, Holmes E *et al.* Metabonomics in toxicology: the COMET project. *Toxicol Appl Pharmacol* submitted (2002).
Nicholls AW, Nicholson JK, Haselden JN, Waterfield CJ. A metabonomic approach to the investigation of drug induced phospholipodosis. *Biomarkers* 5: 410-423 (2000).
Nicholson JK, Buckingham MJ, Sadler PJ. High-resolution proton NMR studies of vertebrate blood and plasma. *Biochem J* 211: 605-615 (1983).
Nicholson JK, Wilson ID. High-resolution proton NMR spectroscopy of biological fluids. *Prog NMR Spectr* 21: 449-501 (1989).
Nicholson JK, Foxall PJD. 750 MHz ^{1}H and ^{1}H-^{13}C NMR spectroscopy of human blood plasma. *Anal Chem* 67: 793-811(1995).
Petricoin EF III, Ardekani AM, Hitt BA *et al.* Use of proteomic patterns in serum to identify ovarian cancer. *Lancet* 395: 572-577 (2002).
Phipps AN, Stewart J, Wright B, Wilson ID. Effect of diet on the urinary excretion of hippuric acid and other dietary-derived aromatics in the rat. A complex interaction between diet, gut microflora and substrate specificity. *Xenobiotica* 28: 527-537 (1998).
Robertson DG, Reily MD, Sigler RE *et al.* Metabonomics: evaluation of nuclear nagnetic resonance (NMR) and pattern recognition technology for rapid *in vivo* screening of liver and kidney toxicants. *Toxicol Sci* 57: 326-337 (2000).
Stevens GJ, Chambers J, Evering W, Deese A. Evaluation of metabonomics technology for liver and kidney toxicants in mice. 41st Meeting of the Society of Toxicology, Nashville (2002).
't Hart BA, Vogels JTWE, Spijksma G *et al.* ^{1}H-NMR spectroscopy combined with pattern recognition analysis reveals characteristic chemical patterns in urines of MS patients and non-human primates with MS-like disease. *J Neurol Sci* submitted (2002).
Wang S, Regnier FE. Proteomics based on selecting and quantifying cysteine containing peptides by covalent chromatography. *J Chromatogr* 924: 345-357 (2001).
Wang Y, Fejzo J, Heiser A *et al.* Application of metabonomics in drug design: can metabonomics and pharmacokinetics (PK) studies be conducted simultaneously in the same set of rats? 41st Meeting of the Society of Toxicology, Nashville (2002).

Chapter 5

METABOLIC PROFILING IN TUMORS BY *IN VIVO* AND *IN VITRO* NMR SPECTROSCOPY

Yeun-Li Chung, Marion Stubbs and John R. Griffiths
Cancer Research UK Biomedical Magnetic Resonance Research Group, Department of Biochemistry and Immunology, St. George's Hospital Medical School, London SW17 ORE, UK

1. INTRODUCTION

Genomic and proteomic information is transforming all fields of cancer research. The increased understanding of the genomics and molecular pathology of cancer also provides a framework for designing new therapeutic strategies. The majority of the genes in most genomes are still of unknown function, and many studies on proteomics are attempting to fill this knowledge gap. Since biochemical information flows from DNA to RNA to protein to function, the role of each gene product in metabolism clearly needs to be studied. A recent theoretical study by ter Kuile and Westerhoff (2001) using the methods of Metabolic Control Analysis demonstrated that there is no general quantitative relationship between mRNA levels and function. Thus a comprehensive study of many metabolites together could become invaluable, and several approaches to this are being developed including metabolite target analysis, metabolite profiling, metabolomics and metabolic fingerprinting as described elsewhere in this book (Fiehn, Chapter 11, see also Fiehn, 2001).

The role of each gene product in metabolism is the link between the genotype and the phenotype (Fiehn, 2002); a single gene mutation may cause alterations of metabolite levels of seemingly unrelated biochemical pathways and this is likely to happen when genes are constitutively over-expressed or anti-sense inhibited. In order to help researchers understand such systems, a comprehensive and quantitative analysis of all metabolites is required, with experimentally robust and reproducible methodologies that aim to include all classes of compounds with high recovery.

So far, most metabolomic research has been performed on plants or microorganisms and relatively little has been published on human metabolomics or on cancer studies. It is important therefore to develop methods for metabolomics that can be used to improve our understanding of cancer and cancer therapy. NMR may play an important role in metabolomic studies, as large arrays and classes of small molecules in cell or tissue extracts can be examined simultaneously in a single experiment without any assumptions as to the types of molecule present. The pattern of concentration changes of these small molecules can be used to infer the metabolic pathways that are involved, and hence, clarify the role of a gene with an unknown function. NMR technologies are able to satisfy most of the criteria required for metabolomic studies. These important factors include: robustness, reproducibility, the ability to identify unknown metabolites and the potential to incorporate them into models of theoretical biochemical networks.

The application of NMR spectroscopy in metabolomic studies will also further complement its role in cancer research and cancer therapy. NMR spectroscopy has already been used to study tumor biology and physiology both in man and in animal models *in vivo* to obtain information on tumor diagnosis and monitoring responses following therapy, as well as predicting patient outcomes. It has also been used for *in vitro* studies of cell and tumor extracts (Aboagye *et al.,* 1998; Florian *et al.*, 1996; Griffiths *et al.*, 2002a, 2002b; Leach *et al.*, 1998; McSheehy *et al.*, 1998; Ronen *et al.*, 1999; Stubbs and Griffiths, 1999; Tate *et al.*, 1998; Williams *et al.*, 1998; Wolf *et al.*, 1998).

2. METABOLOMICS

Proton (^{1}H) NMR has been used in a 'proof-of-principle' demonstration of a practical metabolomic screening method (Raamsdonk *et al.*, 2001) in a eukaryotic genome, with the main objective of ultimately predicting the function of 'silent' genes in yeast. The deletion of silent genes causes no noticeable effect on metabolic fluxes or growth and yet these genes account for about 85% of the yeast genome. The effects of changes in enzyme activity on metabolite concentrations can be much larger than their effects on metabolic fluxes so examining metabolite concentrations is likely to be a good method for revealing the role of 'silent genes' (Cornish-Bowden and Cardenas, 2001).

Raamsdonk *et al.* (2001) used ^{1}H NMR to measure the change of concentrations of large numbers of metabolites simultaneously in yeast, caused by deletion of genes with known function. Different strains were used with knockouts of genes coding for various enzymes in the glycolytic and respiratory pathways. Statistical clustering algorithms were used to show the pattern of metabolic changes corresponding to deletions of genes in each metabolic pathway.

The cell lines with the knockout of (different) enzymes involved in the control of the glycolytic enzyme phosphofructokinase had patterns of altered metabolite concentration that clustered together. Cell lines with knockouts of enzymes that associated with the respiratory chain had patterns of altered metabolite concentrations that did not cluster with the glycolytically compromised ones. Once a database of the patterns of metabolic changes associated with many known gene functions is established, it may be possible to use this methodology to identify the metabolic function of unknown genes.

Because NMR has the advantage of obtaining information on large numbers of metabolites without the need for prior knowledge of the study systems this method has great potential for exploring systems with changes in metabolic pathways following gene mutations.

The rest of this chapter details preliminary studies in mammalian tumors where NMR spectroscopy has been used: a) to understand the metabolic changes caused by tumors deficient in a single transcription factor and b) to look at the effects of novel anticancer drug treatments.

2.1 Evaluating the Effect of a Gene Mutation by NMR Spectroscopy in a Mammalian System

A metabolomic study of a mouse hepatoma tumor model either wild type or deficient in hypoxia-inducible factor 1β (HIF-1β) was carried out by Griffiths *et al.* (2002b) using *in vivo* and *in vitro* NMR methodologies and some classical biochemical methods. Unexpected results were found in the *in vivo* NMR and biochemical study that were difficult to explain until *in vitro* NMR studies provided the extra pieces of information to complement the *in vivo* and biochemical data. This illustrates the complementary application of *in vivo* and *in vitro* NMR based metabolic profiling to studies on genetic manipulation in mammalian systems.

The HIF-1 transcription factor is over-expressed in a broad range of cancers. It is activated by hypoxia and regulates many pathways involved in tumor growth. HIF-1 is a heterodimer, consisting of α and β subunits and even though the hypoxic regulation of HIF-1 occurs mainly through the α subunits, a functional HIF-1 complex is unable to form if the cells are HIF-

1β deficient (Wang *et al.*, 1995; Maxwell *et al.*, 1997). HIF-1β deficient hepatoma cells (Hepa c4) were found to be less able to tolerate hypoxia under hypoxic conditions when compared to wild-type hepatoma (Hepa WT) cells (Stratford *et al.*, 1999). HIF-1 activates transcription in 13 different genes encoding glucose transporters and glycolytic enzymes, and thus can coordinate the regulation of the entire glycolytic pathway from glucose uptake to lactic acid production (Dang and Semenza, 1999). When Hepa c4 cells were grown up as solid tumors in mice, a decreased growth rate and reduced or absent expression of HIF-1 activation targets such as transcription of genes encoding the vascular endothelial growth factor (VEGF) and the glucose transporters, GLUT-1 and GLUT-3, were observed (Maxwell *et al.*, 1997; Williams *et al.*, 2002). The metabolic consequences and importance of the glycolytic pathway in intermediary metabolism in these tumors *in vivo* is unknown, but co-ordinated up-regulation of the glycolytic pathway by HIF-1 would be expected to have a significant effect.

The aims of our metabolomic study were therefore: a) to use *in vivo* MRS and MRI methods, complemented by *in vitro* NMR and classical biochemical analysis of tumor extracts to measure physiological and biochemical parameters in wild-type (Hepa WT) and HIF-1β deficient (Hepa c4) mutant cells grown in solid tumors *in vivo* and b) to determine how a deficiency in HIF-1β alters the metabolic phenotype in mice. The measurement of tumor biochemistry *in vivo* by NMR could be an excellent non-invasive method for monitoring changes caused by gene modification because changes in protein expression can alter the concentration of cellular metabolites.

Hepa c4 tumors grew more slowly than Hepa WT tumors, partly because of a longer lag phase. Two NMR relaxation parameters were measured; i) T_2, calculated from spin-echo images, which reflects aspects of the tissue microstructure and water content and showed a small but significant difference between Hepa WT and Hepa c4 tumors and ii) T_2*, which reflects aspects of the tumor vascularity (through the effects of deoxyhemoglobin on magnetic field inhomogeneities) and which showed no difference between the wild type and HIF-1β deficient tumors, suggesting no difference in vascularity.

Because HIF-1 deficiency has been shown to compromise VEGF expression (which promotes angiogenesis) it might have been expected that a difference in vascularity would be observed in Hepa c4 tumors, and indeed vascular differences had been seen in very small tumors (0.05-0.2g, Maxwell *et al.*, 1997). However for *in vivo* NMR experiments it was necessary to use Hepa c4 and Hepa WT tumors that had reached a larger (~0.5g) size than those used by Maxwell *et al.* (1997). Chalkley vascular counting and gross histology (Griffiths *et al.*, 2002b) confirmed that there were no significant differences in either vascularity, vessel density or necrosis in these larger tumors.

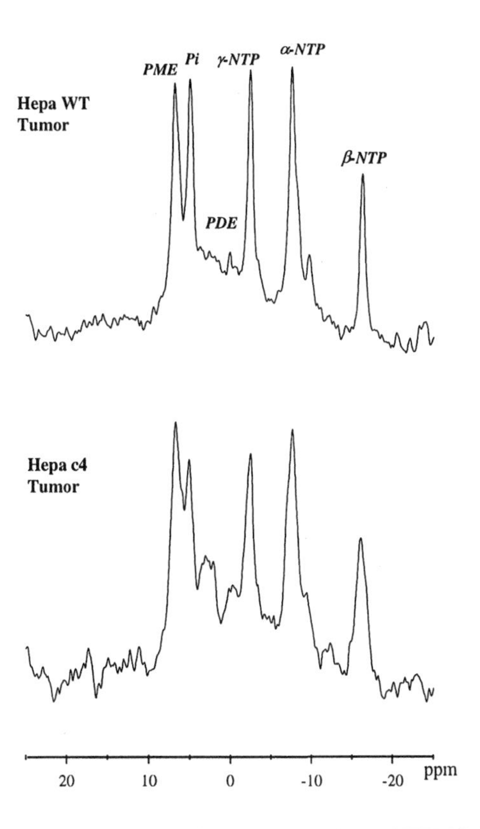

Figure 1. In Vivo ^{31}P NMR spectra of a wild-type (Hepa WT) and HIF-1β deficient (Hepa c4) tumor. Spectra were obtained by ISIS (Image Selected *In Vivo* Spectroscopy) on a 4.7T Varian System. Resonances identified were phosphomonoester (PME), inorganic phosphate (Pi), phosphodiester (PDE) and α-, β-, γ- nucleotide triphosphate (α-, β-, γ- NTP).

Examination of the HIF-1β deficient and wild type tumors by *in vivo* phosphorus (^{31}P) MR (see spectra of a mutant Hepa c4 and Hepa WT tumor in Fig. 1) indicated no differences between the β-phosphate/inorganic phosphate (β-NTP/Pi) ratio, the intracellular pH, the extracellular pH and the phosphomonoester/Pi (PME/Pi) ratio. However the phosphodiester (PDE)/Pi ratio was 3 times higher in the HIF-1β deficient (Hepa c4) tumors compared to the wild type (Hepa WT) tumors (see Table 1). The PDE signal includes catabolites of phospholipid membrane components whereas the PME includes components from precursors of phospholipid membranes.

Table 1. InVivo ^{31}P NMR Measurement of pH and Metabolite Ratios in Wild-Type (Hepa WT) and HIF-1β Deficient (Hepa c4) Tumors (n = 6)

NMR parameter	Hepa WT	Hepa c4
PHi	6.96 ± 0.02	7.03 ± 0.02*
PHe	6.95 ± 0.04	6.96 ± 0.02
β-NTP/Pi	1.06 ± 0.07	0.93 ± 0.17
PME/Pi	0.24 ± 0.02	0.22 ± 0.01
PDE/Pi	0.04 ± 0.01	0.11 ± 0.03*

* for Hepa WT compared with Hepa c4, $p < 0.05$

An *in vitro* enzymatic assay for the ATP content in tumor extracts showed that the level of ATP in the Hepa c4 tumors was about 5 times lower than the Hepa WT tumors and this was reflected in the poorer signal-to-noise ratio and slightly broader spectra observed by NMR in the Hepa c4 tumors (see Fig. 1) compared to the Hepa WT tumor.

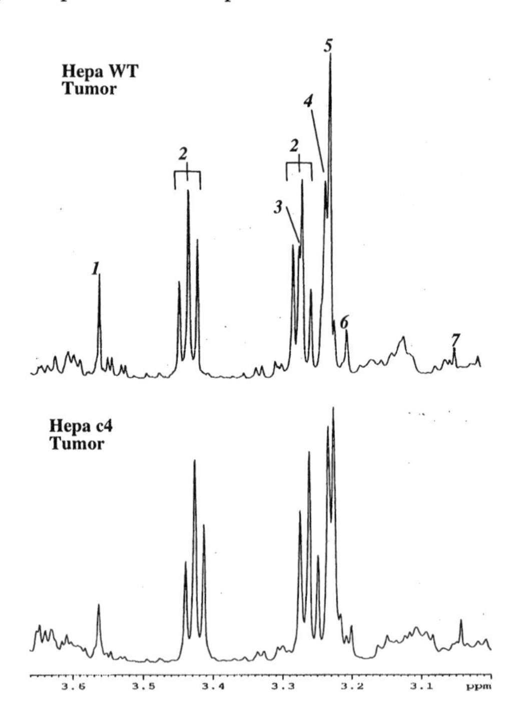

Figure 2. In vitro ^{1}H NMR spectra of a wild-type (Hepa WT) and HIF-1β deficient (Hepa c4) tumor extract. Spectra were obtained from neutralized acid extracts prepared from freeze-clamped tumors on a 500MHz Bruker System. Keys: 1. glycine 2. taurine 3. betaine 4. glycerophosphocholine 5. phosphocholine 6. free choline 7. creatine.

Because these results (similar ATP/Pi ratio, lower ATP content, higher PDE/Pi in the Hepa c4 tumor) were difficult to explain and contrary to what was expected, tumor extracts were examined by *in vitro* ^{1}H NMR. Representative spectra of Hepa c4 and Hepa WT tumor extracts are shown in Fig. 2. Of the 14 signals easily quantified (for details see Griffiths *et al.*, 2002a, 2002b), phosphocholine, free choline, betaine and glycine were found to be significantly lower in the Hepa c4 tumors when compared with Hepa WT (Table 2).

Table 2. In Vitro ^{1}H NMR Measurement of Metabolites in Wild-Type (Hepa WT) and HIF-1β Deficient (Hepa c4) Tumor Extracts (n = 4).

Metabolites	Hepa WT	Hepa c4
Betaine	1.10 ± 0.20	0.40 ± 0.06*
Phosphocholine	3.78 ± 0.65	1.22 ± 0.24*
free choline	0.43 ± 0.08	0.17 ± 0.02*
Glycine	3.93 ± 0.65	1.90 ± 0.38*
Taurine	16.10 ± 3.60	11.90 ± 1.61
Glycerophosphocholine	2.58 ± 0.49	1.55 ± 0.24

* for Hepa WT compared to Hepa c4, $p < 0.05$. Values expressed as μmol/g wet weight.

As mentioned earlier, the most surprising metabolic difference found in the HIF-1β deficient tumors was their much lower ATP content: about 20% of that of Hepa WT cells. Despite this, they had an unchanged ATP/Pi ratio (which reflects the cell's energy status), suggesting that their energy metabolism was not significantly impaired by the deficiency in total ATP concentration. The low ATP content may have been due to the low expression of the gene for HIF-1β disrupting ATP synthesis, probably by preventing the synthesis of adequate amounts of a crucial intermediate. The clue to the explanation for this was provided by the *in vitro* ^{1}H metabolic profile of the two tumor types. Glycine was present at less than half of the concentration in deficient Hepa c4 tumor extracts compared with the Hepa WT tumors. Glycine is formed from 3-phosphoglycerate in the glycolytic pathway via serine. Serine is not resolved and therefore not quantifiable in the MR spectrum but it is a major supplier of one-carbon units for nucleotide biosynthesis (Snell and Fell, 1989). Glycolysis is thus an important precursor for purine synthesis and so the inability of HIF-1β deficient tumors to upregulate glycolysis could prevent newly formed cells from synthesizing sufficient purine moieties to produce the required quantity of ATP.

Glycine can alternatively be formed from choline *via* betaine. Lower levels of phosphocholine, free choline and betaine were also found in the metabolic profile of the Hepa c4 tumor extracts when compared with Hepa WT tumor extracts. These observations would be consistent with a model in which the choline pool was bled off to create glycine and thus ATP, instead of forming phosphocholine and membrane phospholipids (Fig. 3).

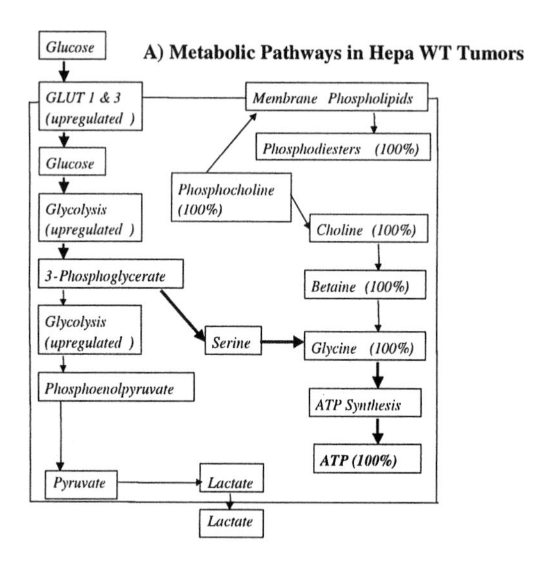

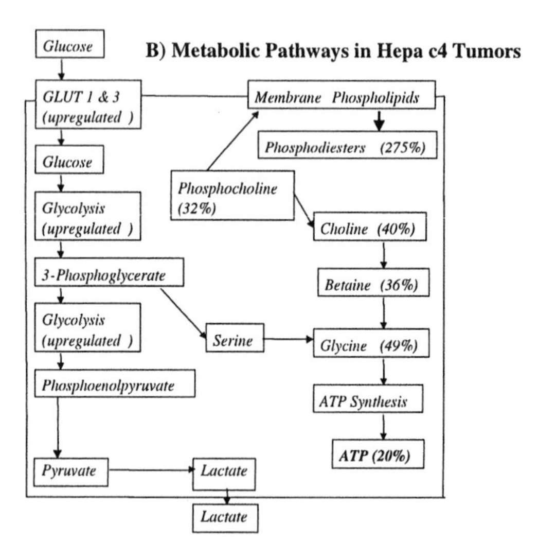

Figure 3. Schematic diagrams of the metabolic pathways in A) wild-type (Hepa WT) and B) HIF-1β deficient (Hepa c4) tumors. Metabolite concentrations in Hepa WT tumors were assumed to be 100% and the percentage metabolite concentrations in Hepa c4 tumors were related to those in the Hepa WT tumor. The percentage values are calculated from concentrations shown in Table 2, except for ATP, which are the mean ATP levels calculated from metabolic imaging and enzymatic essays of the tumor extracts (Griffiths *et al.*, 2002a).

This finding also offers an explanation for the raised PDE in the Hepa c4 tumor; PDEs are catabolites of membrane phospholipids and their higher levels in HIF-1β deficient Hepa c4 tumors might be due the abnormal choline catabolism causing compromised membrane turnover.

In summary, then, it appeared that Hepa c4 tumors failed to upregulate glycolysis because of their deficiency in HIF-1β and therefore could not supply enough 3-phosphoglycerate as a precursor of glycine synthesis. This is in accord with a number of studies that have suggested that cancer cells divert glycolytic metabolites into precursors for anabolic pathways (Eigenbrodt *et al.*, 1992). However it should be pointed out that the amount of glucose used to synthesize new ATP molecules necessary for the growth of a 0.5g tumor is only less than 0.01% of the total glucose used by the tumor. Most of the glucose will be used to generate the energy required for the maintenance of processes such as ion gradients and other processes concerned with tumor growth. Details of these calculations can be found in a forthcoming publication (Griffiths and Stubbs, 2003).

This study has shown that the combined use of both *in vivo* and *in vitro* NMR can provide a powerful tool to measure physiological and biochemical parameters in control and mutant tumors. ^{1}H NMR metabolic profiling of tumor extracts may have a role in studying the altered metabolic phenotype following manipulation of gene function.

2.2 Use of NMR Metabolic Profiling to Identify Changes Associated with Specific Anticancer Drug Targets

The combined use of *in vivo* and *in vitro* NMR metabolic profiling also provides a useful tool for determining the metabolic changes and pathways that are associated with specific drug target mechanisms. Two examples are shown in which different metabolic profiles were obtained for two novel anticancer drugs with different targets and modes of action. Both drugs were tested in a human colorectal xenograft (HT29) grown in mice. The aim of these studies was to identify surrogate markers and pharmacodynamic end points for tumor response to drug treatment and in this regard they were successful (Chung *et al.*, 2002a, 2002b).

2.2.1 17-Allylamino-17-Demethoxygeldanamycin (17-AAG)

17-AAG is an inhibitor of heat shock protein 90 (Hsp90), a molecular chaperone that ensures the correct folding of several proteins associated with tumor growth. Inhibition of Hsp90 results in degradation of oncogenic client proteins by proteasomal degradation. Treatment with 17-AAG over 4 days in these models caused cessation of tumor growth. Examination of the tumors by ^{31}P NMR *in vivo* before and after four days treatment showed a significant reduction in the β-NTP/Pi ratio, a significant increase in both the PME/PDE ratio and the PME/TotP ratio (where TotP is the total phosphorus signal). No significant changes in the control group were found (Chung *et al.*, 2002a).

In vitro analyses of the tumor extracts by ^{1}H NMR from the 17-AAG-treated mice showed a significant increase in the phosphocholine/glycero-phosphocholine ratio and in the levels of phosphocholine and valine when compared with vehicle-controls. Both *in vitro* and *in vivo* NMR technologies revealed changes in metabolites related to membrane phospholipid metabolism (phosphocholine and PME).

Further studies are still needed to fully relate the metabolic changes to the underlying effects of 17-AAG on cancer cell metabolism. These preliminary results, however, clearly reveal patterns of changes that may be useful in identifying the effects of this drug and demonstrate that *in vitro* and *in vivo* NMR may provide a non-invasive surrogate marker of tumor response to 17-AAG and other drugs.

2.2.2 CYC202

CYC202 (R-roscovitine) blocks cell cycle progression through inhibition of cyclin-dependent kinases 1, 2 and 7 and is presently in Phase I clinical trials. (Chung *et al.*, 2002b). Tumor growth was inhibited in CYC202-treated mice when compared with controls. *In vivo* ^{31}P NMR of the tumors showed a significant decrease in the intracellular pH and in the ratio NTP/TotP post-treatment. A significant increase in the Pi/TotP and the Pi/NTP ratios were also observed post-treatment, whereas no significant changes were seen in the vehicle-treated group.

^{1}H MRS of tumor extracts from CYC202-treated mice showed a significant decrease in glycerophosphocholine, glycine and glutamate levels when compared to controls. As with the 17-AAG study the changes observed may be useful in identifying the effects of this drug.

3. CONCLUSION

We have demonstrated that NMR based metabolic profiling has extraordinary potential role in future drug discovery programs. Once a database of changes in metabolite patterns is established it will be possible to correlate them with specific drug target mechanisms. The database can then be used to confirm the mode of drug action and targets of new drugs discovered in the future and to determine pharmacodynamic and efficacy end-points.

REFERENCES

Aboagye EO, Dillehay LE, Bhujwalla ZM, Lee D-J. Hypoxic cell cytotoxin tirapazamine induces acute changes in tumor energy metabolism and pH: a ^{31}P magnetic resonance spectroscopy study. *Radiat Oncol Investigat* 6: 249-254 (1998).

Chung Y-L, Troy H, Banerji U *et al.* The pharmacodynamic effects of 17-AAG on HT29 xenografts in mice monitored by magnetic resonance spectroscopy. *Proc Am Assoc Cancer Res* 43: 73 (2002a).

Chung Y-L, Troy H, Judson IR *et al.* The effects of CYC202 on tumors monitored by magnetic resonance spectroscopy. *Proc Am Assoc Cancer Res* 43: 336 (2002b).

Cornish-Bowden A, Cardenas ML. Functional genomics. Silent genes given voice. *Nature* 409, 571-572 (2001).

Dang CV, Semenza GL. Oncogenic alterations of metabolism. *Trends Biochem Sci* 24: 68-72 (1999).

Eigenbrodt E, Reinacher M, Scheefer-Borchel U *et al.* Double role for pyruvate kinase type M_2 in the expansion of phosphometabolite pools found in tumor cells. *Crit Rev Oncogenesis* 3: 91-115 (1992).

Fiehn O. Combining genomics, metabolome analysis, and biochemical modelling to understand metabolic networks. *Compar Funct Genom* 2: 155-168 (2001).

Fiehn O. Metabolomics - the link between genotypes and phenotypes. *Plant Mol Biol* 48: 155-171 (2002).

Florian CL, Preece NE, Bhakoo KK, Williams SR. Characteristic metabolic profiles revealed by ^{1}H NMR spectroscopy for three types of human brain and nervous system tumors. *NMR Biomed* 8: 253-264 (1996).

Griffiths JR, Tate AR, Howe FA, Stubbs M. Magnetic resonance spectroscopy of cancer - practicalities of multi-centre trials and early results in non-Hodgkins lymphoma. *Eur J Cancer* 38: in press (2002a).

Griffiths JR, McSheehy PMJ, Robinson SP *et al.* Metabolic changes detected by *in vivo* magnetic resonance studies of HEPA-1 wild-type tumor deficient in hypoxia-inducible factor-1β (HIF-1β): evidence of an anabolic role for the HIF-1 pathway. *Cancer Res* 62: 688-695 (2002b).

Griffiths JR, Stubbs M. Opportunities for studying cancer by metabolomics: preliminary observations on tumors deficient in hypoxia-inducible factor 1. *Adv Enzyme Reg* in press (2003).

Leach MO, Verrill M, Glaholm J *et al.* Measurements of human breast cancer using magnetic resonance spectroscopy: a review of clinical measurements and a report of localized ^{31}P measurements of response to treatment. *NMR Biomed* 11: 314-340 (1998).

Maxwell PH, Dachs GU, Gleadle JM *et al.* Hypoxia-inducible factor 1 modulates gene expression in solid tumors and influences both angiogenesis and tumor growth. *Proc Natl Acad Sci USA* 94: 8104-8109 (1997).

McSheehy PMJ, Robinson SP, Ojugo ASE *et al.* Carbogen breathing increases 5-fluorouracil uptake and cytotoxicity in hypoxic murine RIF-1 tumors: a magnetic resonance study *in vivo*. *Cancer Res* 58: 1185-1194 (1998).

Raamsdonk LM, Teusink B, Broadhurst D *et al.* A functional genomics strategy that uses metabolome data to reveal the phenotype of silent mutations. *Nature Biotechnol* 19: 45-50 (2001).

Ronen SM, DiStefano F, McCoy CL *et al.* Magnetic resonance detects metabolic changes associated with chemotherapy-induced apoptosis. *Br J Cancer* 80: 1035-1041 (1999).

Snell K, Fell DA. Metabolic control analysis of mammalian serine metabolism. *Adv Enzyme Reg* 30: 13-32 (1989).

Stratford IJ, Patterson AV, Dachs GU *et al.* Hypoxia-mediated gene expression. In *Tumor Hypoxia.* Vaupel P, Kelleher DK (Ed) pp. 107-113, Wissenschaftliche Verlagsgesellschaft, Stuttgart (1999).

Stubbs M, Griffiths JR. Monitoring cancer by magnetic resonance. *Br J Cancer* 80: 86-94 (1999).

ter Kuile BH, Westerhoff HV. Transcriptome meets metabolome: hierarchical and metabolic regulation of the glycolytic pathway. *FEBS Lett* 500: 169-171 (2001).

Tate AR, Griffiths JR, Martinez-Perez I *et al.* Towards a method for automated classification of ^{1}H MRS spectra from brain tumors. *NMR Biomed* 11: 177-191 (1998).

Wang GL, Jiang BH, Rue EA, Semenza GL. Hypoxia-inducible factor 1 is a basic-helix-loop-PAS heterodimer regulated by cellular O_2 tension. *Proc Natl Acad Sci USA* 92: 5510-5514 (1995).

Williams KJ, Telfer BA, Airley RA *et al.* A protective role for HIF-1 in response to redox manipulation and glucose deprivation: implications for tumorigenesis. *Oncogene* 21: 282-290 (2002).

Chapter 6

RAMAN SPECTROSCOPY FOR WHOLE ORGANISM AND TISSUE PROFILING

Sarah Clarke[1] and Royston Goodacre[1,2]
[1]Institute of Biologial Scicences University of Wales, Abersytwyth, SY23 3DD, UK [2]Department of Chemistry, University of Manchester Institute of Science, PO Box 88, Sackville St., Manchester M60 1QD, UK.

1. INTRODUCTION AND THEORY

The inelastic light scattering process that has become eponymously known as Raman scattering was first observed in 1928 by the Indian physicist Chandrasekhara Venkata Raman and reported in *Nature* (Raman and Krishnan, 1928). Raman scattering spectra, like infrared (IR) absorption spectra, originate from an exchange of energy between photons and vibrational or rotational motions in molecules (Nelson, 1985). Raman scattering is a light scattering phenomenon in which an incident photon beam of well-defined wavelength (a monochromatic laser) is scattered by molecules. While most of the radiation is scattered elastically (Rayleigh scattering), a small fraction of the photons are modified as an irradiated molecule undergoes a vibrational transition (Adar *et al.*, 1997).

Raman spectra are acquired by irradiating a sample with an intense laser source. The interaction of laser light with the molecule produces scattering of three types namely: Rayleigh, Stokes and Anti-Stokes scattering (Fig. 1 for diagrammatic details). Scattering involves a momentary distortion of electrons, followed by re-emission of the radiation as an irradiated chemical bond returns to its ground electronic state (Sanford *et al.*, 2001). When light interacts with molecules, most of the incident photons are scattered from that matter with an identical wavelength. This process is called Rayleigh scattering or elastic scattering (Colthup *et al.*, 1990; Ferraro and Nakamoto,

1994; Maquelin, 2002). When there is a net loss of energy due to the creation of a molecular vibration, the result is termed Stokes; when there is a net gain in energy due to absorption the result is termed an Anti-Stokes process (Fig. 2).

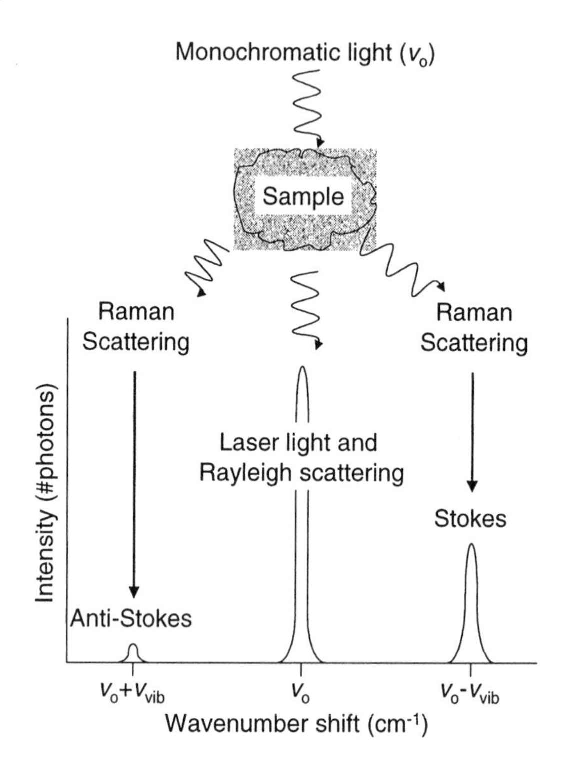

Figure 1. The Raman Effect.

Modern Raman instruments plot the intensity of scattered light versus the difference between its frequency and the frequency of the exciting line (*vide infra*). This frequency difference is called the 'Raman shift' and Raman data are usually given as wavenumber (cm^{-1}) shifts from the incident radiation (Stevenson and Vo-Dinh, 1996).

The vibrational Raman spectrum acquired provides a 'fingerprint' of the molecule and in this sense, the information obtained is similar to IR spectroscopy, another vibrational spectroscopic technique that measures absorption. Raman spectroscopy, like IR, profits as an analytical technique due to the fact that the molecular vibrational signature is a highly characteristic and structurally sensitive fingerprint of the molecule. Raman and IR spectroscopy are considered to be, in many cases, complementary to each other as compounds that tend to produce weak signatures with one of these techniques tend to produce strong signatures with the other.

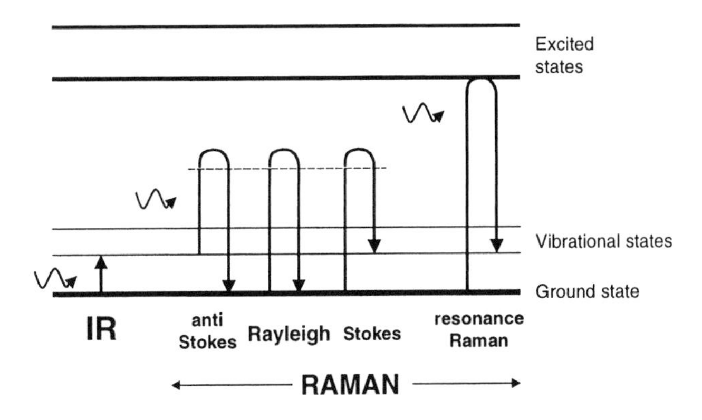

Figure 2. Vibrational States.

There are several important advantages of Raman spectroscopy over other absorption techniques, namely:

1. Water is an ideal solvent as its Raman spectrum is weak. It does not absorb laser irradiation. Neither is it heated by the beam, nor is the Raman scattered light attenuated which would cause distortion in the relative intensities in the spectrum in a way that would preclude quantitative work.
2. There are important molecular functional groups, which are inactive or very weak in absorption processes. For example, the homopolar diatomic gases are only active in Raman spectroscopy and Raman spectra provide much better sensitivity to carbon-carbon bonds (Adar *et al.*, 1997).

2. METHODS

2.1 Raman Lasers

Today Raman instruments use lasers with wavelengths in the near infra-red (NIR), visible and ultra violet (UV) parts of the spectrum along with various methods of enhancement including resonance Raman spectroscopy, surface enhanced Raman spectroscopy (SERS), and surface enhanced resonance Raman spectroscopy (SERRS).

Visible light is often the most convenient for Raman spectroscopy and it yields good sensitivity. However, a significant fraction of samples fluoresce with visible excitation, and better signal-to-noise ratios can be obtained with NIR (750-1064 nm), albeit at some cost in sensitivity and acquisition time.

In the case of Fourier transform (FT) Raman (usually at 1064 nm with a Nd:YAG (neodymium:yttrium aluminum garnet) laser), detection limits are often much higher than for visible Raman. This is because water may absorb both the exciting laser radiation at 1064 nm and the Raman scattered light. It is therefore often necessary to co-add many hundreds of spectra to produce high-quality data (particularly from biological systems), and acquisition times are frequently 15-60 min. This sensitivity loss, however, is compensated by a major decrease in fluorescence interference and the experimenter must decide whether fluorescence reduction or sensitivity is more important for a given application (McCreery, 1996).

The choice of excitation wavelength generally depends on three factors:

1. Whether or not it causes significant fluorescence in the analyte.
2. The attenuation of the light in the optical fibers at the excitation wavelength.
3. The type of detector used to measure the spectrum.

Generally speaking, the best excitation wavelength is the shortest one that does not cause significant fluorescence (Angel and Myrick, 1990) (see also Fig. 3). A shorter excitation wavelength is desirable because the Raman scattering efficiency is directly proportional to the fourth power of the excitation frequency and therefore inversely proportional to the fourth power of the wavelength. Therefore there is a trade-off between the increased Raman scattering efficiency and decreased Raman scattered light (Roberts *et al.*, 1991). FT-Raman with a wavelength of 1064 nm is often used for biological material due to the fact that at this excitation region only a low fluorescence background is observed (Urlaub *et al.*, 1998). The use of 1064 nm excitation circumvents two major drawbacks of Raman spectroscopy, namely fluorescence and photodecomposition (Sato *et al.*, 2001). However, as detailed above, the speed at which spectra can be collected increases. A diode laser excitation at 785 nm suppresses fluorescence from most samples but penetrates water well (Williams *et al.*, 1994; Williams, 1994). This laser is very sensitive with very low noise and therefore data acquisition is fast. However many biological samples still fluoresce at this wavelength.

The *normal* Raman effect is however very weak, since typically only 1 in every 10^8 photons exchange energy with a molecular bond vibration (Colthup *et al.*, 1990), the rest of the photons being Rayleigh scattered. Consequently data acquisition for spectra from samples with only a modest concentration of determinand with a suitably high signal-to-noise ratio still often takes 5-15 min. By contrast, in UV resonance Raman (UVRR) spectroscopy (and in resonance Raman spectroscopy, generally) the inelastic light scattering occurs when the sample is excited with a frequency of light that is within the molecular absorption bands of the sample (Asher, 1993a, 1993b, 1997), in particular (*e.g.*) double and triple bonds, and aromatic ring vibrations. Excitation of this type is in resonance with the electronic transition and yields Raman scattering that is *resonance enhanced.* This resonance enhancement gives signals that are selective for the chromophore.

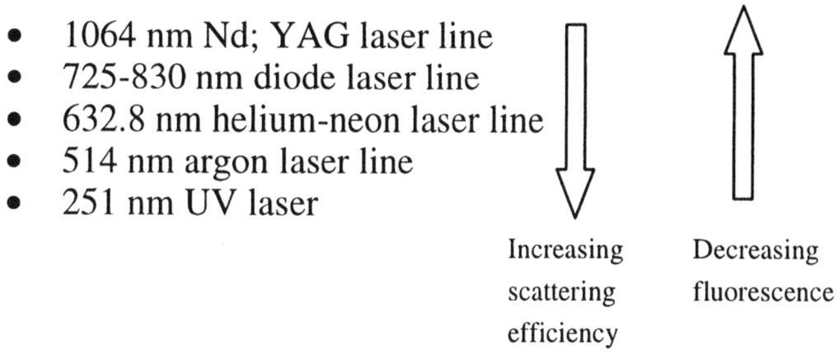

- 1064 nm Nd; YAG laser line
- 725-830 nm diode laser line
- 632.8 nm helium-neon laser line
- 514 nm argon laser line
- 251 nm UV laser

Increasing scattering efficiency

Decreasing fluorescence

Figure 3. Commonly Used Excitation Wavelengths.

The enhancement factor of resonance Raman scattering compared to normal Raman scattering can be as high as 10^8 (Munro and Asher, 1996), but is typically 10^3-10^4. UVRR spectroscopy uses selective excitation in the UV absorption bands of molecules to produce the spectra of particular analytes and chromophoric segments of macromolecules; thus two main advantages of this method are its selectivity and sensitivity (which equates to very rapid measurement times, in the order of a few seconds). Note that while few molecules have visible absorption bands nearly everything absorbs in the deep UV region (Asher, 1993a). UVRR has been used to monitor biological materials; for example in diagnostic cancer studies using a wavelength at 251 nm (Boustany *et al.*, 1999). However, because of the powerful laser used, the application of UVRR spectroscopy requires cells to be suspended in a liquid medium in order to avoid damage due to heating (Nelson *et al.*, 1992a) and from photochemical effects arising from strong absorption of UV radiation by nucleic acids and proteins (Maquelin *et al.*, 2000). However, relatively simple re-circulation sampling systems can greatly reduce this 'burning' phenomenon (Chen *et al.*, 1993).

2.2 Enhancement Techniques

Enhancement effects that can be used along with these wavelengths include resonance Raman, SERS and SERRS. When using resonance Raman (as mentioned above) the energy of the excitation light is tuned close to an electronic transition of the molecule being interrogated and vibrations coupled to the molecules' excited electronic state are preferentially enhanced. This increases the detection sensitivity for compounds present in low concentrations.

SERS is another enhancement technique that can be coupled with resonance Raman, and which combines the structural information content of Raman with ultra-sensitive detection limits. For increased Raman signals the target molecule has to be attached to SERS active substrates (Kneipp *et al.*, 1999) whereby the molecules are adsorbed on or near roughened coinage metals such as silver, copper and gold (Moskovits, 1985). The enhancement is observed when samples are excited in close proximity to roughened metal surfaces (particularly silver) (Pal *et al.*, 1995) and enhancements of up to 10^3-10^6 for some analytes have been recorded, along with a marked reduction in fluorescence (Dou *et al.*, 1997). Recently interest has grown considerably in various applications of SERS (Zhelyaskov *et al.*, 1995) and this method has received considerable attention in biomedical fields as well as in basic biological science (Nabiev and Manfait, 1993). However, if the target analyte is not in close proximity to the metal then SERS does not occur and specialized chemistry is needed (Brown *et al.*, 2001; McAnally *et al.*, 2002).

2.3 Detectors

There are three commonly used types of detectors; photomultipliers, photodiodes and charge-coupled devices (CCD) (Dereniak and Crowe, 1984) and two basic types of spectral analysis instruments; dispersive and interferometric. The cooled CCD is a multi-channel device which has exceptional sensitivity and very low intrinsic noise (dark current), so that the signal-to-noise ratio is improved by at least 2 orders of magnitude (compared with an uncooled CCD) and data acquisition is correspondingly fast (Chase, 1994). CCDs are typically used with lasers in the visible and UV, whilst interferometers are used in the NIR. For further details on detectors used in Raman spectroscopy see McCreery (1994).

3. APPLICATIONS

Raman spectroscopy has been used in the process industries for many years but recent advances in Raman instrumentation have allowed the use of Raman to measure biological materials including cancerous cells (Manoharan *et al.*, 1996; Venkatakrishna, 2001), bacterial cells (Maquelin *et al.*, 2000; Choo-Smith *et al.*, 2001; Kirschner *et al.*, 2001) and tissue.

Outside of *metabolic* fingerprinting, UVRR has emerged as a very promising technique for studying biomolecular structure and function, and in particular for protein structural studies (Efremov *et al.*, 1991; Kitagawa,

1992; Chi and Asher, 1999; Holtz *et al.*, 1999; Lednev *et al.*, 1999), whilst the characterization of (A- and B-) DNA structures has also attracted attention (Gfrorer *et al.*, 1993; Prescott *et al.*, 1984; Mukerji *et al.*, 1996). Within microbiology, UVRR has been used to analyze the DNA and protein constituents of viruses (Overman and Thomas, 1998; Wen and Thomas, 1998; Kaminaka *et al.*, 1999). However, this review concentrates at the exploitation of Raman spectroscopy for metabolite analysis and these applications will be detailed below.

3.1 Raman Analysis of Metabolites in Body Fluids

Determination of blood analyte(s) concentrations is important for diagnosis and clinical management of many medical conditions (Hanlon *et al.*, 2000). In the hospital medical personnel often make diagnoses based on the chemical composition of body fluids, such as blood and urine (Qu *et al.*, 1999). Knowledge of blood metabolite composition is of crucial importance for understanding body metabolism (Wang *et al.*, 1993). Many researchers have investigated methods of measuring blood constituents (analytes) optically, and spectroscopic methods are nowadays frequently employed for applications in clinical chemistry. In many cases the underlying goal is to develop a non-invasive technique for monitoring one or more analytes. In other situations, simpler or more accurate, off-line *in vitro* measurements are sought and interest is strong particularly for assessments of diabetic patients (Berger *et al.*, 1999). Raman spectroscopy is particularly suitable for implementing the advantages of optical blood analysis techniques (Hanlon *et al.*, 2000).

Important categories of metabolites in body fluids, such as glucose (see Fig. 4 for Raman spectrum of glucose) are accessible, for example, by a NIR Raman spectroscopic measurement without sample treatment using reagentless, fast and readily automated *in vitro* multi-component assays. Furthermore, research activities concentrate on non-invasive devices for self-monitoring of blood glucose or for monitoring in intensive care and surgery. *In vivo* sensing and regulation is also necessary for patients with disorders of their carbohydrate metabolism, particularly diabetes mellitus (Heise *et al.*, 2000). Attempts have been made to determine blood metabolite composition indirectly by analyzing body secretions or excretions, such as tears, saliva and urine. Certain correlations between the blood metabolite levels and aqueous humor have been found to exist and metabolite analysis of various concentrations, including glucose, lactate, ascorbate, pyruvate, and urea have been conducted using the 514.5 nm line argon ion laser at a power of 200 mW (Wang *et al.*, 1993). Experiments show that Raman can be used to identify and to detect various metabolites in solutions. At concentrations of interest (1.0 wt %) these metabolite molecules are found to scatter light linearly, proportional to their concentrations.

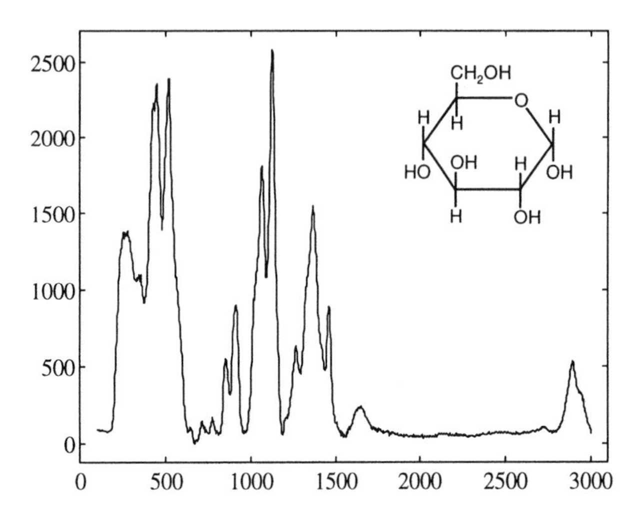

Figure 4. A Raman spectrum of 40% glucose dissolved in H_20. The spectrum was collected for 60 s using a NIR 780-nm diode laser with power at the sampling point typically at 80 mW.

In previous research Berger *et. al.* (1996) used NIR Raman spectroscopy and multivariate calibration methods to extract analyte concentrations from aqueous samples and single-donor doped whole blood (Berger *et al.*, 1997), in both instances at high concentrations. More recently the Raman approach has been extended to demonstrate *in vitro* measurements of analytes in serum and whole blood samples (Berger *et al.*, 1999). Using a 830 nm diode laser, calibration models for glucose, cholesterol, urea and other analytes were developed using partial least squares (PLS) regression methods. It was demonstrated that Raman spectra of serum and whole blood can be acquired and analyzed by use of PLS to extract concentration predictions of many important blood analytes, including glucose, total cholesterol, triglycerides, urea, total protein and albumin (Berger *et al.*, 1999).

Sato's group used Raman to study blood and hemoglobin using 514.5, 720 and 1064 nm laser excitations in an attempt to deal with detailed quantitative analysis including band assignments and investigations of the resonance effect (Sato *et al.*, 2001).

Anti-Stokes is potentially a very attractive method for obtaining Raman spectra from fluorescent materials. This method is intrinsically free from fluorescence, does not depend on the excitation wavelength and does not need complicated instruments. However, the intensity of the Raman anti-

Stokes scattering effect is much weaker than Stokes Raman (Colthup *et al.*, 1990). Dou *et. al.* (1997) have used this method for the quantitative analysis of metabolites in urine including glucose, acetone and urea using a wavelength of 514.5 nm. This is very important from a medical viewpoint, because abnormal values suggest diabetes and/or disease in the liver and kidney. Therefore there is a desire to develop a routine technique, which makes rapid measurements of multiple metabolite concentrations.

This group has also used the excitation wavelength of 820 nm to measure Raman spectra of urine metabolites including glucose, acetone, urea, and creatine artificially. Human urine obtained by micturition was subjected to the spectral measurements within one hour without any pre-treatment. Glucose, acetone, urea, and creatine were then added prior to urine analysis as the first step of the development of the equipment. It has been found that the system provides high correlation coefficients between the concentrations of metabolites and Raman intensities and low detection limits even for urine (Dou *et al.*, 1996).

Knowledge of the metabolite concentrations in an athlete can help make decisions about the level of training that is needed. The physical conditions of an athlete under physical activity can be determined by measuring the metabolite concentrations in blood and muscles. For example the level of lactic acid in the muscles of the body is strongly increased during intense activity. The development and characterization of a method for determining the concentration of metabolites in blood in a non-invasive way is very important in the field of clinical analysis. Raman spectroscopy has been used to evaluate the lactic acid concentrations in blood and muscle using a low power laser beam through the skin and the Raman radiation emitted by the metabolites in the blood are recorded (Pilotto *et al.*, 2001).

3.2 Cancer Diagnostics

Cancer is the second leading cause of death in the USA and occurs in almost all organs of the body. Early detection and prognosis play important roles in clinical management. Raman spectroscopy has the potential to identify markers associated with malignant changes, and biochemical changes may provide new and quantitative information for use in tumour classification (Manoharan *et al.*, 1996).

Vibrational spectroscopic techniques such as Raman spectroscopy can provide highly structural information content on the biological material and allow monitoring of small changes within the molecular structure indicating early stages of disease (Kneipp *et al.*, 1999; Yazdi *et al.*, 1999).

In physiological systems, biomedical changes often precede or cause the onset of disease. Present methods of cancer detection are time consuming and subjective. Optical spectroscopic methods, on the other hand, are usually extremely sensitive, quite specific, and highly objective and can be made non-invasive even for *in vitro* applications. Since cancer is a disease of the cell, it may appear that information on the extracellular matrix may not be of much importance. But the matrix plays a very important and complex role in regulating the behaviour of cells in contact with it, such as influencing their development, migration, proliferation, and metabolic functions (Ullas *et al.*, 1999) and attempts are being made to monitor the matrix for cell disease.

UVRR using an excitation of 251nm has been used to study mucosal surfaces which may provide spatial biochemical information that can help locate regions where cellular metabolites differ from normal. The main objective of this study was to explore the potential of UVRR spectroscopy for the spatial analysis of nucleotide distribution and amino acid/ nucleotide ratios in normal and neoplastic colon tissues. Alterations in DNA metabolism can also be important in identifying early neoplastic changes. Detectable UVRR changes are expected to result from changes in DNA metabolism, proliferation, and changes in DNA content in tumours (Boustany *et al.*, 1999).

Various groups are at present researching the potential of Raman to detect diseased tissues and a review written by Manoharan *et al.* (1996) looks at the use of Raman to analyze diseased biological tissues.

3.3 Characterization of Microorganisms

Using conventional methods, clinical laboratory identification of pathogens is expensive, time consuming and labour intensive as for typical mesophilic organisms approximately 24 hours of incubation on selective media are required to obtain isolated colonies. An additional 12 to 24 hours is then needed for organism identification and susceptibility testing which may further delay administration of the most appropriate narrow-spectrum antibiotic to treat the patient. Therefore for routine microbial identification purposes there is a requirement for methods that would have minimum sample preparation, would analyze samples directly (*i.e.* would not require reagents), would be rapid, automated, and (at least relatively) inexpensive. Raman spectroscopy is one possible candidate solution that is under investigation in several laboratories.

Early Raman studies concentrated on UVRR for characterisation of bacteria (Manoharan *et al.*, 1990; Nelson and Sperry, 1991; Nelson *et al.*, 1992b; Manoharan *et al.*, 1993). The UVRR spectra were largely dominated by the aromatic amino acids tyrosine and tryptophan and nucleic acids which absorb UV-radiation well. Another study investigated *Bacillus* sporulation

(Ghiamati *et al.*, 1992) and found that the spore biomarker dipicolinic acid was also found to be significantly resonance enhanced.

Recently work has concentrated on normal Raman spectroscopy and this has compared very favourably with IR spectroscopy and mass spectrometry for identification of clinical bacterial isolates associated with urinary tract infection (Goodacre *et al.*, 1998). These authors used 780 nm excitation and spectra were collected from bacterial slurries after axenic growth and then analyzed by artificial neural networks. Due to its confocal nature, and in order to speed up diagnosis of pathogens, colonies have been analyzed directly by Raman spectroscopy. In a study by Maquelin and colleagues (2000), a 830 nm laser was used and it was shown that it was possible to obtain Raman spectra directly from microbial microcolonies which had developed after only 6 h of culturing on solid culture medium. However, due to the limited thickness of microcolonies, some of the underlying culture medium was sampled together with the bacteria and this needed to be compensated for mathematically.

The small laser beam volume allows the possibility of heterogeneity within colonies to be investigated. Therefore the same group (Choo-Smith *et al.*, 2001) examined various positions within colonies cultured for 6, 12, and 24 hr. They found that there was little spectral difference in 6 hr microcolonies, but that older cultures exhibited a significant amount of heterogeneity. It was suggested that higher levels of glycogen accumulated on the surface and more RNA was found in deeper layers. Finally, Schuster *et al.* (2000) have shown that even individual cells can be analyzed by Raman spectroscopy and demonstrated this with *Clostridium beijerinckii*.

3.4 Other Applications

With an emerging technology such as Raman spectroscopy, in its many different guises there are many different potential application areas. For example, pefloxacin is an antimicrobial agent with an excellent activity against various bacteria. SERS has been used to gain a better understanding of the mechanism of inhibition of DNA gyrase by pefloxacin using vibrational study (Lecomte *et al.*, 1995).

Other research using Raman spectroscopy to study metabolites include the use of FT-Raman by Urlaub *et al.* (1998) to study the production of naphthylisoquinoline alkaloids, a secondary metabolite produced by *Anastrocladus heyneanus*, a tropical liana. These studies could create the basis for improving the synthetic production of alkaloids to increase pharmacological effectiveness. Studies (Barthus and Poppi, 2001) have also been carried out using FT-Raman to determine total unsaturation of fats and oils.

Shaw *et al.* (1999) describe an on-line monitoring system of the glucose fermentation by yeast to ethanol using a dispersive Raman spectrometer with a 780 nm diode laser, after on-line filtration of the biomass. McGovern and colleagues (2002) have also used the same Raman system to analyze a diverse range of unprocessed, industrial fed-batch fermentation broths containing the gibberellic acid producing fungus *Gibberella fujikuroi*. These fermentation broths were analyzed directly without *a priori* chromatographic separation and gave comparable values with laborious and time consuming HPLC methods.

Raman spectroscopy has also been used to detect endolithic lichen pigments (Wynn-Williams *et al.*, 1997) and to distinguish between biodegradative oxalic acid and metal oxalates (Edwards *et al.*, 1991). Chemical analysis of extracts of *Xanthoria elegans*, *X. parietina* and *X. mawsonii* studied here identified the major secondary metabolite parietin, and other polyphenolic compounds including erythrin, lecanoric acid and atranorin. These compounds could therefore be used as key indicator molecules for Raman spectroscopy of specimens *in vivo* (Edwards *et al.*, 1998).

4. CONCLUSION

Raman spectroscopy is an emerging technology and NIR, visible and UV Raman show great potential for the monitoring of metabolites, along with enhancement techniques such as resonance Raman, SERS and SERRS. This chapter shows that Raman has been used to monitor effectively blood analytes, which will be very beneficial in the medical field particularly for diabetic patients. It has also been shown to be a useful technique for monitoring biochemical changes of tissue cells, an important step towards the chemical understanding of cancer diagnostics. We believe that as more 'biologists' investigate Raman spectroscopy as a complementary tool to IR absorbance spectroscopy the full extent of the importance of this metabolic fingerprinting method will be realised.

REFERENCES

Adar F, Geiger R, Noonan J. Raman spectroscopy for process/ quality control. *Appl Spectr* 32: 45-101 (1997).

Angel MA, Myrick ML. Wavelength selection for fiber optic Raman spectroscopy. *Appl Optics* 29: 1350-1352 (1990).

Asher SA. UV resonance Raman spectroscopy for analytical, physical and biophysical chemistry - Part 1. *Anal Chem* 65: 59A-66A (1993a).

Asher SA. UV resonance Raman spectroscopy for analytical, physical and biophysical chemistry - Part 2. *Anal Chem* 65: 201A-210A (1993b).
Asher SA, Munro CH, Chi ZH. UV lasers revolutionize Raman spectroscopy. *Laser Focus World* 33: 99-109 (1997).
Barthus RC, Poppi RJ. Determination of the total unsaturation in vegetable oils by Fourier transform Raman spectroscopy and multivariate calibration. *Vibrat Spectr* 26: 99-105 (2001).
Berger AJ, Itzkan I, Feld MS. Feasibility of measuring blood glucose concentration by near-infrared Raman spectroscopy. *Spectrochim Acta* 53: 287-292 (1997).
Berger AJ, Koo T-W, Itzkan I *et al.* Multicomponent blood analysis by near-infrared Raman spectroscopy. *Appl Optics* 38: 2916-2926 (1999).
Berger AJ, Wang Y, Feld MS. Rapid, noninvasive concentration measurements of aqueous biological analytes by near-infrared Raman spectroscopy. *Appl Optics* 35: 209-212 (1996).
Boustany NN, Crawford JM, Manoharan R *et al.* Analysis of nucleotides and aromatic amino acids in normal and neoplastic colon mucosa by ultraviolet resonance Raman spectroscopy. *Lab Investigat* 79: 1201-1214 (1999).
Brown R, Smith WE, Graham D. Synthesis of a benzotriazole phosphoramidite for attachment of oligonucleotides to metal surfaces. *Tet Lett* 42: 2197-2200 (2001).
Chase B. A new generation of Raman instrumentation. *Appl Spectr* 48: 14A-19A (1994).
Chen XG, Lemmon DH, Bormett RW, Asher SA. Convenient microsampling system for UV resonance Raman spectroscopy. *Appl Spectr* 47: 248-249 (1993).
Chi ZH, Asher SA. Ultraviolet resonance Raman examination of horse apomyoglobin acid unfolding intermediates. *Biochemistry* 38: 8196-8203 (1999).
Choo-Smith LP, Maquelin K, van Vreeswijk T *et al.* Investigating microbial (micro)colony heterogeneity by vibrational spectroscopy. *Appl Environ Microbiol* 67: 1461-1469 (2001).
Colthup NB, Daly LH, Wiberly SE. *Introduction to Infrared and Raman Spectroscopy.* Academic Press, New York (1990).
Dereniak EL, Crowe DGiR. *Optical Radiation Detectors.* John Wiley and Sons, New York (1984).
Dou X, Yamaguchi Y, Yamamoto H *et al.* Quantitative analysis of metabolites in urine using a highly precise, compact near-infrared Raman spectrometer. *Vibrat Spectr* 13: 83-89 (1996).
Dou X, Yamaguchi Y, Yamamoto H *et al.* Quantitative analysis of metabolites in urine by anti-Stokes Raman spectroscopy. *Biospectroscopy* 3: 113-120 (1997).
Edwards HGM, Farwell DW, Seaward MRD, Giacobini C. Preliminary Raman microscopic analyses of a lichen encrustation involved in the biodeterioration of Renaissance frescoes in central Italy. *Internat Biodeteriorat Biodegrad* 27: 1-9 (1991).
Edwards HGM, Holder JM, Wynn-Williams DD. Comparative FT-Raman spectroscopy of *Xanthoria* lichen-substratum systems from temperate and Antartic habitats. *Soil Biol Biochem* 30: 1947-1953 (1998).
Efremov RG, Feofanov AV, Nabiev IR. Quantitative treatment of UV resonance Raman spectra of biological molecules - application to the study of membrane-bound proteins. *Appl Spectr* 45: 272-278 (1991).
Ferraro JR, Nakamoto K. *Introductory Raman Spectroscopy.* Academic Press, London (1994).
Gfrorer A, Schnetter ME, Wolfrum J, Greulich KO. Double and triple helices of nucleic acid polymers, studied by UV resonance Raman spectroscopy. *Berichte der Bunsen-Gesellschaft-Phys Chem Chem Phys* 97: 155-162 (1993).
Ghiamati E, Manoharan R, Nelson WH, Sperry JF. UV resonance Raman spectra of *Bacillus* spores. *Appl Spectr* 46: 357-364 (1992).

Goodacre R, Timmins ÉM, Burton R *et al.* Rapid identification of urinary tract infection bacteria using hyperspectral, whole organism fingerprinting and artificial neural networks. *Microbiol* 144: 1157-1170 (1998).

Hanlon EB, Manoharan R, Koo T-W *et al.* Prospects for *in vivo* Raman spectroscopy. *Phys Med Biol* 45: R1-R59 (2000).

Heise HM, Bittner A, Marbach R. Near-infrared reflectance spectroscopy for noninvasive monitoring of metabolites. *Clin Chem Lab Med* 38: 137-145 (2000).

Holtz JSW, Holtz JH, Chi ZH, Asher SA. Ultraviolet Raman examination of the environmental dependence of bombolitin I and bombolitin III secondary structure. *Biophys J* 76: 3227-3234 (1999).

Kaminaka S, Imamura Y, Shingu H *et al.* Studies of bovine enterovirus structure by ultraviolet resonance Raman spectroscopy. *J Virol Meth* 77: 117-123 (1999).

Kirschner C, Maquelin K, Pina P *et al.* Classification and identification of enterococci: a comparative phenotypic, genotypic, and vibrational spectroscopic study. *J Clin Microbiol* 39: 1763-1770 (2001).

Kitagawa T. Investigation of higher order structures of proteins by ultraviolet resonance Raman spectroscopy. *Prog Biophys Mol Biol* 58: 1-18 (1992).

Kneipp K, Kneipp H, Itzkan I *et al.* Surface-enhanced Raman scattering: a new tool for biomedical spectroscopy. *Curr Sci* 77: 915-924 (1999).

Lecomte S, Moreau NJ, Manfait M *et al.* Surface-enhanced Raman spectroscopy investigation of fluoroquinoline/ DNA/ DNA gyrase/ Mg^{2+} interactions: Part 1. adsorption of pefloxacin on colloidal silver-effect of drug concentration, electrolytes, and pH. *Biospectroscopy* 1: 423-436 (1995).

Lednev IK, Karnoup AS, Sparrow MC, Asher SA. Alpha-helix peptide folding and unfolding activation barriers: a nanosecond UV resonance Raman study. *J Am Chem Soc* 121: 8074-8086 (1999).

Manoharan R, Ghiamati E, Chadha S *et al.* Effect of cultural conditions on deep UV resonance Raman spectra of bacteria. *Appl Spectr* 47: 2145-2150 (1993).

Manoharan R, Ghiamati E, Dalterio RA *et al.* UV resonance Raman spectra of bacteria, bacterial spores, protoplasts and calcium dipicolinate. *J Microbiol Meth* 11: 1-15 (1990).

Manoharan R, Wang Y, Feld MS. Histochemical analysis of biological tissues using Raman spectroscopy. *Spectrochim Acta Part A*: 215-249 (1996).

Maquelin K. *Confocal Raman Microspectroscopy. A Novel Diagnostic Tool in Medical Microbiology*, Erasmus University, Rotterdam (2002).

Maquelin K, Choo-Smith L-P, van Vreeswijk T *et al.* Raman spectroscopic method for identification of clinically relevant microorganisms growing on solid culture medium. *Anal Chem* 72: 12-19 (2000).

McAnally G, McLaughlin C, Brown R *et al.* SERRS dyes. Part I. Synthesis of benzotriazole monoazo dyes as model analytes for surface enhanced resonance Raman scattering. *Analyst* 127: 838-841 (2002).

McCreery RL. CCD array detectors for multichannel Raman spectroscopy. In *Charge Transfer Devices in Spectroscopy*. Sweedler J, Ratzlaff K, Denton M (Ed) pp. 227-229, VCH, New York (1994).

McCreery RL. Instrumentation for dispersive Raman Sspectroscopy. In *Modern Techniques in Raman Spectroscopy*. Vol. 1. Laserna JJ (Ed) pp. 41-72, John Wiley and Sons, Chichester (1996).

McGovern AC, Broadhurst D, Taylor J *et al.* Monitoring of complex industrial bioprocesses for metabolite concentrations using modern spectroscopies and machine learning: application to gibberellic acid production. *Biotechnol Bioeng* 78: 527-538 (2002).

Moskovits M. Surface-enhanced Raman spectroscopy. *Rev Mod Phys* 57: 783 (1985).
Mukerji I, Shiber MC, Fresco JR, Spiro TG. A UV resonance Raman study of hairpin dimer helices of d(A-G)(10) at neutral pH containing intercalated dA residues and alternating dG tetrads. *Nucleic Acids Res* 24: 5013-5020 (1996).
Munro CH, Asher SA. UV lasers light the way for novel spectroscopy. *Photonics Spectra* 30: 118-120 (1996).
Nabiev I, Manfait M. Industrial applications of the surface-enhanced Raman-spectroscopy. *Rev Industr Fr Petrol* 48: 261- 285 (1993).
Nelson WH. *Instrumental Methods for Rapid Microbiological Analysis*. VCH Publishers (1985).
Nelson WH, Manoharan R, Sperry JF. UV resonance Raman studies of bacteria. *Appl Spectr Rev* 27: 67-124 (1992a).
Nelson WH, Manoharan R, Sperry JF. UV resonance Raman studies of bacteria. *Appl Spectr Rev* 27: 67-124 (1992b).
Nelson WH, Sperry JF. UV resonance Raman spectroscopic detection and identification of bacteria and other microorganisms. In *Modern Techniques for Rapid Microbiological Analysis*. Nelson WH (Ed) pp. 97-143, VCH Publishers, New York (1991).
Overman SA, Thomas GJ. Novel vibrational assignments for proteins from Raman spectra of viruses. *J Raman Spectr* 29: 23-29 (1998).
Pal A, Stokes DL, Alarie JP, Vo-Dinh T. Selective surface-enhanced Raman spectroscopy using a polymer coated-substrate. *Anal Chem* 67: 3154-3159 (1995).
Pilotto S, Pacheco MTT, Silveira L *et al.* Analysis of near-infrared Raman spectroscopy as a new technique for a transcutaneous non-invasive diagnosis of blood components. *Lasers Med Sci* 16: 2-9 (2001).
Prescott B, Steinmetz W, Thomas GJ. Characterisation of DNA structures by laser Raman spectroscopy. *Biopolymers* 23: 235-256 (1984).
Qu JY, Wilson BC, Suria D. Concentration measurements of multiple analytes in human sera by near-infrared laser Raman spectroscopy. *Appl Optics* 38: 5491-5498 (1999).
Raman CV, Krishnan KS. A new type of secondary radiation. *Nature* 121: 501 (1928).
Roberts MJ, Garrison AA, Kercel SW, Muly EC. Raman spectrocopy for on-line, real-time, multi-point industrial chemical analysis. *Process Control Quality* 1: 281-291 (1991).
Sanford CL, Mantooth BA, Jones BT. Determination of ethanol in alcohol samples using a modular Raman spectrometer. *J Chem Ed* 78: 1221-1224 (2001).
Sato H, Chiba H, Tashiro H, Ozaki Y. Excitation wavelength-dependent changes in Raman spectra of whole blood and hemoglobin: comparison of the spectra with 514.5-, 720-, and 1064-nm excitation. *J Biomed Optics* 6: 366-370 (2001).
Schuster KC, Reese I, Urlaub E *et al.* Multidimensional information on the chemical composition of single bacterial cells by confocal Raman microspectroscopy. *Anal Chem* 72: 5529-5534 (2000).
Shaw AD, Kaderbhai N, Jones A *et al.* Noninvasive, on-line monitoring of the biotransformation by yeast of glucose to ethanol using dispersive Raman spectroscopy and chemometrics. *Appl Spectr* 53: 1419-1428 (1999).
Stevenson CL, Vo-Dinh T. Signal expressions in Raman spectroscopy. In *Modern Techniques in Raman Spectroscopy*. Vol. 1. Laserna JJ (Ed) pp. 1-39, John Wiley and Sons, Chichester (1996).
Ullas G, Sudhaker SN, Gopalakrishna K *et al.* Laser Raman spectroscopy: some clinical applications. *Curr Sci* 77: 908-914 (1999).
Urlaub E, Popp J, Keifer W *et al.* FT-Raman investigation of alkaloids in the liana *Ancistrocladus heyneanus*. *Biospectroscopy* 4: 113-120 (1998).

Venkatakrishna K, Kurien J, Pai KM *et al.* Optical pathology of oral tissue: a Raman spectroscopy diagnostic method. *Curr Sci* 80: 665-669 (2000).
Wang SY, Hasty CE, Watson PA *et al.* Analysis of metabolites in aqueous solutions by using laser Raman spectroscopy. *Appl Optics* 32: 925-929 (1993).
Wen ZQ, Thomas GJ. UV resonance Raman spectroscopy of DNA and protein constituents of viruses: Assignments and cross sections for excitations at 257, 244, 238, and 229 nm. *Biopolymers* 45: 247-256 (1998).
Williams KPJ, Pitt GD, Batchelder DN, Kip BJ. Confocal Raman microspectroscopy using a stigmatic spectrograph and CCD detector. *Appl Spectr* 48: 232-235 (1994).
Williams KPJP, Pitt GD. Smith,BJE, Whitley A. Use of a rapid scanning stigmatic Raman imaging spectrograph in the industrial environment. *J Raman Spectr* 25: 131-138 (1994).
Wynn-Williams DD, Edwards HGM, Russell NC. Moisture and habitat structure as regulators for microalgal colonists in diverse Antartic terrestrial habitats. In *Ecosystem Processes in Antartic Ice-Free Landscapes*. Howard-Williams C, Lyons B, Hawes I (Ed) pp. 77-78, Balkema Press, Rotterdam (1997).
Yazdi Y, Ramanujam N, Lotan R *et al.* Resonance Raman spectroscopy at 257 nm excitation of normal and malignant cultured breast and cervical cells. *Appl Spectr* 53: 82-85 (1999).
Zhelyaskov VR, Milne ET, Hetke JF, Morris MD. Silicon substrate microelectrode array for surface-enhanced Raman spectroscopy. *Appl Spectr* 49: 1793-1795 (1995).

Chapter 7

METABOLIC FINGERPRINTING WITH FOURIER TRANSFORM INFRARED SPECTROSCOPY

David I. Ellis[1], George G. Harrigan[2] and Royston Goodacre[1,3]
[1]Institute of Biological Sciences, University of Wales, Aberystwyth, SY23 3DD, UK [2]Global HTS, Pharmacia Corporation, 700 Chesterfield Parkway, Chesterfield, MO 63198, USA
[3]Department of Chemistry, University of Manchester Institute of Science and Technology, PO Box 88, Sackville St, Manchester M60 1QD, UK

1. INTRODUCTION

Fourier transform infrared (FT-IR) spectroscopy is a rapid, reagent-less, non-destructive, analytical technique whose continuing development is resulting in manifold applications in the biosciences. The principle of FT-IR lies in the fact that when a sample is interrogated with light (or electromagnetic radiation), chemical bonds absorb at specific wavelengths and vibrate in one of a number of ways. These absorptions/vibrations can then be correlated to the bonds or functional groups of molecules. Because of its chemical information content and spectral richness (defined as numbers of clearly defined peaks) the major wavenumber region of interest is the mid-infrared, usually defined as 4000-600 cm^{-1} (see Table 1). The infrared spectra of proteins, as a prominent example, exhibit strong amide I absorption bands at 1653 cm^{-1} associated with characteristic stretching of C=O and C-N and bending of the N-H bond (Stuart, 1997).

In FT-IR spectroscopy, any sample can be simultaneously interrogated with electromagnetic radiation corresponding to the mid-IR region and an absorbance spectrum generated. This can be considered a “fingerprint” characteristic of any chemical substance or biological sample (Stuart, 1997). The rapidity of this technique cannot be overstressed as an IR spectrum can be collected in seconds (with minimal sample preparation). This compares to hours for many other analytical methods, some of which often require chemical derivatization steps (*e.g.* gas chromatography). However, FT-IR spectroscopy does have some disadvantages. A problematic issue is that the IR absorption of water is very intense, This can be addressed in one of several ways including 1) dehydration of samples, 2) electronic subtraction of the water signal or 3) application of attenuated total reflectance (ATR) spectroscopy (Schmitt and Flemming, 1998; Ellis *et al.*, 2002).

Table 1. Characteristic Band Assignments in Biological Samples

Region	Wavenumber (cm^{-1})	Band assignments
	3400	OH
	2956	CH_3 asymmetric stretch
Fatty acid region	2920	CH_2 asymmetric stretch
	2870	CH_3 symmetric stretch
	2850	CH_2 symmetric stretch
	1745/1735	C=O stretch (fatty acid esters)
Amide region	1705	C=O stretch (esters, carboxylic groups)
	1652-1648	amide I, (C=O) different conformations
	1550-1548	amide II (NH, C-N)
	1460-1454	CH_2 bend
Mixed region	1400-1398	C-O bend (carboxylate ions)
	1310-1240	Amide III (C-N)
	1240	P=O (phosphate)
	1222	P=O
	1114	C-O-P, P-O-P
Polysaccharide region	1085	sugar ring vibrations
	1052	C-O, C-O-C (polysaccharide)

For biological applications, the mid-IR region can be broken down into spectral windows of interest (*vide infra* for FT-IR spectra) where strong absorption bands can be directly related to specific components. These spectral regions of biological interest include 3050-2800 cm^{-1} where fatty acids predominate; the amide region (1750-1500 cm^{-1}) ascribed primarily to proteins and peptides; the 'mixed region' (1500-1250 cm^{-1}) ascribed to carboxylic groups of proteins, free amino acids and polysaccharides, and finally the polysaccharide region (1200-900 cm^{-1}) (Schmitt and Flemming, 1998). Other spectral regions of interest include 1250-1200 cm^{-1} relevant to RNA, DNA and phospholipid content (Udelhoven *et al.*, 2000).

2. MICROBIOLOGICAL APPLICATIONS

FT-IR spectra of microbes are derived from a composite of different vibrational modes representing a diverse range of biomolecules present in cell-walls, membranes, extracellular polymeric substances and the cytoplasm (Udelhoven *et al.*, 2000) and this information has been exploited to great effect. Naumann and co-workers (Helm *et al.*, 1990; Naumann *et al.*, 1991, 1995; Kirschner *et al.*, 2001) have pioneered research on the rapid, accurate and reproducible identification of bacteria to the subspecies level as well as studies into the structure and composition of intact, living bacterial cells. These studies have subsequently led to further developments such as the ability to identify living colonies of bacteria directly from agar using IR microspectroscopy (Lang and Sang, 1998). Others, including Goodacre and colleagues, have also been active in this area with studies on the differentiation of clinically relevant microorganisms such as *Streptococcus* and *Enterococcus* spp. (Goodacre *et al.*, 1996), urinary tract infection bacteria (Goodacre *et al.*, 1998) and closely related *Candida* spp. (Timmins *et al.*, 1998) (see Fig. 1 for an example of sampling methodology utilized by this group).

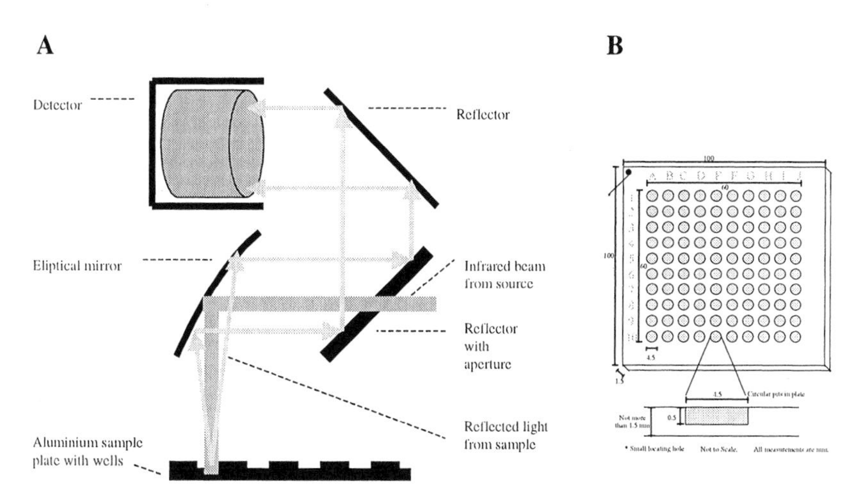

Figure 1. Schematic of sampling technique developed by Goodacre and co-workers. A = sampling method, B = 100 well sampling plate.

As FT-IR spectroscopy is increasingly established as a powerful analytical tool within microbiology, its value to other biomedical areas is now being realized. For example, FT-IR spectroscopy is now demonstrating value in the detection and diagnosis of disease or dysfunction, an application eloquently termed "infrared pathology" (Diem *et al.*, 1999). The following section will concentrate on a selected number of significant diagnostic and biomarker studies dealing primarily with human diseases and disorders through the spectroscopic analysis of tissues, cells or body fluids.

3. INFRARED PATHOLOGY

3.1 Tissue Analysis in Cancer and Other Diseases

Since the development of antibiotics, cancer has eclipsed microbial disease as the major threat to human longevity and quality of life. Early diagnosis of cancerous and pre-cancerous states would therefore be extremely beneficial and increased application of FT-IR spectroscopy in this area is evident. However, level of activity in a field is no measure of the quality of research and doubts have been expressed (Diem *et al.*, 1999) as to whether reported FT-IR analyses in cancer research are supported by sufficiently validated pathological data on assayed biological samples. Nonetheless, several studies have reported changes in FT-IR spectral patterns attributable to variations in cell composition during disease progression. Detailed analyses of spectral variations within the cell cycle have also been investigated (Benedetti *et al.*, 1997, 1998; Chiriboga *et al.*, 1998a, 1998b; Boydston-White *et al.*, 1999; Diem *et al.*, 1999, 2002) as well as spectral variations between different cell types (Haaland *et al.*, 1997). Such analyses indicate that 'normal' cells can be discriminated from cancerous cells as spectral features attributable to DNA are enhanced in cancerous samples (Diem *et al.*, 1999), a perhaps not entirely profound finding. Changes in glycogen concentration associated with different stages of maturation in cervical cancers (Chiriboga *et al.*, 1998a, 1998b) have also been reported.

The most frequent form of cancer in children and adults below the age of 30 is leukemia. Several investigations into the diagnosis of leukemia have been undertaken by FT-IR analysis of normal and leukemic lymphocytes (*e.g.* Liu *et al.*, 1998, 2001; Boydston-White *et al.*, 1999; Liu and Mantsch, 2001). However, although differentiating between normal and leukemic lymphocytes is routine by standard biochemical procedures, differentiating between the many subforms or clones of leukemia is slow, labor intensive and requires specialized knowledge. Schultz *et al.* (1996, 1997) investigated chronic lymphocytic leukemia (CLL), the most common form of leukemia in Western Europe and North America. In comparing CLL cells to normal cells major spectral differences were observed in the region primarily ascribed to absorptions from the DNA backbone (1250-1200 cm^{-1}). Differences were observed in the amide region and a reduction lipid content was noted. Furthermore, it was also possible to separate CLL cells into a number of subclusters based on differences in DNA content and it was suggested that this may provide a useful diagnostic or biomarker tool for staging (progression) of the disease and for multiple clone detection (Schultz *et al.*, 1996, 1997).

Analysis of colorectal adenocarcinoma sections has also been undertaken (Lasch and Naumann, 1998; Lasch *et al.*, 2002) using IR microscopy. This can provide cell-structural information, often presented as false color IR maps, based on biochemical content as determined by selected IR wavenumbers. Results illustrated the exceptional sensitivity of IR microscopy as a diagnostic tool when compared to other pathological analyses. At least one other study has indicated that IR spectroscopy may have potential clinical utility for the early, fast and reagent free assessment of chemotherapeutic efficacy in patients with leukemia (Liu *et al.*, 2001).

The majority of FT-IR studies on cells and tissues exploit the spectral characteristics associated with macromolecules (Benedetti *et al.*, 1997; Diem *et al.*, 1999) and as discussed cancer detection may possible using this approach. Raman spectroscopy has also been applied to the detection of cancerous and pre-cancerous states (Utzinger *et al.*, 2001, Clarke and Goodacre, Chapter 6). Here again, discriminatory factors tend to focus on macromolecules, although differences in cholesterol and lipid species have been observed through this particular technology. In the case of FT-IR, greater spectral richness allows for more detailed metabolite analyses of both cell and tissue content as well as the role of metabolites in discrimination between biological samples (see also Table 2).

Table 2. Comparison of Spectroscopic Whole-Cell and Tissue Fingerprinting Methods

	FT-IR	Raman
Destructive	No	No
Sample size	>50 μg on plates	Slurry in vials
	5 μM diameter for microscope	1 μM diameter for microscope
Typical cell number	Plates: $10^6 - 10^7$	Vials: 10^9 ml^{-1}
	Microscope: aggregates	Microscope: single cells
Typical speed	5-30 s	1-20 min
Automatable	Yes	Yes
Complex data capture	No	Fairly
Typical dimensionality	400-1000 data points	1000-3000+ data points
Data analysis (1= easy to 5 = hard)	2	4

Scrapie is one of a number of related fatal infectious degenerative brain diseases including new variant Creutzfeldt-Jakob disease (vCJD) and bovine spongiform encephalopathy (BSE). These transmissible spongiform encephalopathies (TSE) have gained notoriety in recent years, primarily as a result of BSE outbreaks during the 1990s centered in the UK, followed by concerns about links to CJD in humans (Cousens *et al.*, 2001; Fox, 2001). Studies by (Kneipp *et al.*, 2000, 2002) have utilized IR microspectroscopy to detect disease-associated molecular changes in brain tissue in hamster models of scrapie. The aim was to evaluate the efficacy of FT-IR as a diagnostic or biomarker tool in terminal, early and even preclinical stages of TSE. Histological and molecular differences in TSE-affected nervous tissue are manifold. They include changes in protein expression, composition of membrane systems and alterations in gene expression (Kneipp *et al.*, 2002). The applicability of FT-IR to this problem lies in the fact that it is one of the very few methods that can detect changes in a range of biomolecules simultaneously during a single measurement. In Kneipp *et al.* (2000,2002), hamsters were challenged orally with lethal doses of scrapie and compared to a control population fed normal (non-scrapie) brain homogenate. The molecular alterations in cryosections of three anatomical regions of the brain were studied to determine at which stage IR spectral changes could be observed. Spectra were analyzed at 90 and 120 days post-infection, and at the terminal stages. For identification of brain structure specific spectra, IR images were reconstructed for the two coronal planes of interest for all hamster brains. Results revealed that the most prominent variations between spectra were obtained from scrapie-infected nervous tissue at the terminal stage. However, in the 1060-1040 cm^{-1} region, differences between healthy and infected tissue could be traced back to 90 days post-infection *i.e.* at the preclinical stage. These changes were small but clearly progressed toward the terminal stage until a pronounced shoulder in the spectra appeared at ~1050 cm^{-1}. The changes in this region could be assigned to sugar moieties of nucleic acids, to changed content of metabolic sugar molecules, or to other events yet to be described. In summary, FT-IR spectroscopy demonstrated utility as a method for diagnosis of TSE pathogenesis and could also be developed into a rapid post-mortem and hopefully early diagnostic screening method.

Although the FT-IR derived biomarkers for cancer and scrapie diagnosis focus primarily on macromolecules, the discerning ability of FT-IR is such that biomarker research can be extended to small molecule metabolites. In a study into the effect of the *in vivo* administration of carrageenan, a potent inflammatory agent, FT-IR spectroscopy was used to examine spectral features characteristic of various tissue types (brain, testes, liver) from rats and to attempt to discover new biomarkers for oxidative stress and/or inflammation (Perromat *et al.*, 2001). Difference spectra (carrageenan groups minus control groups) and hierarchical cluster analysis on first and second derivative data, demonstrated clear biochemical differences between tissue not exposed to carrageenan and tissue so exposed.

As already mentioned (and emphasized throughout) FT-IR spectroscopy is a valuable fingerprinting tool owing to its ability to measure proteins, lipids, polysaccharides and other metabolite components simultaneously. Many of the differences observed in this study were evidently attributable to macromolecules. The conformational structure of membrane proteins in liver for example, were changed as illustrated by a large decrease in FT-IR signals attributed to ß-sheet structure. Appreciable depletion of glycogen was also observed in all tissues and there were significant changes in the phosphodiester (DNA) and polysaccharide regions.

Absorption intensities in both the amide I and II regions were increased. However, the increases in the amide II region in exposed brain tissue could also, in principle, be related to increased levels of monoamine (5-hydroxytryptamine, norepinephrine and dopamine) as well as changes in protein structure/content. Such a suggestion would be consistent with known biology and other observed spectral changes such as a decrease in absorption intensity at 1510 cm^{-1} attributable to a decreased level of tyrosine, a common precursor of biological amines. Changes in small molecule metabolites could be detected by monitoring differences in asymmetric vibrations at 2956/2922 cm^{-1} and symmetric vibrations at 2852/2982 cm^{-1}, indicating that carrageenan exposure decreased the ratio of CH_3/CH_2 symmetric band areas. This reflected changes in fatty acid composition and did not implicate CH_3 groups in proteins, as was demonstrated by a low correlation with these ratios and the C=O band in the amide I region. Thus, FT-IR could rapidly demonstrate significant change in the length and concentration of fatty acyl chains on *in vivo* exposure to carrageenan.

An observed increase in intensity at 1170 cm^{-1} in liver from carrageenan exposed rats could be ascribed to O=C-O-C asymmetric vibrations in phospholipids and/or cholesterol esters that are known to be accumulated during inflammation, whilst a decrease in 1150 cm^{-1} could be assigned to decreased glycolipid levels. Brain phospholipids also increased on carrageenan exposure, particularly those containing saccharides (*e.g.* cerebrosides, gangliosides) as demonstrated by an increase in absorption intensities at 1085, 1234 and 1740 cm^{-1}. The absorption increase at 1234 cm^{-1} is also suggestive of increased levels of sphingomyelin. Other discernible changes in brain chemistry including elevated choline levels (1385 cm^{-1}). Thus, this study demonstrated the presence of inflammatory markers and an interpretation of FT-IR spectra in terms consistent with known inflammatory biology (Perromat *et al.*, 2001). The fingerprinting potential highlighted in this study is therefore obvious and extends to other toxicological and biomedical applications.

3.2 Biofluid Analysis and Disease Pattern Recognition

In addition to tissue analyses, FT-IR has been adopted to obtain metabolic fingerprints of a wide variety of biological fluids, and indeed, most 'metabonomic' strategies for biomarker discovery are, in the main, devoted to biofluid analysis. Thomas *et al.* (2000) applied FT-IR spectroscopy to the analysis of follicular fluids from large and small natural luteinized follicles. Analysis of all samples resulted in characteristic, reproducible, FT-IR absorption spectra with recognizable amide I protein vibrations and acyl vibrations from fatty acids. Chemometric analysis, including unsupervised and supervised approaches, demonstrated that fluid from large follicles formed a homogenous closely related cluster, indicating close biochemical similarity, whereas fluid from the small follicles demonstrated greater heterogeneity but were still, in general, distinct from the large follicles. Close inspection of the spectra demonstrated that the differences between and within the follicular fluids did not correlate with measured distributions of steroids, indicating that the FT-IR could reveal other aspects of follicular biochemistry that were important. These differences in the biochemical nature of the fluids may reflect the developmental capacity of the oocyte, suggesting FT-IR could provide a biomarker related to oocyte quality.

A major premise of the metabolic fingerprinting discussed here and alluded to throughout this volume is that an understanding of complex biological systems requires strategies for the acquisition and analysis of datasets comprising hundreds, or potentially thousands, of variables. Applying this approach to biomarker discovery in osteoarthritis (OA), for example, may prove particularly useful. At present there are no diagnostic markers of OA and changes in disease status over a single year are often small and difficult to quantitate. In an analysis of 14 biochemical molecular markers for monitoring OA, Otterness *et al.* (2000) concluded that the magnitude of change in skeletal markers (markers related to changes in cartilage and bone metabolism) was disappointingly small and that this augured poorly for their clinical utility. It was also stated that in spite of statistically significant differences in such biomarkers between the groups of OA patients and the groups of controls, the overlap of patients and controls thwarts the utility of a single skeletal marker to reliably classify individuals (Otterness *et al.*, 2000). In a subsequent report (Otterness *et al.*, 2001) it was demonstrated that there was no correlation between skeletal biomarkers and clinical assessments of OA progression.

Recognition of the complexity of diagnosis in, and the limitations of, single entity markers for osteo- (and rheumatoid) arthritis has encouraged investigations of synovial fluid from the joints of normal and arthritic patients (Eysel *et al.*, 1997). This particular study demonstrated the ability to determine many differences between synovial fluids as a result of disease processes and identified particular IR sub-regions with significant discriminatory power. One of the spectral regions with greatest diagnostic potential was 3500-2800 cm^{-1} a spectral region dominated by CH stretching vibrations of all synovial fluid components (which include components typically found in serum but also the glycosaminoglycans). Discrimination between normal, OA and rheumatoid arthritic patients showed excellent agreement with clinical diagnosis and it was evident that FT-IR spectra of synovial fluids can be used as an aid in the diagnosis of arthritic disorders. Further, FT-IR also offers technical advantages over other possible metabolome data acquisition strategies for synovial fluid. Synovial fluid has a sticky consistency and is not particularly tractable to solubilization, rendering MS difficult. NMR analyses (Damyanovitch *et al.*, 1999) can require as much as 400 μL per sample whereas FT-IR, with appropriate microscopic staging, requires as little as 1 μL (see Fig. 2 for typical FT-IR spectra obtained on these quantities).

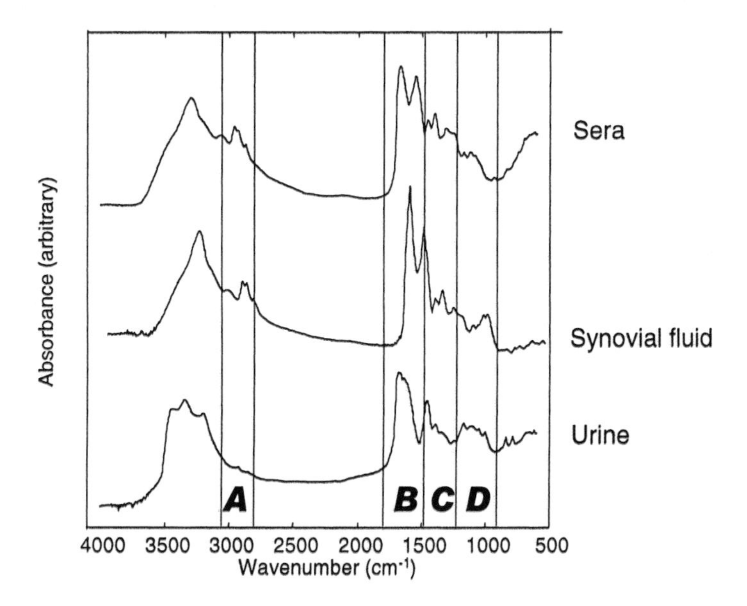

Figure 2. Unpublished data from OA samples showing spectral regions of biological interest A = fatty acid region, B = amide region, C = mixed region, D = polysaccharide region.

It is the case however that sampling of synovial fluid is not necessarily a trivial issue and it would be useful to establish biomarkers for OA status in blood or urine. Work has been undertaken in this area for rheumatoid arthritis (Staib *et al.*, 2001) but not the more challenging OA. We have recently (unpublished) demonstrated that FT-IR and Raman spectroscopy of urine and sera from surgical models of OA in animals can discriminate between control and test subjects. Animal models (Bendele *et al.*, 1999) studied to date include the canine anterior cruciate ligament (ACL) transection model, the rat ACL transection model and the rat meniscectomy model.

Studies have also been undertaken to link FT-IR spectra of sera and urine to a disease related interpretation; this process has been described as disease pattern recognition (DPR). One such study focused on diabetes mellitus (Petrich *et al.*, 2000). Here the full information content of the mid-IR spectra of human blood was recorded and the overall pattern directly related to the donors' disease state. This was achieved by the application of multivariate statistical analyses to the spectra of diabetes type 1, diabetes type 2 and healthy donors. Diabetes type 1 is an autoimmune disorder, where loss of ß-cell function leads to absolute insulin deficiency. Type 2 diabetes is characterized by insulin resistance and/or impaired glucose transport and leads to associated metabolic disorders. Early diagnosis of type 2 diabetes is possible by monitoring blood glucose levels based on a concentration of at least 126 mg/dl (Petrich *et al.*, 2000). However, diabetics have additional metabolic disorders that can, for example, lead to elevated levels of cholesterol and triglycerides. Therefore there are several parameters that can be measured for suitable appraisal of this disease state. Spectroscopic analyses of any body fluid (or tissue) may therefore yield a generalized output more closely correlated to the actual disease state, than an interpretation framed in terms of an individual molecule. Furthermore, a successful interpretation of the spectra using pattern-recognition methods may be possible even when a lack of information (be it partial or complete) exists concerning the underlying molecular components and processes.

In order to illustrate that a quantitative understanding on a molecular basis is feasible, spectra from the above study (Petrich *et al.*, 2000) were correlated successfully with the concentrations of serum glucose, triglycerides and cholesterol, as determined by clinical analyses, by means of partial least squares (PLS) regression. To allow a disease-related interpretation the spectra were also grouped by disease state and the mean spectra for each group calculated. Mathematical classification methods were applied for a more quantitative measure of similarity. Results showed that in supervised classification of any pair of the disease sets specificities and sensitivities of ~80% were achieved within the data set evaluated. This illustrated a clear correlation between a patient's disease state and the mid-IR spectra of that patient's serum. It also illustrates that by applying FT-IR spectroscopy in combination with appropriate multivariate analytical methods, *there is no need for any in-depth knowledge into the underlying metabolic disorder at a molecular level* when using the DPR approach. This approach to diagnostics and biomarker discovery contrasts with the hypothesis-driven research prevalent in the pharmaceutical industry. Other studies employing this approach include Petrich *et al.* (2002) on further diabetes mellitus research and Staib *et al.* (2001) on rheumatoid arthritis, as previously mentioned.

FT-IR has also been advanced as an analytical technique for metabolic profiling of athletes (Petibois *et al.*, 2000a, 2000b, 2002). Studies have primarily been concerned with the analysis of blood, plasma and serum and aimed at monitoring athletes in terms of prevention of overtraining and doping. For example, one study used FT-IR spectroscopy for triglyceride and glycerol concentration measurements in sportsmen taken at rest and after endurance exercises and showed that FT-IR could be applied to routine clinical analyses (Petibois *et al.*, 2002).

Other FT-IR studies of relevance to metabolic profiling include the use of FT-IR in non-invasive multi-component assays of serum (Shaw *et al.*, 1998) and urinary metabolites (Heise *et al.*, 2001, Markus *et al.*, 2001). A study by Shaw and Mantsch (2000), for example, undertook a multi-analyte serum assay that demonstrated measurements of common analytes such as total protein, albumin, urea, glucose, cholesterol and triglycerides. Furthermore, they utilized only the spectral region 4000-2000 cm^{-1}, demonstrating that even a limited spectral region of the mid-IR contains valuable analytical information for target species of interest. The authors of this study also referred to the possible advantages of near infrared (NIR) spectroscopy. These include the fact that remote sampling is possible *via* the use of fiber optics (although it must be stated that this is also possible, but not as common as of yet, with FT-IR). Studies undertaken using NIR spectroscopy include measurement of lactate (Lafrance *et al.*, 2000) and the investigation of metabolites in rat feces (Nakamura *et al.*, 1998). The ability of NIR spectroscopy to quantitate fecal fats was comparable to GC, though it was not nearly as accurate as conventional methods for determining neutral sterol, bile acids and short-chain fatty acids. Its fingerprinting potential however is apparent though mid-IR is generally recognized to be more information rich and to yield a better molecular fingerprint than NIR.

For a historical perspective of vibrational biospectroscopy the reader is directed to Shaw and Mantsch (1999) and for more in-depth reviews of biomedical spectroscopy, Naumann (2001) and Petrich (2001). For a review of the biomedical applications of NIR, see Rempel and Mantsch (1999).

4. CONCLUSION

Rapidity and versatility are two keywords constantly associated with FT-IR spectroscopy. With the continuing development of instrumentation, optical accessories, multivariate analytical techniques, increases in computer processing speeds, FT-IR spectroscopy is assured to be at the forefront of a wealth of new applications devoted to diagnostics and biomarker discovery in both laboratory and clinical settings.

ACKNOWLEDGEMENTS

DIE and RG are indebted to the UK BBSRC (Agri-Food and Engineering and Biological Systems Committee) for financial support. We also thank Margann Wideman for helpful comments on earlier drafts of this manuscript.

REFERENCES

Bendele A, McComb J, Gould T *et al.* Animal models of arthritis: relevance to human disease. *Toxicol Pathol* 27: 134-142 (1999).

Benedetti E, Bramanti E, Papineschi F, Rossi I. Determination of the relative amount of nucleic acids and proteins in leukemic and normal lymphocytes by means of Fourier transform infrared microspectroscopy. *Appl Spectr* 51: 792-797 (1997).

Benedetti E, Bramanti E, Papineschi F, Vergamini P. An approach to the study of primitive thrombocythemia (PT) megakaryocytes by means of Fourier transform infrared microspectroscopy (FT-IR-M). *Cell Mol Biol* 44: 129-139 (1998).

Boydston-White S, Gopen T, Houser S *et al.* Infrared spectroscopy of human tissue. V. Infrared spectroscopic studies of myeloid leukemia (ML-1) cells at different phases of the cell cycle. *Biospectroscopy* 5: 219-227 (1999).

Chiriboga L, Xie P, Yee H *et al.* Infrared spectroscopy of human tissue. I. Differentiation and maturation of epithelial cells in the human cervix. *Biospectroscopy* 4: 47-53 (1998a).

Chiriboga L, Xie P, Yee H *et al.* Infrared spectroscopy of human tissue. IV. Detection of dysplastic and neoplastic changes of human cervical tissue via infrared microscopy. *Cell Mol Biol* 44: 219-229 (1998b).

Cousens S, Smith PG, Ward H *et al.* Geographical distribution of variant Creutzfeldt-Jakob disease in Great Britain, 1994-2000. *Lancet* 357: 1002-1007 (2001).

Damyanovitch AZ, Staples JR, Marshall KW. ^{1}H NMR investigation of changes in the metabolic profile of synovial fluid in bilateral canine osteoarthritis with unilateral joint denervation. *Osteoarthritis Cartilage* 7: 165-172 (1999).

Diem M, Boydston-White S, Chiriboga L. Infrared spectroscopy of cells and tissues: shining light onto a novel subject. *Appl Spectr* 53: 148A-161A (1999).

Diem M, Chiriboga L, Lasch P, Pacifico A. IR spectra and IR spectral maps of individual normal and cancerous cells. *Biopolymers* 67: 349-353 (2002).

Ellis DI, Broadhurst D, Kell DB *et al.* Rapid and quantitative detection of the microbial spoilage of meat by Fourier transform infrared spectroscopy and machine learning. *Appl Environ Microbiol* 68: 2822-2828 (2002).

Eysel HH, Jackson M, Nikulin A *et al.* A novel diagnostic test for arthritis: multivariate analysis of infrared spectra of synovial fluid. *Biospectroscopy* 3: 161-167 (1997).

Fox S. WHO to convene on worldwide risk of BSE and vCJD. *Infect Med* 18: 69-69 (2001).

Goodacre R, Timmins EM, Burton R *et al.* Rapid identification of urinary tract infection bacteria using hyperspectral whole-organism fingerprinting and artificial neural networks. *Microbiology* 144: 1157-1170 (1998).

Goodacre R, Timmins EM, Rooney PJ *et al.* Rapid identification of *Streptococcus* and *Enterococcus* species using diffuse reflectance-absorbance Fourier transform infrared spectroscopy and artificial neural networks. *FEMS Microbiol Lett* 140: 233-239 (1996).

Haaland DM, Jones HDT, Thomas EV. Multivariate classification of the infrared spectra of cell and tissue samples. *Appl Spectr* 51: 340-345 (1997).

Heise HM, Voigt G, Lampen P *et al.* Multivariate calibration for the determination of analytes in urine using mid-infrared attenuated total reflection spectroscopy. *Appl Spectr* 55: 434-443 (2001).

Helm D, Labischinski H, Schallehn G, Naumann D. Classification and identification of bacteria by Fourier-transform infrared spectroscopy. *J Gen Microbiol* 137: 69-79 (1990).

Kirschner C, Maquelin K, Pina P *et al.* Classification and identification of enterococci: a comparative phenotypic, genotypic, and vibrational spectroscopic study. *J Clin Microbiol* 39: 1763-1770 (2001).

Kneipp J, Beekes M, Lasch P, Naumann D. Molecular changes of preclinical scrapie can be detected by infrared spectroscopy. *J Neurosci* 22: 2989-2997 (2002).

Kneipp J, Lasch P, Baldauf E *et al.* Detection of pathological molecular alterations in scrapie-infected hamster brain by Fourier transform infrared (FT-IR) spectroscopy. *Biochim Biophys Acta-Mol Basis Dis* 1501: 189-199 (2000).

Lafrance D, Lands LC, Hornby L, Burns DH. Near-infrared spectroscopic measurement of lactate in human plasma. *Appl Spectr* 54: 300-304 (2000).

Lang PL, Sang SC. The in situ infrared microspectroscopy of bacterial colonies on agar plates. *Cell Mol Biol* 44: 231-238 (1998).

Lasch P, Haensch W, Lewis EN *et al.* Characterization of colorectal adenocarcinoma sections by spatially resolved FT-IR microspectroscopy. *Appl Spectr* 56: 1-9 (2002).

Lasch P, Naumann D. FT-IR microspectroscopic imaging of human carcinoma thin sections based on pattern recognition techniques. *Cell Mol Biol* 44: 189-202 (1998).

Liu KZ, Jia L, Kelsey SM *et al.* Quantitative determination of apoptosis on leukemia cells by infrared spectroscopy. *Apoptosis* 6: 269-278 (2001).

Liu KZ, Mantsch HH. Apoptosis-induced structural changes in leukemia cells identified by IR spectroscopy. *J Mol Struct* 565: 299-304 (2001).

Liu KZ, Schultz CP, Mohammad RM *et al.* Similarities between the sensitivity to 2-chlorodeoxyadenosine of lymphocytes from CLL patients and bryostatin 1 treated WSU-CLL cells: an infrared spectroscopic study. *Cancer Lett* 127: 185-193 (1998).

Markus A, Swinkels DW, Jakobs BS *et al.* New technique for diagnosis and monitoring of alcaptonuria: quantification of homogentisic acid in urine with mid-infrared spectrometry. *Anal Chim Acta* 429: 287-292 (2001).

Nakamura T, Takeuchi T, Terada A *et al.* Near-infrared spectrometry analysis of fat, neutral sterols, bile acids, and short-chain fatty acids in the feces of patients with pancreatic maldigestion and malabsorption. *Int J Pancreatol* 23: 137-143 (1998).

Naumann D. FT-infrared and FT-Raman spectroscopy in biomedical research. *Appl Spectr Rev* 36: 239-298 (2001).

Naumann D, Helm D, Labischinski H. Microbiological characterizations by FT-IR spectroscopy. *Nature* 351: 81-82 (1991).

Naumann D, Keller S, Helm D *et al.* Ft-Ir Spectroscopy and Ft-Raman spectroscopy are powerful analytical tools for the noninvasive characterization of intact microbial-cells. *J Mol Struct* 347: 399-405 (1995).

Otterness IG, Swindell AC, Zimmerer RO *et al.* An analysis of 14 molecular markers for monitoring osteoarthritis: segregation of the markers into clusters and distinguishing osteoarthritis at baseline. *Osteoarthritis Cartilage* 8: 180-185 (2000).

Otterness IG, Weiner E, Swindell AC *et al.* An analysis of 14 molecular markers for monitoring osteoarthritis: relationship of the markers to clinical end- points. *Osteoarthritis Cartilage* 9: 224-231 (2001).

Perromat A, Melin AM, Deleris G. Pharmacologic application of Fourier transform infrared spectroscopy: the *in vivo* toxic effect of carrageenan. *Appl Spectr* 55: 1166-1172 (2001).

Petibois C, Cazorla G, Deleris G. Triglycerides and glycerol concentration determinations using plasma FT-IR spectra. *Appl Spectr* 56: 10-16 (2002).

Petibois C, Deleris G, Cazorla G. Perspectives in the utilisation of Fourier-transform infrared spectroscopy of serum in sports medicine - health monitoring of athletes and prevention of doping. *Sports Med* 29: 387-396 (2000a).

Petibois C, Melin AM, Perromat A *et al.* Glucose and lactate concentration determination on single microsamples by Fourier-transform infrared spectroscopy. *J Lab Clin Med* 135: 210-215 (2000b).

Petrich W. Mid-infrared and Raman spectroscopy for medical diagnostics. *Appl Spectr Rev* 36: 181-237 (2001).

Petrich W, Dolenko B, Fruh J *et al.* Disease pattern recognition in infrared spectra of human sera with diabetes mellitus as an example. *Appl Optics* 39: 3372-3379 (2000).

Petrich W, Staib A, Otto M, Somorjai RL. Correlation between the state of health of blood donors and the corresponding mid-infrared spectra of the serum. *Vibrat Spectr* 28: 117-129 (2002).

Rempel SP, Mantsch HH. Biomedical applications of near-infrared spectroscopy: a review. *Can J Anal Sci Spectr* 44: 171-179 (1999).

Schmitt J, Flemming HC. FTIR-spectroscopy in microbial and material analysis. *Internat Biodeterior Biodegrad* 41: 1-11 (1998).

Schultz CP, Liu KZ, Johnston JB, Mantsch HH. Study of chronic lymphocytic leukemia cells by FT-IR spectroscopy and cluster analysis. *Leukemia Res* 20: 649-655 (1996).

Schultz CP, Liu KZ, Johnston JB, Mantsch HH. Prognosis of chronic lymphocytic leukemia from infrared spectra of lymphocytes. *J Mol Struct* 408: 253-256 (1997).

Shaw RA, Kotowich S, Leroux M, Mantsch HH. Multianalyte serum analysis using mid-infrared spectroscopy. *Ann Clin Biochem* 35: 624-632 (1998).

Shaw RA, Mantsch HH. Vibrational biospectroscopy: from plants to animals to humans. A historical perspective. *J Mol Struct* 481: 1-13 (1999).

Shaw RA, Mantsch HH. Multianalyte serum assays from mid-IR spectra of dry films on glass slides. *Appl Spectr* 54: 885-889 (2000).

Staib A, Dolenko B, Fink DJ *et al.* Disease pattern recognition testing for rheumatoid arthritis using infrared spectra of human serum. *Clin Chim Acta* 308: 79-89 (2001).

Stuart B. *Biological Applications of Infrared Spectroscopy.* John Wiley and Sons, Chichester (1997).

Thomas N, Goodacre R, Timmins EM *et al.* Fourier transform infrared spectroscopy of follicular fluids from large and small antral follicles. *Hum Reprod* 15: 1667-1671 (2000).

Timmins EM, Howell SA, Alsberg BK *et al.* Rapid differentiation of closely related *Candida* species and strains by pyrolysis mass spectrometry and Fourier transform- infrared spectroscopy. *J Clin Microbiol* 36: 367-374 (1998).

Udelhoven T, Naumann D, Schmitt J. Development of a hierarchical classification system with artificial neural networks and FT-IR spectra for the identification of bacteria. *Appl Spectr* 54: 1471-1479 (2000).

Utzinger U, Heintzelman DL, Mahadevan-Jansen A *et al.* Near-infrared Raman spectroscopy for *in vivo* detection of cervical precancers. *Appl Spectr* 55: 955-959 (2001).

Chapter 8

METABOLOMIC ANALYSIS WITH FOURIER TRANSFORM ION CYCLOTRON RESONANCE MASS SPECTROMETRY

Dayan Goodenowe
Phenomenome Discoveries Inc., 204-407 Downey Road, Saskatoon SK S7N 4L8, Canada

1. INTRODUCTION

All organisms contain a finite set of genes. However, the expression of these genes under varying developmental and environmental conditions results in an almost infinite number of possible phenotypes. To properly study and understand an organism we need to know more than its genetic identity (*i.e.* its genome) and how its genome responds to developmental and environmental challenges. We need to know how these responses either benefit or harm the organism. To study properly the events leading to, and the effects resulting from, the expression of a gene we need to be able to characterize the corresponding phenotypic changes in an objective and quantifiable manner. Metabolomics is the science of phenotype analysis through the measurement of the small molecules (metabolites) that interact with, are used by, and are created by, the functionally active proteins of a biological system.

Since the entire gene sequences of some species are now known, gene-chip technology makes it possible to simultaneously monitor and quantitate the changes in expression of every gene within a genome to developmental and environmental changes. Gene-chip technology can therefore be viewed as non-targeted gene expression analysis. It is however, in actuality, a targeted analysis that just happens to contain all possible targets. This is a powerful comprehensive capability, but it was made possible by the fact that the genome is a finite and unitary entity. Other methods such as massively parallel signature sequencing and Serial Analysis of Gene Expression (SAGE) are technologies that also attempt to analyze the transcriptome in a truly non-targeted and unbiased manner.

At the molecular level, the phenotype of a given biological system is manifested in its proteome and metabolome. Since gene expression results in protein synthesis, the proteome is the first and most direct link to gene expression. However, due to the complex interactions of metabolic pathways, it is difficult to predict the effects that changes in expression levels of a given protein will have on the overall cellular processes that it is involved in. The metabolome, on the other hand, is the summation of all metabolic (proteomic) activities occurring in an organism at any given point in time. The metabolome is therefore a direct measure of the overall or end effect of gene expression on the cellular processes of any given biological system. For this reason, the metabolome could prove to be the more powerful of the two phenotypes in actually understanding the effects of gene function and manipulation.

The current bottleneck in functional genomics is objective and quantitative phenotype analysis. Although gene-chip technology has opened our minds to non-targeted analysis methods, it is not possible to directly apply these tools to phenotype analysis due to the fact that not only are there multiple phenotypes, but a virtually infinite number of metabolites and proteins are possible. To be complementary to the current state of genomic analysis, phenotype analysis must eventually be non-targeted in "actuality". In this chapter, our objective is to discuss the requirements of truly non-targeted, comprehensive, metabolome analysis. More specifically, we will discuss how the use of Fourier transform ion cyclotron resonance mass spectrometry (FT-ICR-MS) can attempt to reach some of these goals.

2. FOURIER TRANSFORM MASS SPECTROMETRY

FT mass analysers use magnetic fields and work on the premise that a charged particle in a magnetic field experiences a force at right-angles to its motion which makes it move in a helix (see Marshall and Grosshans, 1991; Marshall, 2000, Marshall and Hendrickson, 2002, for technical discussions and detailed reviews on FT-ICR-MS). Faster ions move in bigger helices, so that the time taken to describe a period of the helix is always the same, proportional to the mass-to-charge ratio (*m/z*). The particle thus operates at its cyclotron frequency, and determining the signal intensity at that frequency yields the number of ions with a given *m/z*. All ions contribute to the overall output at any given instant, and the separation of different masses is performed mathematically by Fourier transformation of the whole signal. Very high resolutions (in excess of 10^6), and high detection efficiencies are achievable. FT-ICR-MS also possesses inherent MS^n capabilities. Technical demands in terms of high-vacuum requirements and difficulty in coupling with high-pressure ion sources, and high cost limit its widespread use. Recent biological applications of FT-ICR-MS, however, do include analyses of changes in phospholipids in mastocytoma cells during degranulation (Ivanova *et al.*, 2001) and in short tandem repeat genotyping (Null and Muddiman, 2001). In studies on mastocytoma cell phospholipids more than 130 glycerophospholipids could be spectrally resolved (at 1 mass unit resolution) and changes initiated by addition of exogenous phospholipases easily monitored at the sub-nanomole level.

Important steps in all metabolomic analyses by FT-ICR-MS (and other approaches) include sample extraction and preparation, metabolite separation and detection, metabolite identification and metabolite quantitation, issues now addressed.

2.1 Metabolite Extraction

A multitude of extraction technologies have been developed over the years to recover all classes of metabolites. However, extracts can be biased by a chosen extraction procedure. In non-targeted metabolomic strategies fractions are generally split into polar and non-polar fractions. However issues such as gaining access to the cellular and sub-cellular compartments containing the metabolites of interest remain important. For example, Griffin and Sang ask (Chapter 3) "in neural tissue can all the glutamate present be extracted by an aqueous extraction procedure when a significant amount of the neurotransmitter is contained within lipid vesicles?" Thus, extraction methods must be tailored to and validated for any matrix being studied. For direct infusion MS methods, such as discussed here, it is also of critical importance that ultra-pure solvents are used as well as only volatile organic buffering molecules.

2.2 Metabolite Separation and Detection

To establish whether levels of a specific metabolite have changed as a result of genetic or environmental perturbation, that metabolite clearly has to be isolated and independently detected. Polar and non-polar extraction methods can be considered a crude form of metabolite separation allowing *e.g.* separation of sugar phosphates from triglycerides.

Polarity and mass are, of course, the two main chemical characteristics exploited in separation technologies; chromatographic techniques exploit differences in polarity between molecules whereas MS exploits differences in mass. Maximal separation can be achieved by using combinations of the above technologies (*e.g.* liquid chromatography-MS). However, the detection methods used in chromatography (*e.g.* ultraviolet (UV), fluorescence, electrochemical) and the ionization methods used in MS (electrospray, atmospheric chemical ionization, electron impact, fast atom bombardment, secondary ion mass spectrometry) all have varying degrees of selectivity and therefore, also introduce bias. A comprehensive, non-targeted metabolome analysis needs to take advantage of these characteristics by balancing different technologies and maximizing both the separation and detection of molecules. With FT-ICR-MS, the combination of ultra-high mass resolution and ionization techniques that cover almost all biological molecules, make it an ideal choice in this area.

MS is both a separation and a detection technology with separation based upon *m/z* ratios, or mass. The first important concept in MS is in establishing the difference between the nominal and accurate mass of a molecule. Only carbon-12 has an accurate mass equal to its nominal mass. This is simply because it has been set as such so we can then use it to compare with all other elemental masses. Therefore *e.g.* hydrogen-1 has a nominal mass of 1 but a true mass of 1.007825 and oxygen-16 has a true mass of 15.99491.

The second important concept in MS is isotopic abundance. Most elements have more than one isotope. This means that a given molecule will exist in nature in many different forms, all of which will have similar biological activity but a different mass. All isotopomers will be treated as a single entity in a chromatographic separation but MS can separate out the different isotopes as different molecules. Indeed, it is imperative that direct infusion, non-chromatographic MS separation method can separate and identify the difference between isotopes and unique molecules.

The last important distinction to recognize is the difference between the average mass and the accurate mass. The accurate mass always refers to the mass of a specific isotope whereas the average mass takes into account the natural abundance and mass of each isotope.

Table 1. Nominal and Accurate Masses of Selected Metabolites

Common Name	Empirical Formula	Nominal mass	Accurate Mass
Allantoin	$C_4H_6N_4O_3$	158	158.043990
Indole-3-acetaldehyde	$C_{10}H_8NO$	158	158.060589
Ethyl heptanoate	$C_9H_{18}O_2$	158	158.130680
Decanol	$C_{10}H_{22}O$	158	158.167065

The point of this discussion is illustrated in Table 1. Metabolites of different empirical formulas may have the same nominal mass but different accurate masses. Each unique empirical formula has a unique mass and with sufficient mass resolution different empirical formulas can be separated from one another. FT-MS is currently the highest resolution MS technology available and is the only technology that can reliably separate such isobaric peaks from one another. Fig. 1 illustrates a high resolution spectra of strawberry fruit extracts where a metabolite change between the green and red stages can be clearly elucidated even though the mass difference between it an its nearest neighbor is only 0.027 Daltons.

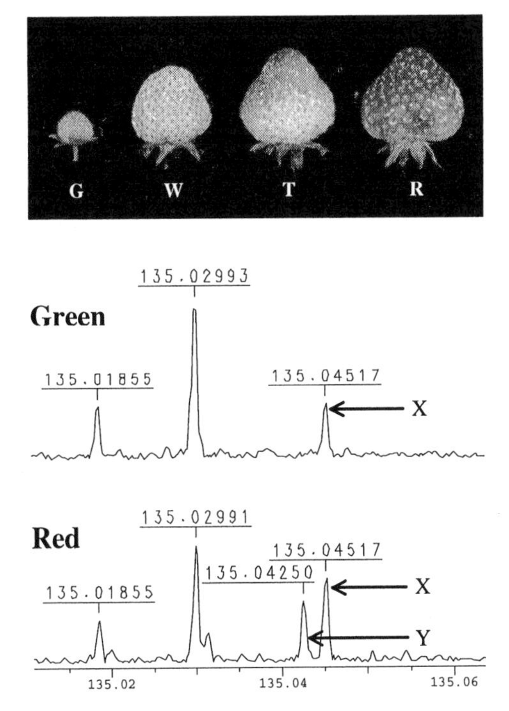

Figure 1. High resolution mass spectra of strawberry extracts.

2.3 Metabolite Identification

It is at the point of metabolite identification that the true differences between a targeted and non-targeted approach become apparent. Classically, non-targeted approaches generated "fingerprints" in which the pattern of metabolite peaks from different biological states can be used to group samples into "like" groups. These fingerprints provided a broad coverage of metabolites but the identity of individual peaks often remains unknown. Nuclear magnetic resonance (NMR) spectroscopy has identified new biomarkers in *e.g.* phospholipidosis (Espina *et al.*, 2001) but generally addresses only high abundance compounds. Phospholipids tend to dominate mass spectra in direct infusion studies of microbes (see Vaidyanathan and Goodacre, Chapter 2).

Classical targeted approaches are used to profile a selected (and identifiable) group of metabolites using methods specifically designed and based upon known standards. Metabolites are traditionally identified based upon their behavior under specific method parameters and not their inherent chemical makeup unlike proteins and genes that can be identified based upon their amino acid or nucleotide sequences, respectively. To be able to identify metabolites in a non-targeted approach the primary identification process must be based upon the intrinsic chemistry of the metabolite and its behavior under a given set of conditions used as supporting information.

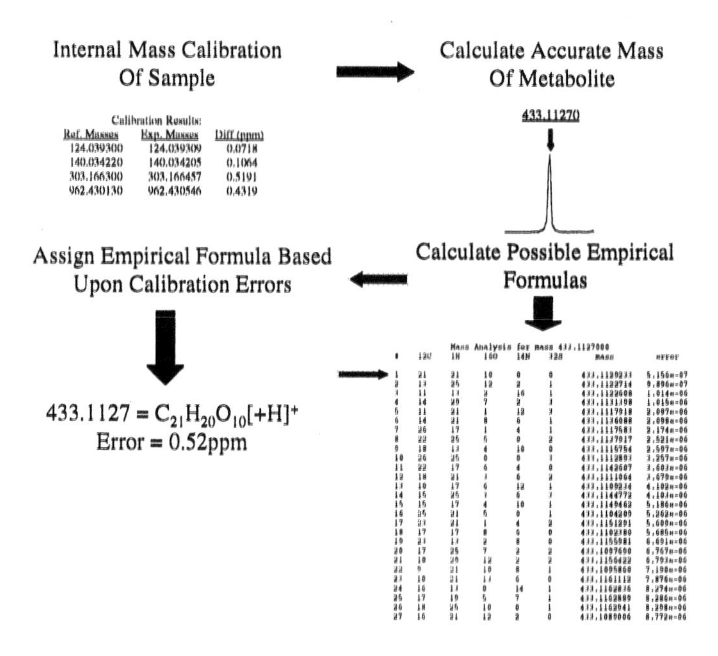

Figure 2. Determining empirical formula from high resolution data.

As has been shown in the previous section, metabolites with different empirical formulas have a different accurate mass. Therefore, if the mass of a metabolite can be accurately determined it should be then be possible to determine the empirical formula. Fig. 2 illustrates this process. The utility of this process is dependent upon both mass accuracy and resolution. In this example, mass accuracy of 1 ppm or less is required to assign a unique empirical formula to the metabolite. With a mass accuracy of 10 ppm or less 37 empirical formulas are possible. Only FT-ICR-MS technology with internal standard calibration is capable of routinely achieving mass accuracies of less than 1 ppm.

Metabolite identification is critical to the utility of metabolomics in functional genomics research; however there are multiple levels of metabolite identification. The FT-ICR-MS approach described allows for non-targeted separation of metabolites and the assignment of an accurate mass tag to each of these metabolites without any *a priori* knowledge of the metabolites. This accurate mass tag can then be used to determine the empirical formula of the metabolite. In this sense it can be viewed as an expansion on the capabilities of the classical non-targeted approach.

The FT-ICR-MS approach is a one-dimensional approach to metabolite identification; one dimensional in that it returns the accurate mass and intensity of the intact metabolite. The major advantage is that it allows for the simultaneous analysis of thousands of metabolites. The major limitation is that all metabolites with the same empirical formula (*i.e.* isomers) will have the same mass and are therefore pooled. Switching from parallel processing of multiple parent ions to the serial analysis of these parents in MS/MS mode allows for additional structural information to be obtained. In MS/MS mode, an individual metabolite is isolated in the FT-ICR-MS and then fragmented by either collision induced dissociation (CID) or by infra-red multi-photon dissociation (IRMPD) where the metabolite is bombarded with either an inert gas or photons, respectively. This approach can be used to differentiate geometric and stereoisomers. In performing MS/MS analysis it is important that the parent ion is resolved from all of the other metabolites prior to fragmentation. If the parent ion is not fully separated, it is not possible to confidently assign the fragment ions generated to a specific parent ion. Currently, FT-ICR-MS is the only technology capable of high resolution parent ion selection for MS/MS. It is generally accepted that accurate mass in combination with NMR and MS/MS data is the best way to assign a three dimensional structure to an "unknown" metabolite.

2.4 Metabolite Quantitation

Targeted and non-targeted approaches are further differentiated at the level of metabolite quantitation. Targeted approaches are typically more robust with lower levels of variability and the quantitation is absolute (*i.e.* sample x has 5.3 ug/g of serine) whereas non-targeted methods are typically less robust with higher levels of variability and the quantitation is relative (*i.e.* sample x has 2.3 times the level of serine vs. sample y). Minimizing variability is critical to reliable quantitation. Errors associated with sampling and metabolite ionization are responsible for the majority of variability observed in MS based metabolomic methods. Sampling errors are most commonly addressed by either pooling samples or by replicate analyses. If time and cost are not limiting both corrective measures should be taken. Variability associated with sampling is common to both targeted and non-targeted methodologies. Variability in metabolite ionization, however, can be dramatically different between targeted and non-targeted methods.

Methods that rely upon external calibration parameters (as is the case with most targeted methods) generate two sets of recovery data; true recovery and observed recovery. True recovery is defined as the amount of analyte that is in the final extract compared to the theoretically determined amount. Observed recovery is the amount of analyte determined to be in the final extract as calculated from a calibration curve generated from known amounts of analyte in pure solvent mixtures. Situations in which the observed recovery is different from the true recovery arise when the instrument response for analytes in pure solvent solutions is different from the instrument response for analytes in the final extract or matrix. These differences are called matrix effects and all MS and LC-MS related methods are susceptible to matrix effects. Matrix effects are instrument and more specifically source-dependent. Essentially the matrix present in a sample extract can affect instrument response by either enhancing or suppressing the ability of the source to generate and/or transfer ions into the mass spectrometer. The two most important factors that are influenced by matrix are: ion generation and desolvation.

In chromatography-based systems the matrix effect is variable with time (*i.e.* different amounts and different kinds of matrix components will elute at different times) and both suppression and enhancement can be observed depending on the elution time. In targeted methods which make use of stable isotope internal standards these effects can easily be monitored and corrected if necessary because stable isotopes will have the same retention time as the endogenous isotope of interest. Matrix effects are much simpler in direct infusion methods as the concentration of matrix is constant and therefore the matrix effect is also constant. More importantly, the effect is constant across all metabolites within the matrix and therefore the matrix effect on all metabolites can be measured by using a small set of internal standards.

Matrix effects require that approaches to quantitation should be based upon a fit for purpose model. Purposes that require high degrees of precision (a 5-20% change in concentration is significant) should use targeted methods with enough replication to perform reliable statistics. However, in most metabolomic studies, where research is primarily interested in what systems have changed, this is excessive. In these systems, it is the assessment of the relative changes of a large number of components that is most important. Here the goal should be to obtain reproducibility such that changes in concentration of 50% or more can be confidently measured.

3. INTERPRETATION OF METABOLOMIC DATA

As increasing amounts of metabolomic data continue to be collected, the ability to store, retrieve and interpret this information will become a rate limiting factor to its efficient use in biological research. It is here that the FT-MS based metabolomic approach described can provide its most significant contribution. The two most important factors in managing and interpreting research data are 1) the ratio of the physical size of the raw data files to the amount of relevant information contained in these files, and 2) the ability to compare, quantitatively and qualitatively, data files from discreet biological conditions and experimental tests.

One of the advantages of targeted methods is that they are focused on a finite set of pre-specified metabolites. This means that however large and complex the generated data files may be during the analysis of these metabolites, the requirements for data interpretation can be reduced to a simple table that quantitates metabolites of interest. It must be noted that such exquisite data reduction is possible only through the confidence expressed in the validation of the analysis method. When confidence in the raw data becomes suspect complete reduction of such data is no longer justified. For this reason 2D protein gels or gene array maps generally continue to be used side by side with the derived processed data. The retention of large data files with sparce information places an enormous stress on computing systems. In fact it is precisely these issues that have pushed computing to such amazing heights. Unfortunately, the requirement of super computers is value-limiting.

In some respects, the processing of metabolomic data is in its infancy when compared to proteomics and genomics. Non-targeted methods based upon chromatography, low resolution MS or NMR rely on the analysis of patterns resulting from the analysis without the ability to reliably reduce this continuous data to discreet data. Van der Greef (Chapter 10) discusses in detail some of the data preprocessing and processing stages currently used in increasing platform reproducibility for metabolomic analyses. Phenomenome Discoveries Inc. believe that the FT-ICR-MS method alluded to here offers a unique ability to analyze metabolites in a non-targeted manner as well as reduce this information to the similar format employed by targeted methods – simple tables.

Our concept is simple. Independent biological extracts are analyzed by high-resolution FT-ICR-MS. Each analysis is internally calibrated. This internal calibration enables not only the monitoring of performance but ensures that all samples are the same, or more accurately, the degree of sameness can be calculated. High resolution ensures that metabolite peaks are resolved from one another. A high degree of mass accuracy ensures that each of these peaks is assigned a consistent mass. A large analysis data file can therefore be reliably converted into a simple text file containing accurate mass and intensity data. This is the first and most important step. From this point forward, multiple data files from different ionization and extraction conditions from a single biological sample can be combined with confidence. Most importantly, the final combined file contains a high density of useful information.

Once confidence in the empirical data from a specific sample is obtained, the focus switches to the real work of science – the interpretation of experimental data to confirm or refute existing hypotheses and the creation of new hypotheses to be further tested.

Metabolomic data is unique from proteomic and genomic data in that the homology between metabolites can be measured–exactly. Metabolites within a pathway are directly related to one another by the chemical transformations of that pathway. The ability to elucidate and resolve these chemical transformations is the most important requirement of metabolomics. Biological systems are in a constant state of dynamic equilibrium. This means that most perturbations are compensated for in some fashion. If a given pathway is disrupted, accumulation of a substrate will not occur unchecked. This substrate may be transported and subsequently modified in some secondary pathway. Some, but not all, of these secondary transformations can be predicted. Successful metabolomics must be able to find and identify these effects.

Phemonenome Inc. believe that an accurate mass metabolomic database approach is the only approach that can comprehensively address these issues. The accurate mass and empirical formula data generated from an FT-ICR-MS experiment offers unparalleled processing power. However, to have value in functional genomics, this data needs to visualized in a meaningful way, specific tools need to be created to draw relationships between metabolites within and between samples and to assist in the identification of new or novel metabolites. This data must ultimately be integrated with other "omic" data.

An independent and comprehensive summary data file (Met-Ex) for each discreet experimental condition is our starting point. This file contains the accurate mass of each detected metabolite, its calculated neutral mass, an extraction and ionization code, its intensity, the putative empirical formula(e) for the metabolite as well as any error associated with this prediction. In addition, the internal standard QA/QC statistics such as measured versus known mass and intensity are included.

4. VISUALIZATION OF CHANGES IN PATHWAYS

To compare metabolic pathway changes between two conditions requires that, 1) the metabolites contained within the pathways are known and, 2) that one can access the intensity values for each of the metabolites. Since these are fixed pathways, the visual map needs to be built only once and then be capable of interfacing with changing data files.

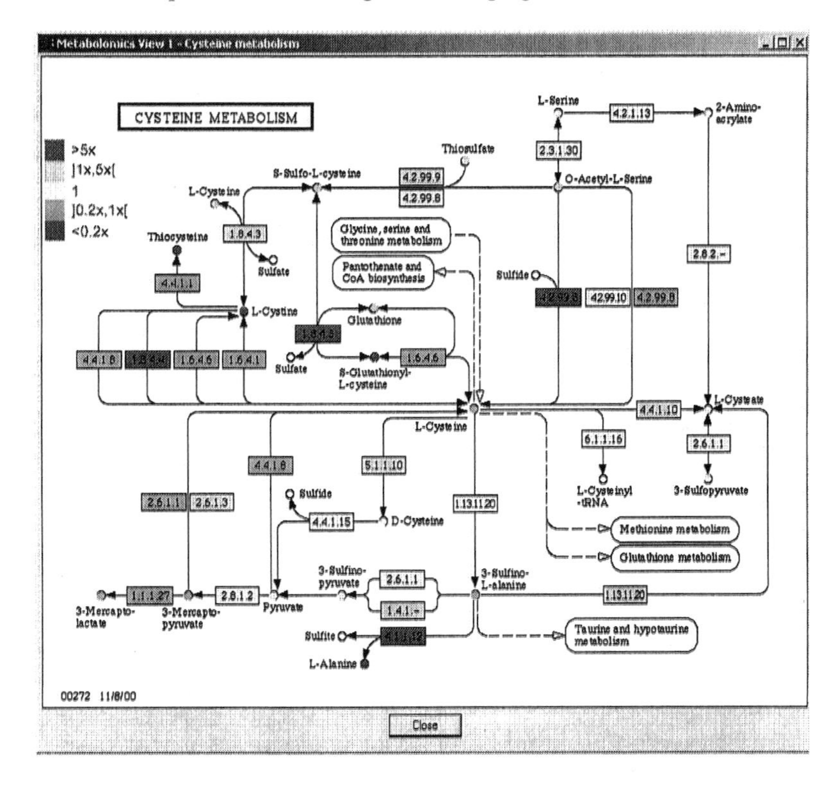

Figure 3. Visualization of metabolite changes in known pathways.

Pedro Mendes' group at the Virginia Bioinformatics Institute, Blacksburg Virginia, in collaboration with Phenomenome Discoveries Inc., has applied this simple logic to the 107 pathways listed in the Kyoto Encyclopedia of Genes and Genomes (KEGG) database (see also Jing Li *et al*, Chapter 16).

As is illustrated in Fig. 3, a fixed metabolic pathway map (*e.g.* cysteine metabolism) can be used as an interface between any two Met-Ex files. All that is required to do is calculate the accurate mass from the empirical formula of the metabolite, search for it in both selected files, convert it to a ratio and then display this ratio using a color coding scheme.

4.1 Visualization of Metabolite Changes Using an Array

Unfortunately the fixed pathway format cannot be used to compare multiple files or to compare metabolites of unknown identity. The metabolite array visualization format is the most powerful and comprehensive visualization tool made possible by the Met-Ex format.

Mean_Mass	RSD (ppm)	Mode	N	24D/C 1h	24D/C 4h	24D/C 8h	24D/C 24	24D/C 48h	24D/C 144h
219.9694	0.23	12	24	2.4	13.9	7.9	17.4	45.0	36.8
214.0396	0.21	12	24	6.0	35.3	20.1	45.7	104.8	81.7
298.0008	0.57	12	21	1.2	4.4	2.5	5.0	11.6	8.0
343.9968	0.67	12	18	1.0	3.6	2.1	3.8	7.5	4.8
260.0449	0.53	12	20	1.0	3.4	2.5	4.7	9.7	7.4
328.0193	0.69	12	16	1.0	2.4	1.6	3.2	5.1	4.6
282.0268	0.66	12	15	1.0	2.1	1.8	2.2	3.8	3.6
910.2484	7.07	22	13	1.0	1.8	1.2	4.2	2.3	1.0
342.0977	0.95	12	10	1.0	1.0	1.9	3.2	4.5	17.5
328.0193	0.69	12	16	1.0	2.4	1.6	3.2	5.3	4.6
303.9303	0.82	12	11	1.0	1.7	1.4	2.4	4.0	2.8
126.0510	0.34	12	12	1.0	1.6	1.0	2.5	3.2	3.7
310.1258	0.76	12	13	1.0	1.0	1.0	2.4	2.0	1.3
875.5676	7.41	22	16	1.0	1.4	1.9	2.3	5.3	3.4
269.9446	0.60	12	22	0.9	1.0	1.0	2.3	5.4	2.8
910.2284	8.96	22	14	1.0	1.7	1.3	2.2	8.4	4.5
349.9616	3.06	12	13	1.0	1.4	1.0	2.1	2.2	1.6
277.0350	0.57	12	7	1.0	1.0	1.0	1.5	5.0	3.1
771.3524	3.08	22	9	1.0	1.2	1.0	1.5	4.1	2.8
333.9492	0.79	12	5	1.0	1.0	1.0	1.7	4.1	1.8
834.9808	2.71	22	8	1.0	1.0	1.0	1.5	3.1	2.4
282.9647	0.47	12	5	1.0	1.0	1.0	1.0	2.9	2.0
349.9599	0.26	22	3	1.0	1.0	1.0	1.0	2.8	2.1
287.9562	1.81	12	5	1.0	1.4	1.0	1.4	2.9	1.7
771.3555	5.28	12	11	1.0	1.3	1.0	1.2	2.6	2.1
341.1318	1.10	12	6	1.0	1.0	1.0	1.0	2.2	3.9
285.9613	0.52	12	8	1.0	1.0	0.6	1.0	0.6	3.2
217.9739	0.36	12	60	1.2	0.9	0.7	1.9	1.1	3.1
253.9582	0.34	12	9	1.0	1.0	1.0	1.0	1.6	3.1
255.9386	0.41	12	14	1.5	1.0	1.0	1.8	1.5	2.6
217.9829	0.56	12	9	1.0	0.9	1.0	1.9	1.6	2.6

Figure 4. The Met-Array representing changes in metabolite concentrations in herbicide treated *Arabidposis thaliana.*

The Met-Array is based upon the observation that if consistent high mass resolution and accuracy is obtained across different samples, when the Met-Ex data files from these samples are combined, metabolites having identical empirical formulae will form a discrete cluster about the accurate mass, whereas metabolites of different empirical formulae will have masses that are significantly different. Each of these clusters represents one line in the array. The average mass and associated statistics for each cluster can then be represented in the first columns of each row of the array. The subsequent columns in each row contain file-specific information (*i.e.* intensity or mass) for each of the files used in the creation of the array. In addition to displaying the absolute values specific to each file, any number of mathematical comparisons can be made between the files selected, for example, the ratio of intensities between any two samples can also be displayed. A zero-fill table is required to insert a number in those cells in which no metabolite was detected. Fig. 4 is an array representing metabolite changes that occur in *Arabidopsis thaliana* after treatment with herbicide. The first column is the average neutral mass of the metabolite cluster, followed by the relative standard deviation of the cluster. The greyed cells represent the ratio of the herbicide treated samples versus control. The array is further clustered to show which metabolites change early and which metabolites change later.

4.2 Metabolite Analysis Tools

Metabolite analysis tools can be differentiated by whether the focus is on identification of the selected metabolite or on relationships between selected metabolites and other metabolites.

The actual identification of a metabolite is a contentious issue. The FT-ICR-MS method described here accurately measures the mass of the metabolite in a given extract. The information available to the scientist to use in predictions of metabolite identity are the theoretically possible empirical formulae that could have such a mass within a given range of error, the chemical class of the metabolite based upon extraction conditions, and the primary functional group of the metabolite, based upon ionization mode. Through evaluation of these data the most probable empirical formula can be predicted and a search made of existing databases. An example of the available information, in an easy-to-interpret interface is displayed in Fig. 5. In this example, there are four putative empirical formulae that could exist for the selected metabolite with an error of less than 1 ppm. Additional information that aids interpretation is that the internal standards of that file had ppm ranges from 0.01 to 0.1 and the external recovery standard had a mass accuracy of 0.23 ppm. This indicates that the correct empirical formula will most likely have an error of less than 0.3 ppm. Furthermore the metabolite in question was analyzed in a mode that corresponds to an aqueous extract analyzed in the negative ion electrospray mode. All of this data strongly suggests that the metabolite has the empirical formula $C_{21}H_{20}O_{11}$. This empirical formula is used to search an online or proprietary database to assist in identification.

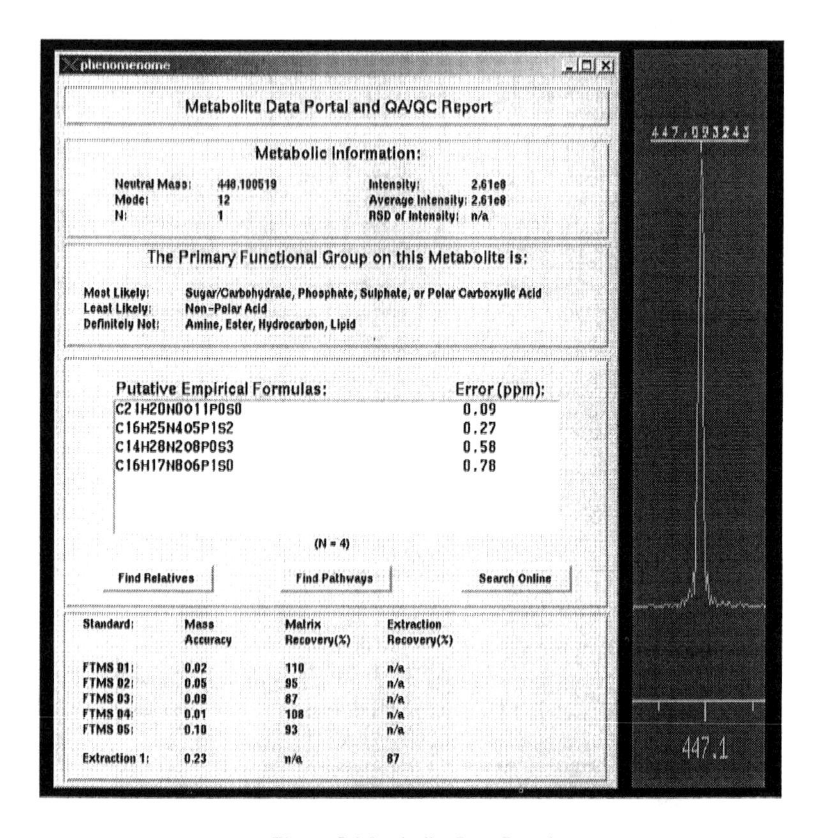

Figure 5. Metabolite Data Portal.

Accurate mass data on all of the expressed metabolites within a given biological extract allow for more elegant analyses to be undertaken. Since we know that metabolic pathways are governed by a finite set of enzymatic reactions and that these enzymatic reactions result in the chemical transformation of a metabolite by either the addition or subtraction of chemical entities, we can search the array for a list of possible related metabolites by simply looking for the associated change in mass. This search can go on for a finite set of iterations or until all possible steps are exhausted. Taken one step further, all discrete metabolic families from all of the metabolites detected can be elucidated.

5. CONCLUSION

In summary, a non-targeted metabolomic analysis using FT-MS to resolve and then ascertain the accurate mass of all expressed metabolites within a biological extract brings the study of metabolomics to the level of comprehensiveness and utility that is currently possible in the areas of proteomics and genomics.

REFERENCES

Espina JR, Shockcor JP, Herron WJ *et al.* Detection of *in vivo* biomarkers of phospholipidosis using NMR-based metabonomics approaches. *Magn Reson Chem* 28: 559-565 (2001).

Ivanova PT, Cerda BA, Horn DM *et al.* Electrospray ionization mass spectrometry analysis of changes in phospholipids in RBL-2H3 mastocytoma cells during degranulation. *Proc Natl Acad Sci USA* 98: 7152-7157 (2001).

Marshall AG, Grosshans PG. Fourier transform ion cyclotron resonance mass spectrometry; the teenage years. *Anal Chem* 63: 215A-229A (1991).

Marshall AG. Milestones in Fourier transform ion cyclotron resonance mass spectrometry technique development. *Int J Mass Spectrom* 200: 331-356 (2000).

Marshall AG, Hendrickson CL. Fourier transform ion cyclotron resonance detection: principles and experimental configurations. *Int J Mass Spectrom* 215: 59-75 (2002).

Null AP, Muddiman DC. Perspectives on the use of electrospray ionization Fourier transform ion cyclotron resonance mass spectrometry for short tandem repeat genotyping in the post-genome era. *J Mass Spectrom* 36: 589-606 (2001).

Chapter 9

STABLE ISOTOPE-BASED DYNAMIC METABOLIC PROFILING IN DISEASE AND HEALTH

Tracer Methods and Applications

László G. Boros[1], Marta Cascante[2] and Wai-Nang Paul Lee[1]

[1]Harbor-UCLA Research and Education Institute, 1124 West Carson Street RB1, Torrance, CA 90502, USA [2]Department of Biochemistry and Molecular Biology, University of Barcelona, Marti/I Franques 1, Barcelona 08028, Catalonia, Spain

1. INTRODUCTION

Substrates, intermediary metabolites and products, the so called "global metabolite pool", constitute a living organism's metabolome. Variations and changes in components of the metabolome reflect adaptation of an organism to its microenvironment, as defined by substrate availability and hormonal milieu, through altered gene expression and through the activation of signaling cascades (Tweeddale *et al.*, 1998). The major regulatory components of cell function, the genome, transcriptome and proteome, ultimately act on the metabolome resulting in the expression of a specific phenotype. The close interactions among these components establish a rationale for integrating functional genomics, proteomics and metabolomics, as a means to study the complete intracellular signal processing apparatus that regulates metabolic adaptation, phenotype and ultimately cellular function (Kell and King, 2000).

Although broad applications of metabolome analyses have been reported in the literature, most studies involve narrowly defined metabolic reactions such as CO_2 release or the measurements of selected metabolites from particular pathways in primitive organisms (Tweeddale *et al.*, 1999, Liu *et al.*, 2000, Yoder and Turgeon, 2001, Oliver *et al.*, 2002, Oliver, 2002, Fiehn, 2002, Aharoni and Vorst, 2002, Eglington *et al.*, 2002). Recent reports on mammalian systems include investigations of diet-dependent changes in the serum metabolome (metabolic serotype) of rat. These studies describe validation of the various separation and detection techniques used for the characterization of serum metabolites (Vigneau-Callahan *et al.*, 2001) and their variations by sex (Shi *et al.*, 2002). Currently there are approximately 250 metabolites in serum that are sufficiently reliable, both analytically and biologically, for potential use in databases that store levels and ratios of serum metabolites in disease and health. This approach however may prove only tenuously effective in metabolic profiling as it cannot determine precursor product relationships or reveal specific synthesis pathways of serum metabolites.

The physiology, biochemistry and morphology of mammalian cells will never be fully understood unless the uptake, flow, intracellular distribution and utilization of substrates are revealed and correlated with gene expression and signal based regulatory events in disease and health states (Tomita 2001). The rapid accumulation of biological data from the genome, proteome, transcriptome and metabolome increasingly allows us to enter potential interactions in developing computer models to investigate the dynamics and levels of organization of mammalian cells from a functional perspective. However, in the case of gene regulation of metabolism, there is not always a quantitative relationship between gene expression, mRNA levels and metabolic response. In fact, regulatory interactions between the genome and metabolome are rather slack as demonstrated by weak correlations between glycolytic flux and the expression of glycolytic enzymes (ter Kuile and Westerhoff, 2001). This indicates that metabolism is rarely regulated by gene expression alone and certainly casts doubts on whether transcriptome and proteome analyses suffice to assess biological function and phenotype modifications, especially when mammalian cells respond to external changes in the environment that primarily alter substrate availability.

The challenges of characterizing the metabolome are several; 1) although metabolite levels can be measured easily and accurately, their precursors and synthesis pathways may remain elusive, 2) in many instances there exist alternative synthesis pathways for the production of a given metabolite and 3) label or tracer systems developed to target a narrow range of metabolites in order to reveal specific synthesis steps do so only with low specificity and efficacy. Current metabolome labeling technologies allow measurements of the accumulation of specific components of pathological importance or measurements of cell replication. There is however less emphasis on studying specific synthesis pathways and the contribution of alternative synthesis branches to the production of biomolecules in response to specific genetic and signaling events. Because gene mutations, and even deletion of certain genes, often yield "silent" phenotypes when alternative pathways contribute to the production of a target metabolite, diseases and drug actions often remain poorly understood or confusing from a molecular point of view (Raamsdonk *et al.*, 2001, Cascante *et al.*, 2002).

It is clear that without appropriate labeling technologies that target a broad range of metabolites, yet address specific metabolic steps and reliably differentiate among alternate synthesis routes, biomarker discovery and gene function analysis and ultimately the understanding of human cell behavior in disease and health will develop only slowly. The genome and proteome already have their own highly specific labeling technologies (*e.g.*, PCR, ^{32}P, precipitating antibodies, immunohistochemistry), and an analogous approach could be developed for metabolome studies. One major difficulty, of course, is that the metabolome contains relatively small molecules with many structural and functional differences. One may therefore expect significant differences in labeling technologies, analysis methods and instrumentation applied in studies of the metabolome. Nevertheless, the need for an isotope-based dynamic labeling methodology to aid biomarker discovery and gene function analysis through the metabolome is irrefutable.

Our chapter describes tracer methods and applications of a stable isotope-based dynamic metabolic profiling approach, broadly applicable for metabolic profiling of various organisms, disease processes, drug actions, drug toxicity testing or diagnostic purposes. The technique utilizes the principles of accumulation, exchange, dilution and rearrangements of specific stable isotope labeled atoms in biomolecules during synthesis and enzymatic modifications, and it is particularly suited for studying healthy and diseased conditions as well as drug effects (Boros *et al.*, 2002a). The instrumentation utilized includes mass spectrometers capable of measuring molecular weight differences in ^{13}C labeled metabolites although nuclear magnetic resonance instruments or isotope ratio mass spectrometers, which determine isotope composition accurately in biomolecules can also be utilized. Separation methods using liquid or gas phase chromatography improve identification and analytical purity of metabolites for mass analyses. Indeed, based on current separation technologies and the available mass selective detectors it is not very challenging to apply stable isotope labels and reveal their positional distribution in the metabolome; *the real challenge is to find the appropriate tracer to gain the most biologically relevant information* and generate databases for comparative functional gene analyses and biomarker discovery. As will be demonstrated herein the single [1,2-$^{13}C_2$]glucose tracer represents one example with broad labeling potential and specific rearrangements of ^{13}C in various carbon positions in metabolites. It can thus provide reliable specificity in measuring pathway activities and can differentiate between alternative synthesis pathways.

The potential applications of specifically labeled stable isotope precursors are immense and include practically all areas of biomedical research and diagnostics. The stable-isotope based metabolic profiling technology, as described herein, can also be used to screen natural compounds with beneficial metabolic actions and thus aid the pharmaceutical drug development process by reducing the trial and error approach for the selection of new parent drugs for laboratory testing and further development.

As the metabolome ultimately absorbs and assimilates all genetic and signaling events in the chain of molecular interactions in living organisms the description of how metabolic profiling can be applied to uncover new biomarkers, study gene function and determine signaling events through specific changes in the metabolome will greatly assist further developments in academic research and mechanism based drug development efforts for the industry. Yet, the greatest promise of the stable isotope-based dynamic metabolic profiling technology is to integrate the metabolome into the functional genomics concept of biomedical research in the new millennium.

2. THE ISOTOPE LABELED METABOLOME: HISTORIC OVERVIEW AND CURRENT STATE OF THE ART

The utilization of various isotopes for studying organic compounds started in 1929 when the American chemist William Giauque discovered that oxygen is a mixture of 3 isotopes, ^{16}O, ^{17}O and ^{18}O. Shortly after that, Harold Urey discovered deuterium (1932). The unstable, radioactive isotope of carbon, ^{14}C, was then discovered in 1940. This became a pivotal research tool in studies of metabolic processes by Melvin Calvin, who used it to uncover the sequence of events in photosynthesis. Two years later Willard Frank Libby introduced the radioactive carbon-14 method of dating ancient objects, facilitating the broad use of carbon isotopes in many research fields including medicine, anthropology, biochemistry and physiology.

Early studies that utilized stable isotope labeled precursors and their incorporation into various intermediates were performed to study catabolic reactions in primitive organisms (Henderson *et al.*, 1967, Saur *et al.*, 1968). These studies clearly demonstrated that isotope labeled precursors are biologically active in bacteria (Flaumenhaft *et al.*, 1970), mammalian cells (Katz *et al.*, 1975) and plants (Uphaus and Katz, 1976), and that they offer great advantages in the study of complex biochemical networks. Several isotope designs and substrates have now been developed for studies of glycogen synthesis (Katz *et al.*, 1991, Katz and Lee, 1991), gluconeogenesis (Katz *et al.*, 1993; Wykes *et al.*, 1998; Mao *et al.*, 2002), glycolysis and the pentose cycle (Katz *et al.*, 1966; Katz and Rognstad, 1967), the tricarboxylic acid (TCA) cycle (Chance *et al.*, 1983), amino acid/protein synthesis (Papageorgopoulus *et al.*, 1999) and fatty acid metabolism (Fernandez *et al.*, 1996, Wadke *et al.*, 1973). These techniques all utilize specific isotope tracer designs for labeling precursors in order to determine synthesis and turnover rates (Previs *et al.*, 1994). Although metabolic processes are inherently complex and there have been disputes about the accuracy and efficacy of uniformly ^{13}C labeled glucose or lactate for metabolic studies, especially that for gluconeogenesis (Mao *et al.*, 2002), these techniques are, in general, appropriate for isotope based metabolic profiling once their limitations and applicability have been determined.

One major advantage of the stable isotope approach compared to the radiating isotope tracer technique is that stable isotopes can readily be used in animals or humans with little or no risk to the organism. This has brought about great advancements in the determination of proliferation and turnover rates of various cell pools in human diseases using *in vivo* labeling in, for example, patients with AIDS (Macallan *et al.*, 1998; Neese *et al.*, 2001). The label design for *in vivo* studies is uniformly labeled glucose, lactate, acetate or glycerol and the goal is to introduce as many tracer carbons as possible into the metabolome for accurate and sensitive measurements of DNA synthesis, cell proliferation, fatty acid synthesis or gluconeogenesis (Previs and Brunengraber, 1998; Lee *et al.*, 2000, Bassilian *et al.*, 2002). These studies, again, do not attempt to determine specific synthesis steps but rather specific metabolite synthesis rates or their buildup in health and disease.

Another main advantage of the stable isotope approach is the possibility of determining the *positional distribution* of tracer-derived labeled carbons within newly formed intermediary metabolites. Therefore "smart" stable isotope tracers with broad labeling capabilities for determining net metabolite synthesis and turnover rates, but which effect specific rearrangements into specific carbon positions, reflective of intermediate synthesis patterns can offer further benefit to academic and pharmaceutical research. Such a labeling system would reflect the dynamic nature of interconnected metabolic pathways demonstrating how stable isotope labeled carbons of substrates and metabolites accumulate, exchange, dilute and recycle within the metabolome. This, in turn, could reveal the "metabolic history" and exact precursor product relationships in intermediates that advance through consecutive synthesis steps in order to produce macromolecules with biological function.

Needless to say, metabolic profiling will best serve the discovery process only if it is reasonably priced, so that experiments can be carried out repeatedly to yield statistically relevant quality data. New efforts in metabolomics should adequately address these points and utilize knowledge from previous efforts in order to accelerate the discovery of new reliable metabolic biomarkers and to reveal gene function.

2.1 Stable Isotope-Based Dynamic Metabolic Profiling Technology using the [1,2-$^{13}C_2$]Glucose Tracer

As discussed above, desirable features of a comprehensive metabolic profiling technique for studies of biological functions of biomarkers, genes, signals and drugs in the functional genomics era include cost effective broad labeling potential, yet specific labeling patterns imprinted by individual metabolic reactions into intermediates and products contained by the metabolome. Fig. 1 demonstrates major metabolite pools of living organisms that provide intermediate precursors for nucleotide, fatty acid and amino acid synthesis; these metabolic products are key structural and functional biomolecules in health and disease. The dynamics of metabolic networks is evident by the connectivities of major metabolic pathways, their sharing of ^{13}C labeled substrates and their passing on of ^{13}C labeled products through consecutive reactions. In principle, the [1,2-$^{13}C_2$]glucose tracer readily labels glycolytic intermediates, pentose cycle metabolites, glycogen, the TCA cycle metabolite pool, amino acids of the non-essential glycogenic type as well as the non-essential saturated and unsaturated fatty acid pools including palmitate, stearate and oleate. The loss, positional rearrangements, dilution and distribution of ^{13}C in these metabolome constituents allow us to quantitate absolute and relative anabolic and catabolic pathway activities and to determine their dynamic interactions in health or their interruptions in disease. One main advantage of this technology is that it can also be used to generate a dynamic metabolic profile database where the characteristics and effects of metabolic biomarkers, gene function, drug treatments and signaling events are stored by recorded stable isotope distribution patterns in common metabolites of major macromolecule synthesis routes (Fig. 1). Such a searchable database for sharing by academia and industry should employ a standard tracer design that produces consistent label patterns in intermediates of structural molecules (proteins and lipids), cell proliferation related molecules (RNA and DNA) and secretory molecules (hormones, enzymes and proteins) of a variety of mammalian cells under differing conditions.

The excess accumulation of the ^{18}Fluor tracer attached to deoxyglucose (^{18}DOG) is the only known reliable biomarker of increased metabolic activity in mammalian cells that correlates with transformation, malignancy, growth rates and response to anti-cancer therapies (Stokkel *et al.*, 2001 Smith, 2001). Although indispensable as a diagnostic tool, this tracer does not permit detailed characterization of the destinations or routes taken by glucose carbons in macromolecule and energy producing processes. It does, however, demonstrate the broad labeling potential of glucose tracers in virtually all mammalian cells. Metabolic profiling must therefore focus on developing tracers that can uncover detailed and specific substrate flow modifications in health and disease.

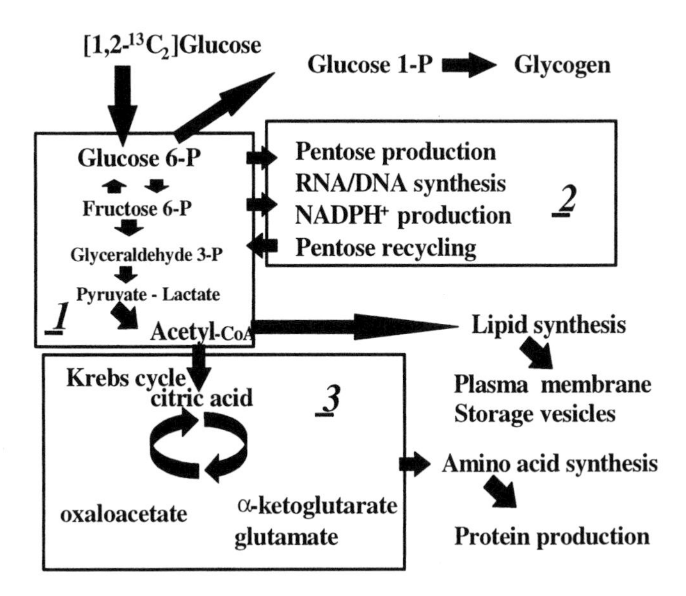

Figure 1. Interconnected metabolic pathways and their dynamic cross labeling by ^{13}C labeled glucose as the precursor. Glucose broadly utilized in mammalian cells readily labels major metabolite pools either as a direct substrate or through carbon exchange. The specificity for metabolic pathway substrate flow measurement is provided by the loss and rearrangements of the label from [1,2-$^{13}C_2$]glucose in various metabolites, intermediates and product pools. 1 glycolysis; 2 pentose cycle; 3 TCA cycle.

In general, metabolic profiling, as defined here, determines specifically labeled glucose flow towards *lactate*, *glycogen*, *glutamate*, nucleic acid *ribose/deoxyribose*, *palmitate*, *stearate* and *oleate* syntheses, as well as the release of CO_2 in cell cultures, animals or humans. Thus, it indicates specific changes in glucose utilization for biomolecule synthesis in glycolysis, the pentose and TCA cycles and in fatty acid synthesis pathways. [1,2-$^{13}C_2$]glucose metabolism produces four classes of isotope-labeled intermediary metabolite species, also called mass isotopomers: *m1*: with one

^{13}C substitution; *m2*: with two ^{13}C substitutions; *m3*: with three ^{13}C substitutions; and *m4*: with four ^{13}C substitutions.

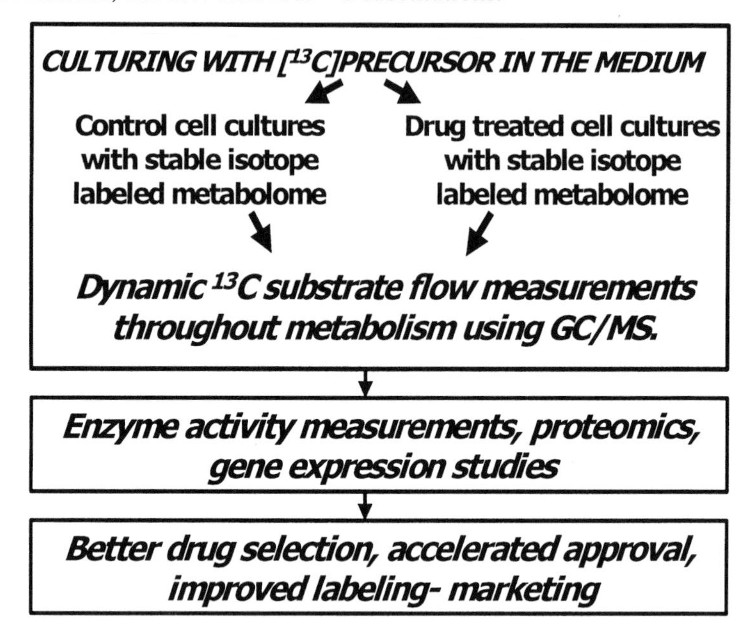

Figure 2. Cell culture study design for stable isotope-based metabolic profiling in the presence of a test drug and the [1,2-$^{13}C_2$]glucose tracer. Increasing doses of test drug are applied to cell cultures for a desired period of time (usually 48 or 72 hours) and the generated metabolic profiles compared with untreated control cultures, which contain only vehicle. Cell pellets and culture media are collected and ^{13}C distribution and rearrangements determined using LC- or GC-MS. Protein analyses can be performed and the expression of metabolic enzymes and their activities monitored as potential biomarkers. The pharmaceutical industry benefits from better drug target selection, more specific information on drug mechanism of action, accelerated approval, improved labeling as a source of information for patients and physicians and ultimately improvements in development and marketing of new drugs.

These isotopomers are readily separated and measured using conventional gas or liquid chromatography/mass spectrometry (MS) techniques. Chemical ionization MS reveals label accumulation into chemically and structurally intact biomolecules, while electron impact ionization reveals positional distributions of the ^{13}C label in mass fragments of the metabolite studied. Measurements can be performed in parallel or subsequently from the same experimental material (culture media, cell pellets or tissue) to obtain comprehensive and comparative metabolic profiles. The scheme of a cell culture experiment for metabolic profiling studies is given in Fig. 2.

Control (untreated) cells are incubated in the presence of the isotope labeled glucose substrate in order to generate a reference metabolic profile based on stable isotope distribution patterns, to which treated test cultures are compared. A single metabolic profiling experiment can reveal a comprehensive metabolic response of adaptive changes to increasing doses of a drug treatment or gene modifications, which are then compared to the reference control cultures by collecting the following information:

1. *Glucose uptake and lactate release (chemistry analysis of the culture media)* are important indicators of the cell's overall aerobic and anaerobic activities and can quantitate intracellular substrate flow values after the cell number of the culture has also been determined.

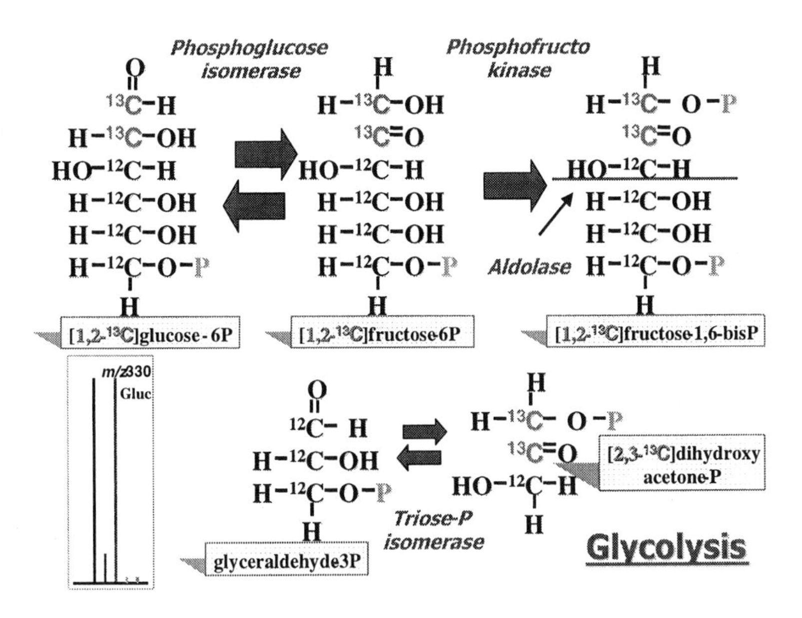

Figure 3. Possible rearrangements of ^{13}C in various metabolites of glycolysis using [1,2-$^{13}C_2$]glucose as the single tracer. Glucose activation via *hexokinase/glucokinase* and the formation of fructose-1,6-bisphosphate maintain the ^{13}C labeled carbons in the 1st and 2nd positions.

2. *Direct glucose oxidation relative to glycolysis (lactate m1/m2 ^{13}C ratios).* <u>Lactate</u> is the main three-carbon product of glycolysis. It is readily secreted into cell culture media, and can therefore be utilized for the measurement of label incorporation into the three-carbon metabolite pool of glycolysis. The possible arrangements of ^{13}C labels from [1,2-^{13}C]glucose to

lactate through glycolysis are shown in Figs. 3 and 4. [1,2-^{13}C]glucose oxidation through the pentose cycle, on the other hand, results in a loss of the first ^{13}C of glucose as shown in Fig. 5. It is easy to see that the ratio between *m1* (recycled lactate from oxidized glucose via the oxidative branch of pentose cycle) and *m2* (lactate produced by the Embden-Meyerhoff-Parnas glycolytic pathway) is indicative of the activity of glucose-6-phosphate dehydrogenase (G6PDH) and glucose recycling in the pentose cycle. During glucose oxidation $^{13}CO_2$ is also released which reflects glucose utilization for energy production in the pentose and TCA cycles. A detailed description of the reactions and calculations can be found elsewhere (Lee *et al.*, 1998). Drug treatments that affect direct glucose oxidation or glycolytic flux are expected to alter glucose ^{13}C label rearrangement in lactate and $^{13}CO_2$ release as shown later.

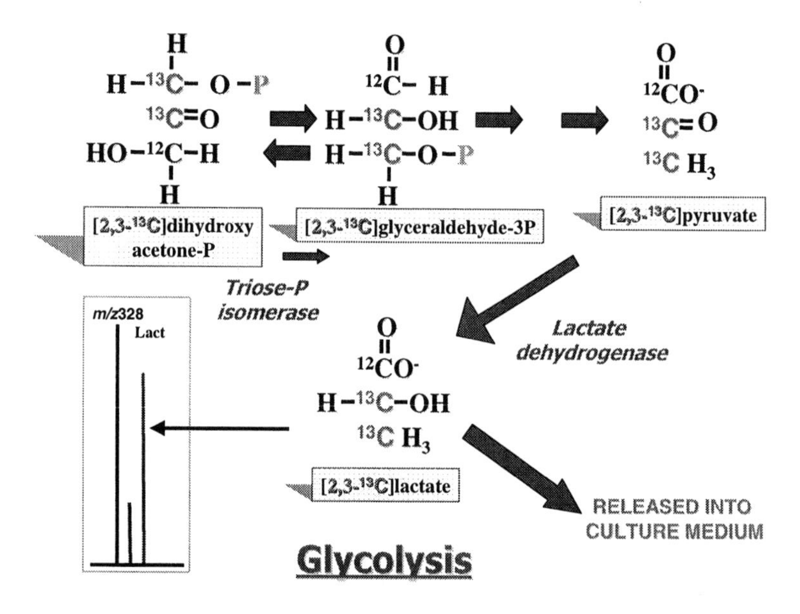

Figure 4. Formation of [2,3-$^{13}C_2$]lactate from [1,2-$^{13}C_2$]glucose through the Embden-Meyerhoff-Parnas pathway. The production by aldolase of the three-carbon metabolites, glyceraldehyde and dihydroxy acetone phosphate, transfers the labeled carbons into the 2nd and 3rd positions of glyceraldehyde (see Fig. 3). There are no subsequent positional changes in terms of ^{13}C labeling by *triose phosphate isomerase* in the three-carbon metabolite pool that undergoes glycolysis. The mass spectrum of lactate produced by glycolysis shows a high *m2* peak at *m/z* 330 as indicated by the arrow.

3. *Direct glucose oxidation/nonoxidative ribose synthesis for RNA/DNA nucleotide syntheses (m1/Σm versus m2/Σm, m3/Σm and m4/Σm).* Ribose and deoxyribose are the building blocks of nucleotides and therefore ^{13}C

incorporation from glucose into RNA ribose or DNA deoxyribose indicates changes in nucleic acid synthesis rates through the respective branches of the pentose cycle. Singly labeled ribose molecules on the first carbon position (*m1*) represent ribose that is produced by direct glucose oxidation through G6PDH (Fig. 5). This ribose can either be incorporated into nucleic acid or returned to glycolysis as shown in Fig. 6. The alternative pathway for ribose synthesis is through the non-oxidative steps of the pentose cycle using glycolytic metabolites. The non-oxidative synthesis of ribose from glucose is controlled by transketolase as shown in Fig. 7.

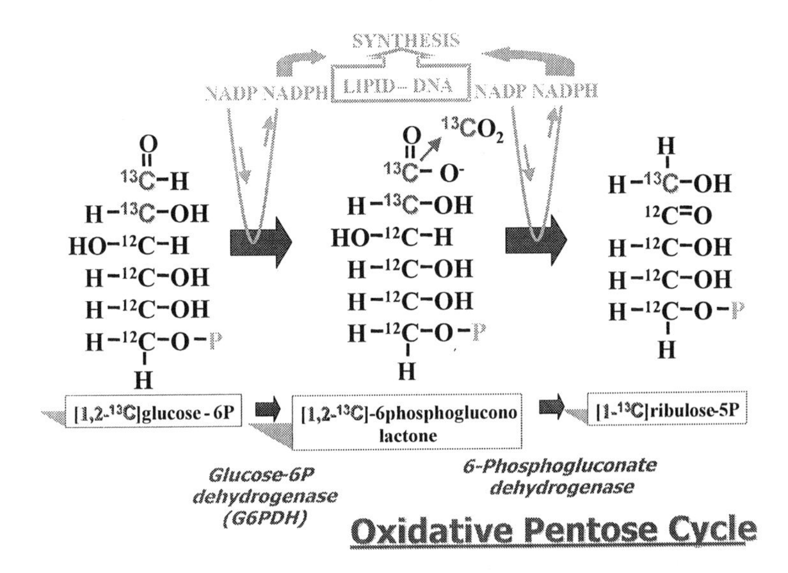

Figure 5. The rearrangement of ^{13}C due to direct [1,2-$^{13}C_2$]glucose oxidation in the pentose cycle. The loss of the first labeled carbon of glucose due to direct oxidation produces ribulose molecules that are labeled only on the 1st position with ^{13}C. During the oxidation of glucose $^{13}CO_2$ is released, which can easily be detected using isotope ratio MS. The reducing equivalent $NADPH^+$ is also produced and can be used in lipid synthesis, DNA nucleotide production or to maintain reductive/oxidative reactions.

There is no net carbon loss throughout the non-oxidative steps of the pentose cycle but there are significant label rearrangements in the five carbon metabolite pool based on participating non-oxidative reactions as shown on Figs. 7, 8 and 9; therefore, ribose molecules labeled on 1,2-, 1,2,4,5- and 1,4,5- carbon positions with ^{13}C (*m2, m3* and *m4*) represent nucleic acid ribose synthesis through the non-oxidative route. The ratio between these isotopomers of ribose/deoxyribose in nucleic acids closely reflects the involvement of a variety of metabolic reactions with great specificity and sensitivity that include glucose oxidation and subsequent non-oxidative reactions (Boros *et al.*, 1997).

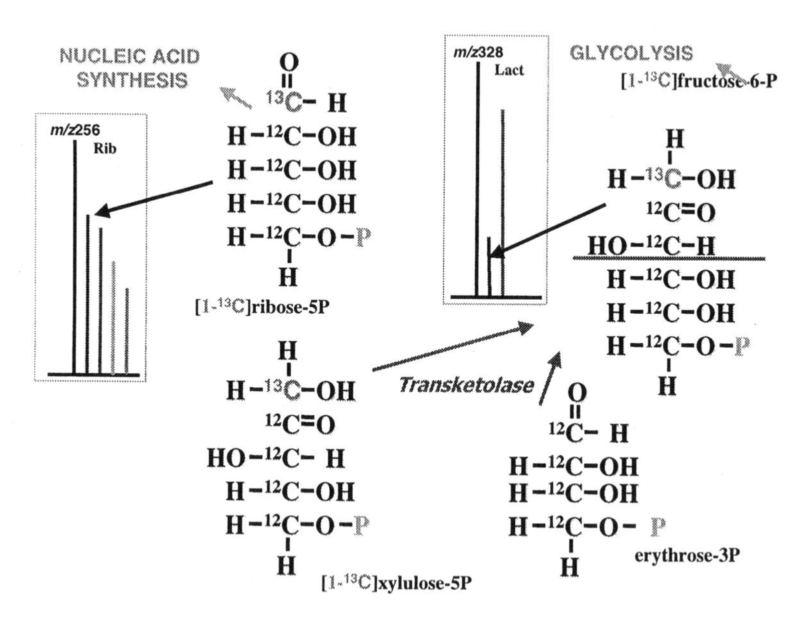

Figure 6. [1-^{13}C]ribose-5P formation in the non-oxidative pentose cycle after [1,2-$^{13}C_2$]glucose oxidation. The non-oxidative steps of the pentose cycle generate intermediates that can be used for nucleic acid synthesis (ribose-5P, as seen in proliferating cells) or recycled back to glycolysis (glyceraldehyde-3P and fructose-6P, as seen in non-proliferating/resting cells). Lactate that is produced from [1-^{13}C]fructose-6P contains only 1 ^{13}C substitution on the 3rd carbon position (*m/z* 329), therefore the mass spectral ratio of *m1/m2* in lactate reflects direct glucose oxidation and pentose recycling relative to the glycolytic flux. Nucleic acid ribose that contains one label produces spectral peak at *m/z* 257 indicating glucose oxidation and ribose/deoxyribose synthesis through the oxidative branch of the pentose cycle.

These reactions have high significance for the testing of novel and effective cancer treatment modalities during *de novo* nucleic acid synthesis and cell growth. Glucose oxidation and oxidative ribose synthesis are parts of the normal cell development process where reducing equivalents are also generated that can be used for reductive synthesis of deoxyribose and fatty acids. Transformed cells switch to non-oxidative nucleic acid ribose synthesis, a potential metabolic biomarker in cancer, as shown later.

4. *Glucose oxidation in the pentose and TCA cycles ($^{13}CO_2/^{12}CO_2$ ratios).* $^{13}CO_2$ release is a reliable marker of glucose oxidation (Figs. 5, 10). $^{13}CO_2$ production from [1,2-^{13}C]glucose takes place in both the pentose and TCA cycles and it is measured as part of the metabolic profiling processes to determine the rate of glucose oxidation versus anabolic glucose utilization. Decreased glucose oxidation with increased glucose uptake is always a reliable sign of increased glucose-based anabolic activities as seen in transformed cells.

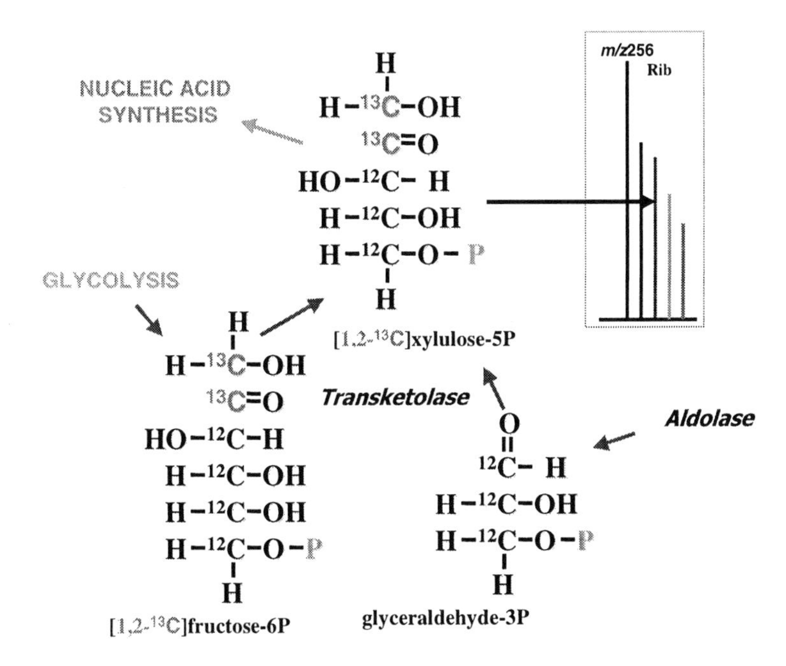

Figure 7. The formation of [1,2-$^{13}C_2$]xylulose-5P through the non-oxidative reactions of the pentose cycle. Mammalian cells are able to synthesize ribose-5P directly *via* the non-oxidative pentose cycle reactions *via* several reactions. This process allows the unrestrained production of ribose-5P, independent of available NADP. Nucleic acid ribose/deoxyribose molecules that are synthesized from glycolytic metabolites by aldolase and transketolase produce the *m/z* 258 spectral peak as shown by the dark arrow.

5. *TCA cycle anaplerotic flux (glutamate m2/m1 ratios).* Glutamate, a non-essential amino acid, is partially produced from mitochondrial α-ketoglutarate, which is a central intermediate of the TCA cycle. Glutamate is readily released into the culture medium after synthesis, which represents one of the routes for glucose carbon utilization. Therefore, label incorporation from glucose into glutamate is a good indicator of TCA cycle anabolic metabolism for amino acid synthesis instead of glucose oxidation

(Fig. 10) (Lee *et al.*, 1996, Leimer *et al.*, 1977). Anaplerosis refers to the reactions that allow the entry of carbon into the TCA cycle intermediate pools other than *via* citrate synthase. Any carbon that enters the cycle as acetyl-CoA is oxidized to carbon dioxide and water; any carbon that enters the citric acid cycle *via* an anaplerotic pathway is not oxidized, but must be disposed of by some other route. Glutamate dehydrogenase is one possible route that converts α-ketoglutarate to glutamate. Some other reactions include pyruvate carboxylation, transamination reactions or propionate carboxylation.

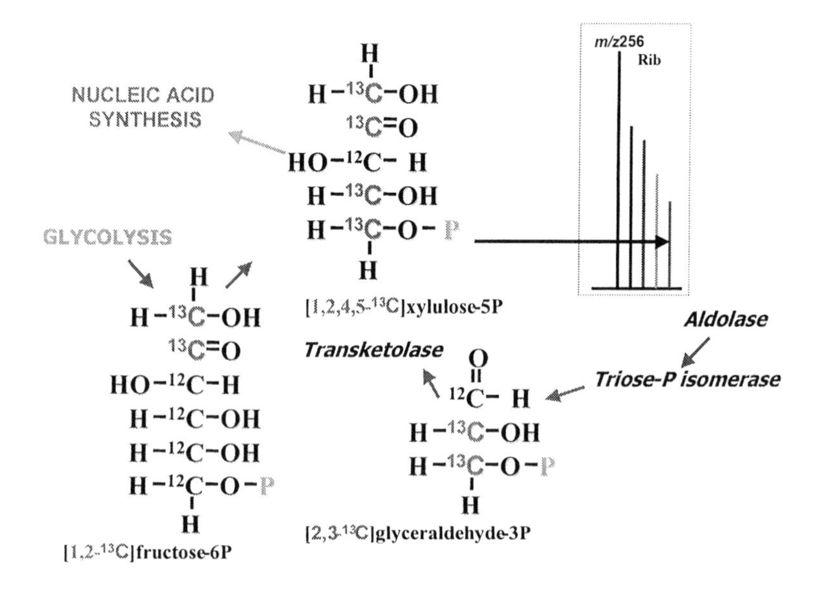

Figure 8. The formation of [1,2,4,5-$^{13}C_3$]xylulose-5P from [1,2-$^{13}C_2$]glucose through the non-oxidative reactions of the pentose cycle. Nucleic acid ribose/deoxyribose molecules that are synthesized from glycolysis metabolites by aldolase, triose phosphate isomerase and transketolase produce the *m/z* 260 spectral peak as shown by the dark arrow.

6. *De novo fatty acid palmitate synthesis (^{13}C enrichment in palmitate).* Fatty acid synthesis is also strongly dependent on glucose carbons through the formation of acetyl-CoA *via* pyruvate dehydrogenase. The incorporation of ^{13}C from [1,2-$^{13}C_2$]glucose gives key information about the fraction of *de novo* lipogenesis, chain elongation and desaturation which as well as glucose carbon contribution to acetyl-CoA for fatty acid synthesis (Fig. 11) (Lee *et al.*, 1995).

7. *Chain elongation into stearate (^{13}C enrichment in stearate and the isotope distribution ratios in palmitate and stearate).* Stearate is a long chain non-essential fatty acid that can be produced from dietary or *de novo*

synthesized palmitate by fatty acid elongase. The activity of this enzyme can be predicted by the accumulation of ^{13}C from glucose and the ratios of *m2/m4* in palmitate and stearate, respectively (Lee *et al.*, 2000).

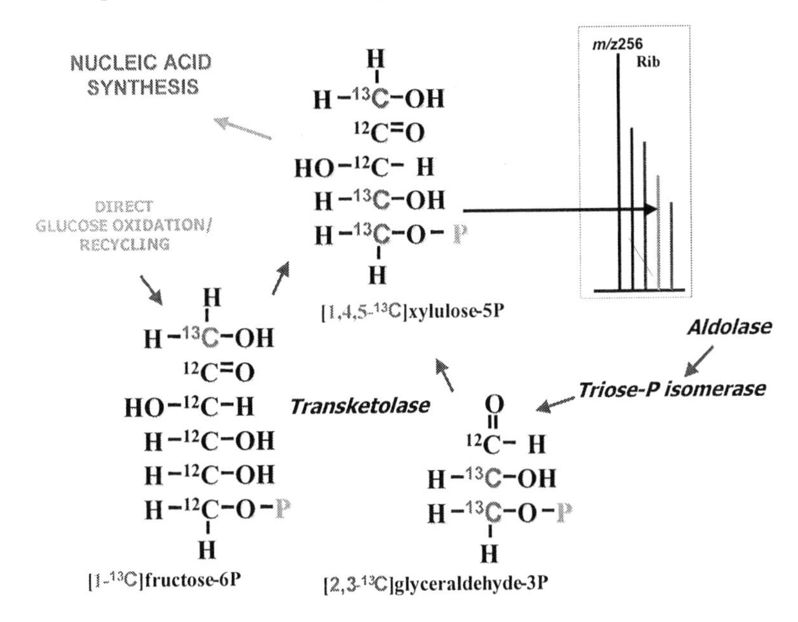

Figure 9. The formation of [1,4,5-$^{13}C_4$]xylulose-5P from [1,2-$^{13}C_2$]glucose through the non-oxidative reactions of the pentose cycle. Nucleic acid ribose/deoxyribose molecules are synthesized utilizing recycled intermediates of the oxidative pentose cycle, which results in the loss of ^{13}C from the 1st carbon position of glucose, by aldolase, triose phosphate isomerase and transketolase produce the *m/z* 259 spectral peak as shown by the dark arrow.

8. *Acetyl-CoA glucose carbon enrichment (palmitate m1/m2 ^{13}C ratios).* <u>Acetyl-CoA</u> ^{13}C enrichment closely reflects glucose carbon utilization for de novo fatty acid synthesis, chain elongation and desaturation that affect cell function and structure by several mechanisms. Acetyl units participate in the assembly of short and long chain non-essential fatty acids and they can also be used for energy production in the TCA cycle. The rate of glucose utilization for *de novo* fatty acid synthesis is an important indicator of lipid anabolic reactions and cell differentiation in response to various signaling events and anti-cancer drug treatments as shown later.

9. *Direct and indirect glycogen synthesis (glycogen m1/m2 ratios).* <u>Glycogen glucose</u> makes up the substrate storage compartment of the metabolome to release and mobilize glucose for various metabolic reactions. Glycogen can be synthesized from glucose-1-phosphate directly or through the pentose cycle indirectly where the loss of the first label occurs (Fig. 5).

Glycogen *m2* to *m2 of* medium glucose ratios indicate therefore the rate of direct and indirect glycogen synthesis in various metabolic states of the cell (Katz *et al.*, 1989).

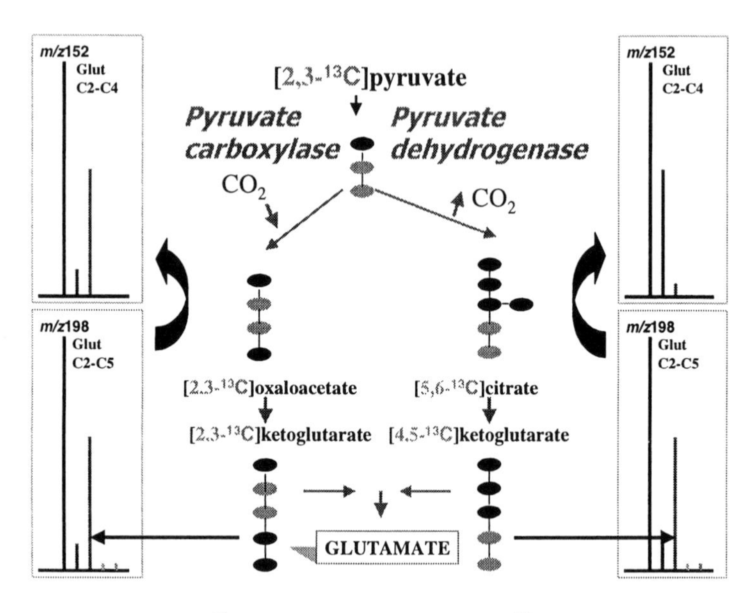

Figure 10. Formation of ^{13}C labeled glutamate from [1,2-$^{13}C_2$]glucose through pyruvate dehydrogenase and pyruvate carboxylase in the TCA cycle. Glucose carbons readily label TCA cycle metabolites because the first two carbons of glucose provide the 2nd and 3rd carbon of pyruvate that assume positions as the 1st and 2nd carbons in acetate by pyruvate dehydrogenase (PDH). When the ^{13}C tracer enriched acetyl molecules enter the TCA cycle carbon pool through citrate synthesis they assume positions as [4,5-$^{13}C_2$]ketoglutarate. The alternative entry of pyruvate into the TCA cycle is *via* pyruvate carboxylase that results in the formation of [2, 3-$^{13}C_2$]ketoglutarate. Glutamate electron impact analysis of the *m/z* 152 (C2-4) and *m/z* 198 (C2-5) fragments clearly reveals these labeling patterns by the ratios of *m1* and *m2*. Releasing the 5th carbon of glutamate by electron impact yields *m1* if the ^{13}C labeled carbons reside on the 4th and 5th carbon positions in glutamate due to pyruvate dehydrogenase activity, however there are no spectral changes if the ^{13}C carbons reside on the 2nd and 3rd positions as shown by the curved arrows. Note that ^{13}C-labeled carbon positions derived from [1,2-$^{13}C_2$]glucose are shown in grey, while ^{12}C native-labeled carbon positions are black here and in subsequent figures.

10. Gluconeogenesis (ratios of [1,2-$^{13}C_2$]glucose and other recombined isotope labeled glucose species in the media). Specialized cells like hepatocytes are capable of producing glucose from various three-carbon metabolites, lipids or amino acids. The rate of gluconeogenesis can be indirectly measured using the ratios between the original [1,2-$^{13}C_2$]glucose label pattern and any other ^{13}C labeled glucose species that appear in the culture medium, including excess naturally labeled glucose that indicates gluconeogenesis of amino acid or lipid source. The study of metabolic profiles using stable isotopes in cell cultures in response to cell transforming agents and cancer growth-controlling compounds provides good examples of metabolic profiling and reveals how growth signaling, genetic and metabolic processes connect and how substantially metabolic pathway flux influences cell growth. The utility of the stable isotope methodology in studying mammalian cell physiology in the context of signaling events and the function of genetic background is one of the new advances in metabolome research.

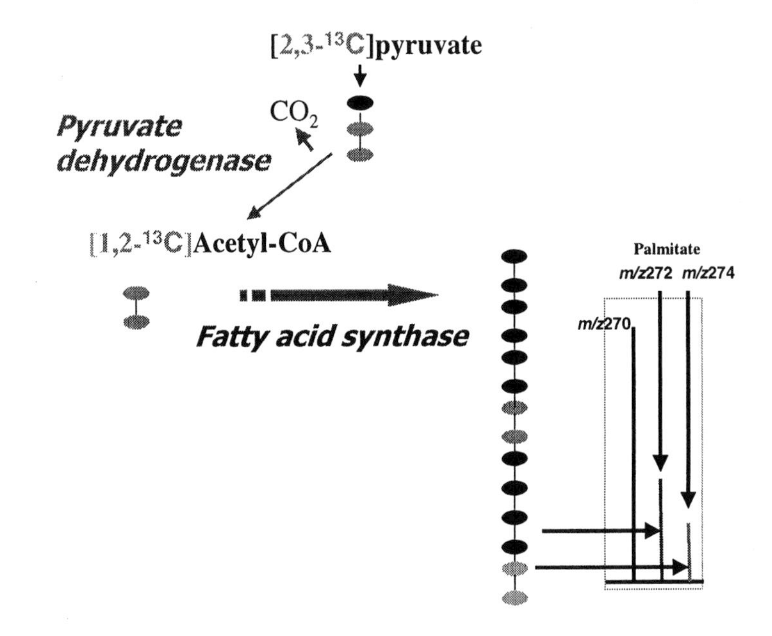

Figure 11. Formation of ^{13}C labeled acetyl-CoA through pyruvate dehydrogenase for fatty acid synthesis. Glucose carbons readily label fatty acids because the first two carbons of glucose provide the 2nd and 3rd carbon of pyruvate that assume positions as the 1st and 2nd fatty acid pool and produce palmitate measured at *m/z* 272 and even multiple increments depending on the extent of ^{13}C enrichment.

As we demonstrate below, effective anti-cancer therapeutics limit carbon flow toward nucleic acid synthesis and shift glucose toward oxidation or fatty acid synthesis through specific metabolic reactions, making these reactions and enzymes suitable as new biomarkers for screening potential drugs to treat cancer and to determine if cell transformation takes place.

One can expect entire new dimensions in biomarker discovery and gene function analysis by exploring the stable isotope tracer labeled metabolome as a new window to study how cellular metabolism correlates with genotypic and phenotypic modifications.

2.2 The Stable Isotope Labeled Metabolome and the Effects of Growth-Promoting and Growth-Inhibiting Drugs

The stable isotope labeled metabolome of selected intermediates using natural ^{13}C labeled glucose as the substrate of myeloid leukemia cells is shown in Fig. 12.

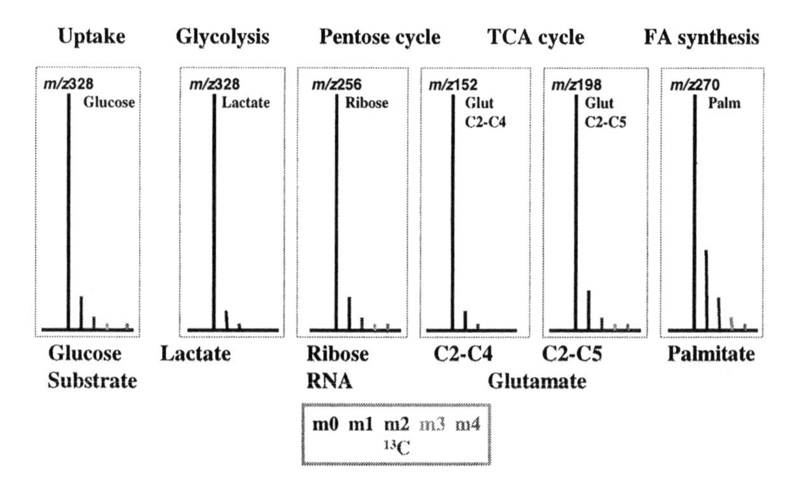

Figure 12. The stable isotope labeled metabolome using natural ^{13}C labeled glucose as the substrate of K562 myeloid leukemia cell metabolism. Biomolecules of the mammalian cell metabolome are readily labeled by natural ^{13}C abundance. Therefore, metabolites of glucose become naturally labeled by ^{13}C where the unlabeled species (*m0*) are black and 1, 2, 3 and 4 ^{13}C substitutions (*m1*, *m2*, *m3* and *m4*) are shown consecutively. The highest peak is the unlabeled species in all naturally labeled metabolites and the distribution of stable isotope labeled positional topomers, also called isotopomers, follow a binominal distribution based on the number of carbons in the molecule. As natural isotope labeling occurs both in control and drug treated metabolomes the corresponding natural ^{13}C labeled spectra have to be subtracted from the spectrum of each metabolite generated by the subsequent [1,2-$^{13}C_2$]glucose tracer experiment.

These biomolecules are naturally labeled by ^{13}C, which is ~1% of all carbon atoms in the Earth's atmosphere. Therefore, all metabolites of glucose generated by a series of enzymatic reactions such as lactate, ribose, deoxyribose, glutamate and palmitate are naturally labeled by ^{13}C as shown by the mass/charge (*m/z*) ratios. Therefore, natural compounds also form isotopomers that need to be subtracted. In natural small molecular weight compounds the highest peak is always the uniformly ^{12}C (unlabeled) species and the distribution of stable isotope labeled positional topomers, the so called isotopomers, follow the binominal distribution based on the number of carbons in the molecule, as described elsewhere (Katz *et al.*, 1989).

The stable isotope labeled metabolome of myeloid leukemia cells using 50% [1,2-$^{13}C_2$]glucose as the tracer for 72 hours culturing shows a dramatically altered profile (Fig. 13).

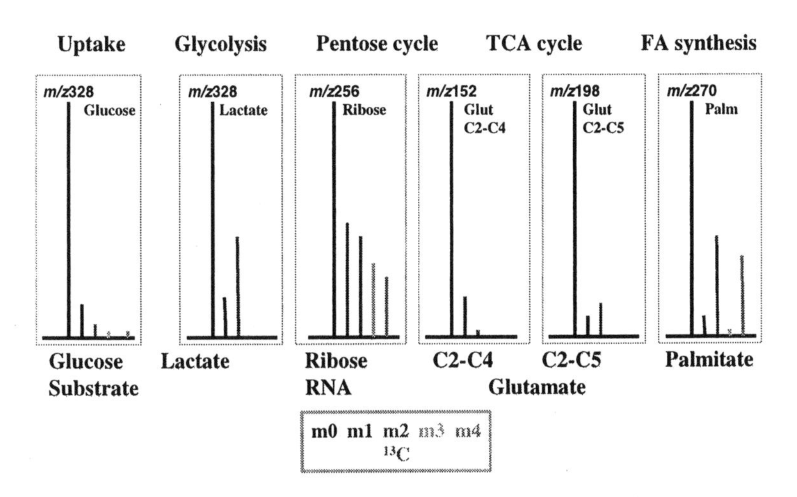

Figure 13. The stable isotope labeled metabolome using [1,2-$^{13}C_2$]glucose as the substrate of K562 myeloid leukemia cells for 72 hours. Adding [1,2-$^{13}C_2$]glucose to the culture medium results in a significant change in the ratios of ^{13}C labeled isotopomers in the metabolome, after subtracting natural ^{13}C enrichment. As expected, *m2* lactate indicate a significant increase in glucose uptake because glycolysis is intense in tumor cells. The majority of ^{13}C label from [1, 2-$^{13}C_2$]glucose accumulates in RNA and fatty acid. Surprisingly there is little label accumulation into the TCA cycle metabolite glutamate. The particular ratios among *m1*, *m2*, *m3* and *m4* ^{13}C labeled isotopomers indicate specific enzymatic activities and substrate flow/redistribution among major metabolic pathways and their products.

As expected, *m2* lactate with 2 ^{13}C substitutions shows a significant increase as the results of glucose uptake and glycolytic activity. The majority of ^{13}C label from [1,2-$^{13}C_2$]glucose accumulates in the RNA and fatty acid fractions of the tumor cell metabolome. In general, the ratios among ^{13}C labeled isotopomers indicate specific enzymatic activities and substrate redistribution among major metabolic pathways through the respective branches of metabolism. Overall changes in the metabolome in terms of total ^{13}C accumulation, known as Σm_n, and the ratios of the various ^{13}C isotopomers serve as the basis of determining specific pathway activities as the function of total ^{13}C label accumulation. This, in turn, produces the quantitative and comparable matrices of metabolic profiles of mammalian cells, their metabolic response to growth promoting or growth controlling signaling events, genetic changes or drug treatments, which can be used to build the stable isotope-based dynamic metabolic profiles database that describe metabolic profile and cell function mathematically *via* the stable isotope labeled metabolome.

3. METABOLIC PROFILES OF TRANSFORMED CELLS MANIFEST BY INCREASED NON-OXIDATIVE PENTOSE CYCLE CARBON FLOW

Cancer progression is enhanced by the autonomous growth promoting tyrosine kinase signaling ligand, transforming growth factor-β (TGF-β_2) (Hojo *et al.*, 1999). This process primarily depends on non-oxidative glucose conversion into ribose as the end-result of this signaling pathway (Boros *et al.*, 2000). Similarly, the carcinogen pesticide isofenphos directs glucose carbon flow toward nucleic acid ribose synthesis in myeloid cells, resulting in a proliferative phenotype (Boros and Williams, 2001). Metabolic profile changes include an increase in the ^{13}C accumulation into RNA and DNA with an increase in all (especially *m2*, *m3* and *m4*) isotopomers of ribose indicating intense non-oxidative synthesis, while there is a significant decrease in fatty acid ^{13}C incorporation as shown in Fig. 14.

Cell-transforming agents uniformly induce carbon flow changes consistent with increased pentose cycle metabolism and the diversion of glucose carbons towards fatty acid synthesis with limited lipid and TCA cycle metabolism. Therefore, the primary direction of glucose carbon flow in transformed human cells is toward nucleic acid ribose synthesis. The phenotypic consequences include rapidly proliferating, poorly differentiated cells with aneuploidy.

Molecular genetic studies indicate that metabolic enzyme gene expression patterns support increased carbon flow toward nucleic acid synthesis that requires the unremitting expansion of six, five and three-carbon phospho-metabolite pools via the oncogenic expression of the dimer M_2 isoform of pyruvate kinase (M_2PK) (Zwerschke *et al.*, 1999).

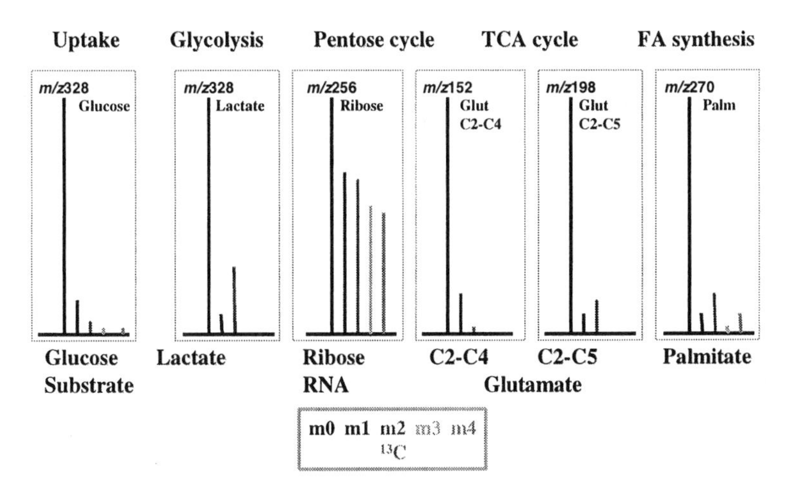

Figure 14. The stable isotope labeled metabolome of K562 myeloid leukemia cells in response to the transforming effects of isofenphos pesticide using [1,2-$^{13}C_2$] glucose as the tracer. The carcinogen pesticide isofenphos directs glucose carbon flow toward nucleic acid ribose synthesis in myeloid cells, resulting in a proliferative phenotype. Metabolic profile changes include an increase in the ^{13}C accumulation into RNA with an increase in all labeled species, especially *m2, m3* and *m4* of ribose, indicating intense non-oxidative ribose synthesis. Please note the significant decrease in fatty acid ^{13}C incorporation. Cell-transforming agents that induce these effects include TGF-β in lung epithelial carcinoma cells. The metabolic phenotype of tumor cells forces glucose substrate carbon flow toward nucleic acid ribose synthesis, a profile representing the metabolic substrate distribution biomarker for cell transformation.

The oncogenic transition of metabolic gene expression lowers the affinity of substrate phosphoenolpyruvate for pyruvate kinase decreasing glycolytic flux but expanding all glucose phosphate intermediary metabolites in tumor cells that promote nucleotide synthesis in the non-oxidative branch of the pentose cycle. The expression of M_2PK has consistently been observed in kidney (Oremek *et al.*, 1999), lung (Schneider *et al.*, 2000), gastrointestinal (Hardt *et al.*, 2000), and breast malignancies (Luftner *et al.*, 2000). Based on this correlation between the metabolic enzyme expression profile and metabolic substrate flow changes in tumor cells and it is evident that metabolic profile changes of intense glucose utilization primarily for nucleic acid ribose synthesis through the non-oxidative steps of the pentose cycle can be considered a reliable metabolic biomarker of cell transformation. When the stable isotope-based metabolic profile database is generated, this profile indicates cell transformation and provides reference values of isotope ratios in the metabolome to determine the level of transformation of other mammalian cells when comparing the transforming effects of growth hormones, pesticides, food additives, gene mutations and signaling events.

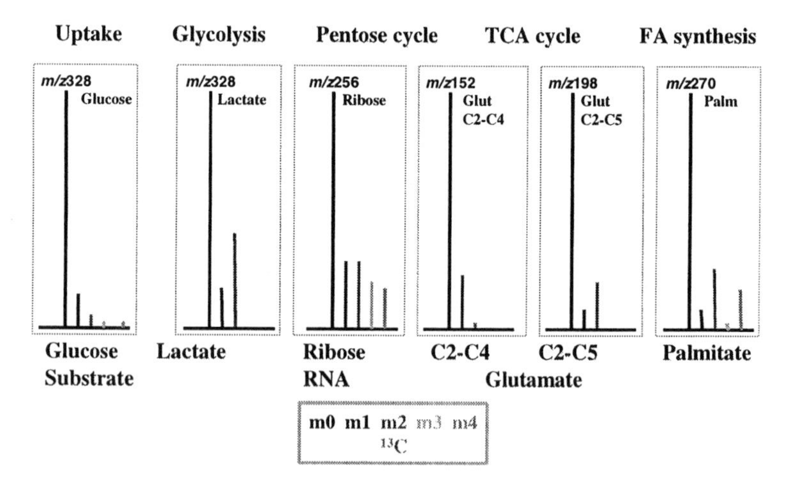

Figure 15. The stable isotope labeled metabolome of K562 myeloid leukemia cells in response to the new effective anti-leukemia drug Gleevec (STI57, imatinib mesylate) using [1,2-$^{13}C_2$] glucose as the tracer. Gleevec effectively controls glucose carbon flow toward nucleic acid ribose synthesis in myeloid cells, resulting in cell cycle arrest. Metabolic profile changes include a decrease of ^{13}C accumulation into RNA, due to limited oxidative and non-oxidative ribose synthesis. There is a significant increase in direct glucose oxidation and recycling in the pentose cycle as shown by increased *m1/m2* ratio in lactate. This provides more reducing equivalents for cell differentiation through increased glycolytic substrate flow and decreased nucleic acid ribose synthesis, which limits cell proliferation. The metabolic profile shown here is common in other anti-cancer treatment modalities. Thus, this labeling pattern can serve as a substrate flow metabolic biomarker of effective cancer treatment.

3.1 Decreased Pentose Cycle Carbon Flow and Increased Fatty Acid Synthesis as Cell Differentiating and Growth Controlling Mechanisms

The expression of a constitutively active tyrosine kinase signaling protein construct as a result of the re-alignment of the breakpoint cluster region and Ableson leukemia virus proto-oncogene sequences (Bcr/Abl) (Gyger *et al.*,

1985) is the basis of oncogenic transformation of myeloid cells in chronic myeloid leukemia (CML). This construct stimulates glucose transport in multipotent hematopoietic cells (Bentley *et al.*, 2001).

The inhibition of this constitutively active Bcr/Abl tyrosine kinase by the anticancer drug STI571 (Gleevec) has been shown to effectively inhibit glucose utilization towards nucleic acid synthesis and leukemia cell proliferation (Boren *et al.*, 2001). At doses comparable to those used in the treatment of myelogenous leukemia, STI571 suppresses hexokinase and G6PDH activities in K562 myeloid leukemia cells. As a result, STI571 limits pentose cycle carbon flow toward nucleic acid ribose synthesis, as depicted in Fig. 15 as the metabolic profile of K562 cells in response to STI571 treatment.

Similar substrate flow modifications, namely decreased glucose carbon flow toward nucleic acid synthesis has also been observed in cancer cells after treatment with other agents including phytochemicals such as genistein (Boros *et al*, 2001) and natural hormones such as dehydroepiandrosterone sulfate (Boros *et al.*, 1997). Novel cancer treatment modalities, which effectively limit carbon flow in the pentose cycle by re-directing carbon substrate flow toward oxidation or fatty acid synthesis introduce a biomarker metabolic profile of effective cancer treatment *via* limited nucleotide macromolecule synthesis. This metabolic biomarker in tumor cells likely predicts disease progression and clinical outcome, therefore, these cell growth limiting metabolic profiles can in the future be used to screen and test new anti-cancer treatment modalities for effectiveness and mechanism of anti-proliferative actions for the pharmaceutical industry.

4. THE ROLE OF METABOLIC CONTROL ANALYSIS IN BIOMARKER AND DRUG DISCOVERY

Based on metabolic profiling data and results from molecular genetic studies it is evident that high levels of pentose cycle enzymes are associated with cell aging, uncontrolled cell proliferation and prolonged cell survival (Orgel, 1973). Pentose cycle enzymes provide the necessary ribose substrate for *de novo* nucleic acid synthesis as well as the reducing equivalent NADPH for deoxyribose nucleotide and fatty acid syntheses. A shortage in glucose carbon flow toward nucleic acid synthesis as well as decreased NADPH production are reliable biomarkers for cell cycle withdrawal and apoptosis in rapidly dividing cells (Tian *et al*, 1999, Tuttle *et al.*, 2000). In order to understand why pentose cycle substrate flow controlling enzymes

(G6PDH and transketolase) have such a strong influence on cell physiology and how they control cell growth and transformation, one needs to consider the concept of Metabolic Control Analysis (MCA). MCA provides a quantitative description of how changes in system properties, such as metabolic fluxes or cell growth, are related with changes in system components such as gene expression or enzyme activities (Schuster *et al.*, 1999). MCA and metabolic profiling are closely related fields as substrate flow changes are the combined results of gene expression, enzyme protein synthesis, enzyme activity changes and substrate availability and redistribution. These factors constitute integrated regulatory mechanisms of cell function, and they can only be understood as integrated elements of a cell's complex genetic, signaling and metabolic architecture. For example, the growth control coefficient of transketolase in the *Ehrlich*'s tumor model was recently reported to be high (Comin-Anduix *et al.*, 2001) and this enzyme is now ranked as a new promising specific biomarker for cancer as well as an anticancer drug target (Cornish-Bowden, 1999). In other words, enzyme biomarkers and targets for new anti-cancer therapies have to be those that demonstrate strong control properties over substrate flow for nucleic acid ribose synthesis. This is, of course, also a crucial criterion in drug development efforts in which the efficacy of potential anti-proliferative drugs is determined by metabolic screening.

5. METABOLIC PROFILING IN BIOMARKER AND DRUG DISCOVERY

Drug discovery programs currently place considerable emphasis on discovering novel anti-cancer agents that inhibit growth through modulation of specific signal transduction pathways or genetic targets. The use of metabolic profiling to discover new biomarkers and drug targets emphasizes critical enzymatic steps that control substrate flow during transformation of mammalian cells or in various disease states. Metabolite profiling can thus identify new biomarkers that indicate the disruption of unique metabolic networks in diseased cells. New targets can also be explored based on the novelty and specificity of stable isotope-based dynamic metabolic profiling. Thus, the ability to monitor the destinies of glucose carbons within cancer cells elevates this technology above, for example, ^{18}DOG uptake based biomarker strategies because it can highlight specific intracellular sites throughout metabolism as targets for therapeutic intervention. Specific biomarkers and drug target sites can only be identified by "smart" and comprehensive labeling of the metabolome. The efficacy of such an approach can be evaluated by MCA and the use of direct metabolic enzyme inhibitor drugs. A combined study of signaling, genetic and metabolic events allows us to better define metabolic processes in mammalian cell growth and death and in health and disease (Cascante *et al.*, 2002, Boros *et al.*, 2002a, 2002b, 2002c). Biomarkers for human diseases will therefore be expanded beyond the known genetic and signaling patterns by newly described metabolic profile patterns in the near future. Characterization of these biomarkers will supplement current testing methods and facilitate discovery new disease mechanisms, genetic function and drug effects.

6. CONCLUSION

Although molecular genetic studies can anticipate changes in metabolism, they cannot fully reveal whether metabolic enzymes are active and their substrates present. Thus metabolic profiling using stable isotope labelled substrates is clearly needed for the understanding of genotype-phenotype correlation. The metabolome represents the final event of phenotype modifications by the adjustment of substrate flow. Metabolic profiling therefore provides vital information regarding cell transformation and phenotypic modification beyond the scope of signal transduction and genetic studies. The ability to both identify biomarkers, targets and fully establish whether a candidate drug actually produces the desired metabolic effect makes metabolic profiling a valuable new tool in mechanism-based drug discovery efforts. The application of stable isotope-based dynamic metabolic profiling also provides a new business model for the pharmaceutical industry by facilitating more comprehensive and mechanism-based evaluation of the drug selection and drug testing procedures. This will improve drug approval, drug labeling and marketing.

REFERENCES

Aharoni A, Vorst O. DNA microarrays for functional plant genomics. *Plant Mol Biol* 48: 99-118 (2002).

Bassilian S, Ahmed S, Lim SK *et al.* Loss of regulation of lipogenesis in the Zucker diabetic rat. II. Changes in stearate and oleate synthesis. *Am J Physiol Endocrinol Metab* 282: E507-513 (2002).

Bentley J, Walker I, McIntosh E *et al.* Glucose transport regulation by p210 Bcr-Abl in a chronic myeloid leukaemia model. *Br J Haematol* 112: 212-215 (2001).

Boren J, Cascante M, Marin S *et al.* Gleevec (STI571) influences metabolic enzyme activities and glucose carbon flow toward nucleic acid and fatty acid synthesis in myeloid tumor cells. *J Biol Chem* 276: 37747-37753 (2001).

Boros LG, Puigjaner J, Cascante M *et al.* Oxythiamine and dehydroepiandrosterone inhibit the nonoxidative synthesis of ribose and tumor cell proliferation. *Cancer Res* 57: 4242-4248 (1997).

Boros LG, Torday JS, Lim S *et al.* Transforming growth factor β_2 promotes glucose carbon incorporation into nucleic acid ribose through the nonoxidative pentose cycle in lung epithelial carcinoma cells. *Cancer Res* 60: 1183-1185 (2000).

Boros LG, Bassilian S, Lim S, Lee WN. Genistein inhibits nonoxidative ribose synthesis in MIA pancreatic adenocarcinoma cells: a new mechanism of controlling tumor growth. *Pancreas* 22: 1-7 (2001).

Boros LG, Williams RD. Isofenphos induced metabolic changes in K562 myeloid blast cells. *Leukemia Res* 25: 883-890 (2001).

Boros LG, Cascante M, Paul Lee WN. Metabolic profiling of cell growth and death in cancer: applications in drug discovery. *Drug Discov Today* 7: 364-372 (2002a).

Boros LG, Lee WN, Go VL. A metabolic hypothesis of cell growth and death in pancreatic cancer. *Pancreas* 24: 26-33 (2002b).

Boros LG, Lee WN, Cascante M. Imatinib and chronic-phase leukemias. *N Engl J Med* 347: 67-68 (2000c).

Cascante M, Boros LG, Comin-Anduix B *et al.* Metabolic control analysis in drug discovery and disease. *Nature Biotechnol* 20: 243-249 (2002).

Chance EM, Seeholzer SH, Kobayashi K, Williamson JR. Mathematical analysis of isotope labeling in the citric acid cycle with applications to ^{13}C NMR studies in perfused rat hearts. *J Biol Chem* 258: 13785-13794 (1983).

Comín-Anduix B, Boren J, Martinez S *et al.* The effect of thiamine supplementation on tumor proliferation: a metabolic control analysis study. *Eur J Biochem* 268: 4177-4182 (2001).

Cornish-Bowden A. Metabolic control analysis in biotechnology and medicine. *Nature Biotechnol* 17: 641-643 (1999).

Eglinton JM, Heinrich AJ, Pollnitz AP *et al.* Decreasing acetic acid accumulation by a glycerol overproducing strain of *Saccharomyces cerevisiae* by deleting the ALD6 aldehyde dehydrogenase gene. *Yeast* 19: 295-301 (2002).

Fernandez CA, Des Rosiers C, Previs SF *et al.* Correction of ^{13}C mass isotopomer distributions for natural stable isotope abundance. *J Mass Spectrom* 31: 255-262 (1996).

Fiehn O. Metabolomics-the link between genotypes and phenotypes. *Plant Mol Biol* 48: 155-171 (2002).

Flaumenhaft E, Uphaus RA, Katz JJ. Isotope biology of ^{13}C. Extensive incorporation of highly enriched ^{13}C in the alga *Chlorella vulgaris*. *Biochim Biophys Acta* 215: 421-429 (1970).

Guo ZK, Lee WN, Katz J, Bergner AE. Quantitation of positional isomers of deuterium-labeled glucose by gas chromatography/mass spectrometry. *Anal Biochem* 204: 273-282 (1992).

Gyger M, Perreault C, Belanger R, Bonny Y. Chronic myeloid leukemia with an unusual simple variant translocation: t(22;22)(q13;q11). *Acta Haematol* 73: 193-195 (1985).

Hardt PD, Ngoumou BK, Rupp J *et al.* Tumor M_2-pyruvate kinase: a promising tumor marker in the diagnosis of gastro-intestinal cancer. *Anticancer Res* 20: 4965-4968 (2000).

Henderson TR, Dacus JM, Crespi HL, Katz JJ. Deuterium isotope effects on beta-galactosidase formation by *Escherichia coli*. *Arch Biochem Biophys* 120: 316-321 (1967).

Hojo M, Morimoto T, Maluccio M *et al.* Cyclosporine induces cancer progression by a cell-autonomous mechanism. *Nature* 397: 530-534 (1999).

Katz J, Landau BR, Bartsch GE. The pentose cycle, triose phosphate isomerization, and lipogenesis in rat adipose tissue. *J Biol Chem* 241: 727-740 (1966).

Katz J, Rognstad R. The labeling of pentose phosphate from glucose-^{14}C and estimation of the rates of transaldolase, transketolase, the contribution of the pentose cycle, and ribose phosphate synthesis. *Biochemistry* 6: 2227-2247 (1967).

Katz J, Wals PA, Lee WN. Determination of pathways of glycogen synthesis and the dilution of the three-carbon pool with [U-^{13}C]glucose. *Proc Natl Acad Sci USA* 88: 2103-2107 (1991).

Katz J, Wals PA, Golden S, Rognstad R. Recycling of glucose by rat hepatocytes. *Eur J Biochem* 60: 91-101 (1975).

Katz J, Lee WN, Wals PA, Bergner EA. Studies of glycogen synthesis and the Krebs cycle by mass isotopomer analysis with [U-^{13}C]glucose in rats. *J Biol Chem* 264: 12994-13004 (1989).

Katz J, Lee WN. Application of mass isotopomer analysis for determination of pathways of glycogen synthesis. *Am J Physiol* 261: E332-336 (1991).

Katz J, Wals P, Lee WN. Isotopomer studies of gluconeogenesis and the Krebs cycle with ^{13}C-labeled lactate. *J Biol Chem* 268: 25509-25521 (1993).

Kell DB, King RD. On the optimization of classes for the assignment of unidentified reading frames in functional genomics programmes: the need for machine learning. *Trends Biotechnol* 18: 93-98 (2000).

Lee WN, Byerley LO, Bassilian S *et al.* Isotopomer study of lipogenesis in human hepatoma cells in culture: contribution of carbon and hydrogen atoms from glucose. *Anal Biochem* 226: 100-112 (1995).

Lee WN, Edmond J, Bassilian S, Morrow JW. Mass isotopomer study of glutamine oxidation and synthesis in primary culture of astrocytes. *Dev Neurosci* 18: 469-477 (1996).

Lee WN, Boros LG, Puigjaner J *et al.* Mass isotopomer study of the nonoxidative pathways of the pentose cycle with [1,2-$^{13}C_2$]glucose. *Am J Physiol* 274: E843-851 (1998).

Lee WN, Bassilian S, Lim S, Boros LG. Loss of regulation of lipogenesis in the Zucker diabetic (ZDF) rat. *Am J Physiol Endocrinol Metab* 279: E425-432 (2000).

Leimer KR, Rice RH, Gehrke CW. Complete mass spectra of N-TAB esters of amino acids. *J Chromatogr* 141: 121-144 (1977).

Liu X, Ng C, Ferenci T. Global adaptations resulting from high population densities in *Escherichia coli* cultures. *J Bacteriol* 182: 4158-4164 (2000).

Luftner D, Mesterharm J, Akrivakis C *et al.* Tumor type M_2 pyruvate kinase expression in advanced breast cancer. *Anticancer Res* 20: 5077-5082 (2000).

Macallan DC, Fullerton CA, Neese RA *et al.* Measurement of cell proliferation by labeling of DNA with stable isotope-labeled glucose: studies *in vitro*, in animals, and in humans. *Proc Natl Acad Sci USA* 95: 708-713 (1998).

Mao CS, Bassilian S, Lim SK, Lee WN. Underestimation of gluconeogenesis by the [U-$^{13}C_6$]glucose method: effect of lack of isotope equilibrium. *Am J Physiol Endocrinol Metab* 282: E376-85 (2002).

Neese RA, Siler SQ, Cesar D *et al.* Advances in the stable isotope-mass spectrometric measurement of DNA synthesis and cell proliferation. *Anal Biochem* 298: 189-95 (2001).

Oliver DJ, Nikolau B, Wurtele ES. Functional genomics: high-throughput mRNA, protein, and metabolite analyses. *Metab Eng* 4: 98-106 (2002).

Oliver SG. Functional genomics: lessons from yeast. *Philos Trans R Soc Lond B* 357: 17-23 (2002).

Oremek GM, Teigelkamp S, Kramer W *et al.* The pyruvate kinase isoenzyme tumor M_2 (Tu M_2-PK) as a tumor marker for renal carcinoma. *Anticancer Res* 19: 2599-2601 (1999).

Orgel LE. Ageing of clones of mammalian cells. *Nature* 243: 441-445 (1973).

Papageorgopoulos C, Caldwell K, Shackleton C *et al.* Measuring protein synthesis by mass isotopomer distribution analysis (MIDA). *Anal Biochem* 267: 1-16 (1999).

Previs SF, Ciraolo ST, Fernandez CA *et al.* Use of [6,6-$^2H_2$]glucose and of low-enrichment [U-$^{13}C_6$]-glucose for sequential or simultaneous measurements of glucose turnover by gas chromatography-mass spectrometry. *Anal Biochem* 218: 192-196 (1994).

Previs SF, Brunengraber H. Methods for measuring gluconeogenesis *in vivo*. *Curr Opin Clin Nutr Metab Care* 1: 461-465 (1998).

Raamsdonk LM, Teusink B, Broadhurst D *et al.* A functional genomics strategy that uses metabolome data to reveal the phenotype of silent mutations. *Nature Biotechnol* 19: 45-50 (2001).

Rais B, Comin B, Puigjaner J *et al.* Oxythiamine and dehydroepiandrosterone induce a G_1 phase cycle arrest in *Ehrlich*'s tumor cells through inhibition of the pentose cycle. *FEBS Lett* 456: 113-118 (1999).

Saur WK, Peterson DT, Halevi EA *et al.* Deuterium isotope effects in the fermentation of hexoses to ethanol by *Saccharomyces cerevisiae*. II. A steady-state kinetic analysis of the isotopic composition of the methyl group of ethanol in an isotopic mirror fermentation experiment. *Biochemistry* 7: 3537-3546 (1968).

Schneider J, Morr H, Velcovsky HG *et al.* Quantitative detection of tumor M_2-pyruvate kinase in plasma of patients with lung cancer in comparison to other lung diseases. *Cancer Detect Prev* 24: 531-535 (2000).

Schuster S, Dandekar T, Fell AD. Detection of elementary flux modes in biochemical networks: a promising tool for pathway analysis and metabolic engineering. *Trends Biotechnol* 17: 53-60 (1999).

Shi H, Vigneau-Callahan KE, Shestopalov AI *et al.* Characterization of diet-dependent metabolic serotypes: proof of principle in female and male rats. *J Nutr* 132: 1031-1038 (2002).

Shi H, Vigneau-Callahan KE, Shestopalov *et al.* Characterization of diet-dependent metabolic serotypes: primary validation of male and female serotypes in independent cohorts of rats. *J Nutr* 132: 1039-1046 (2002).

Smith TA. The rate-limiting step for tumor [^{18}F]fluoro-2-deoxy-D-glucose (FDG) incorporation. *Nucl Med Biol* 28: 1-4 (2001).

Stokkel MP, Draisma A, Pauwels EK. Positron emission tomography with 2-[^{18}F]-fluoro-2-deoxy-D-glucose in oncology. Part IIIb: Therapy response monitoring in colorectal and lung tumours, head and neck cancer, hepatocellular carcinoma and sarcoma. *J Cancer Res Clin Oncol* 127: 278-285 (2001).

ter Kuile BH, Westerhoff HV. Transcriptome meets metabolome: hierarchical and metabolic regulation of the glycolytic pathway. *FEBS Lett* 500: 169-171 (2001).

Tian WN, Braunstein LD, Apse K *et al.* Importance of glucose-6-phosphate dehydrogenase activity in cell death. *Am J Physiol* 276: C1121-1131 (1999).

Tomita M. Whole-cell simulation: a grand challenge of the 21st century. *Trends Biotechnol* 19: 205-210 (2001).

Tuttle S, Stamato T, Perez ML, Biaglow J. Glucose-6-phosphate dehydrogenase and the oxidative pentose phosphate cycle protect cells against apoptosis induced by low doses of ionizing radiation. *Radiat Res* 153: 781-787 (2000).

Tweeddale H, Notley-McRobb L, Ferenci T. Effect of slow growth on metabolism of *Escherichia coli*, as revealed by global metabolite pool ("metabolome") analysis. *J Bacteriol* 180: 5109-5116 (1998).

Tweeddale H, Notley-McRobb L, Ferenci T. Assessing the effect of reactive oxygen species on *Escherichia coli* using a metabolome approach. *Redox Rep* 4: 237-241 (1999).

Uphaus RA, Katz JJ. Culture of isotopically substituted plants of pharmacological importance: conservation and recycling of stable isotope substrates. *J Pharm Sci* 65:1096-1099 (1976).

Vigneau-Callahan KE, Shestopalov AI *et al.* Characterization of diet-dependent metabolic serotypes: analytical and biological variability issues in rats. *J Nutr* 131: 924S-932S (2001).

Wadke M, Brunengraber H, Lowenstein JM *et al.* Fatty acid synthesis by liver perfused with deuterated and tritiated water. *Biochemistry* 12: 2619-2624 (1973).

Wykes LJ, Jahoor F, Reeds PJ. Gluconeogenesis measured with [U-^{13}C]glucose and mass isotopomer analysis of apoB-100 amino acids in pigs. *Am J Physiol* 274: E365-376 (1998).

Yoder OC, Turgeon BG. Fungal genomics and pathogenicity. *Curr Opin Plant Biol* 4: 315-312 (2001).

Zwerschke W, Mazurek S, Massimi P *et al.* Modulation of type M_2 pyruvate kinase activity by the human papillomavirus type 16 E7 oncoprotein. *Proc Natl Acad Sci USA.* 96: 1291-1296 (1999).

Chapter 10

THE ROLE OF METABOLOMICS IN SYSTEMS BIOLOGY

A New Vision for Drug Discovery and Development

Jan van der Greef[1,2,3], Eugene Davidov[1], Elwin Verheij[2], Jack Vogels[2], Rob van der Heijden[3], Aram S. Adourian[1], Matej Oresic[1], Edward W. Marple[1] and Stephen Naylor[1]

[1]Beyond Genomics Inc., 40 Bear Hill Rd., Waltham, MA 02451, USA [2]TNO Pharma, PO Box 2215 2031CE, Zernikedreff 9, 2333 CK, Leiden, Netherlands [3]Leiden University/Amsterdam Center for Drug Research, Division of Analytical Sciences, NL-2300 RA Leiden, Netherlands

1. INTRODUCTION

The recent completion of major milestones of the Human Genome Project has demonstrated the impact of the "omics revolution" on modern research in the life sciences (International Human Genome Sequencing Consortium, 2001; Venter *et al.*, 2001). In turn, the enormous degree of complexity inherent in genomic information has revealed the limitations of purely genomic investigation. Recent research has extended to the study of the proteome, defined as the total protein complement encoded for by a genome. The enormity of this effort becomes evident when it is considered that the estimated 30,000 - 40,000 genes in the human are predicted to yield as many as 1 million distinct proteins due to processes such as transcriptional splicing and post-translational modifications. As such, proteomic analysis has stimulated dramatic technological advances for protein quantification, characterization, and identification (Gygi *et al.*, 1999; Geng *et al.*, 2000; Regnier *et al.*, 2000; Barnes and Clemmer, 2001; McLuckey *et al.*, 2001; Valentine *et al.*, 2001).

Yet an under-appreciated but essential and complementary element necessary for a coherent understanding of both genome and proteome, and by extension biomolecular function, is the metabolome. Along with its counterparts in the "omics revolution," metabolomics has emerged as a significant and compelling discipline in life sciences research (Fig. 1). The study of metabolites can provide both a comprehensive signature of the physiological state of an organism as well as unique insights into specific biochemical processes. Monitoring the metabolome as a function of time allows the study of the temporal dynamics of normal homeostasis as well as the effects of perturbations or stimuli on a system. The importance of metabolites in biological control and communication, and as building blocks and energy transporters makes the metabolome particularly suited to phenotypic and dynamic profiling.

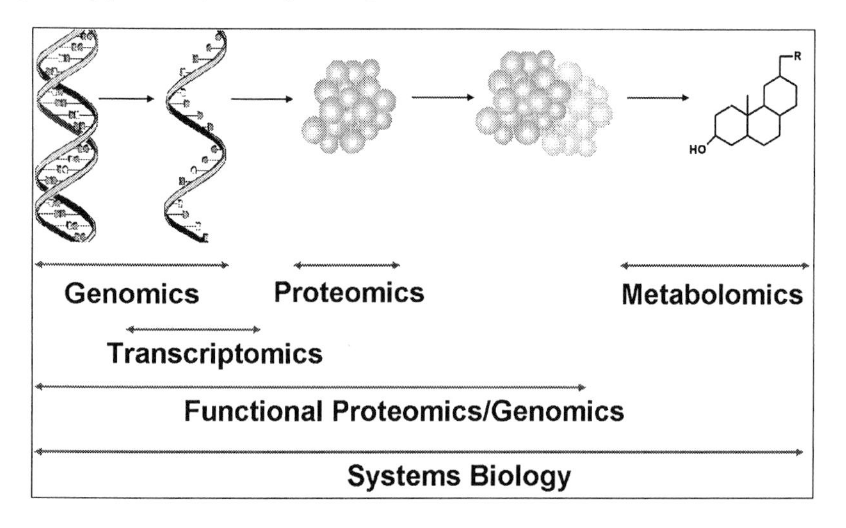

Figure 1. The "omics" nomenclature.

The analysis of the metabolome can be dramatically more complex than that of the genome or the proteome. This is due, not only to the number of molecules which constitute the metabolome, but also to the sophistication required to identify and map their structural and compositional diversity. The basic building blocks of genes and transcription elements (*i.e.* nucleotides) and proteins (*i.e.* amino acids) have provided a focus for the development of rapid and robust analytical procedures for facile identification of these biopolymers. In contrast, such tools are not as readily available for comprehensive metabolite identification and analysis. Moreover the concentration range to be covered in metabolomics can also be extremely large, presenting the researcher with considerable technical challenges.

Despite these challenges, metabolomics plays a key role in the convergence of the "omics" disciplines within an integrative biological framework. Such a "Systems Biology" approach aims to understand phenotypic variation, to assemble comprehensive data and models of cellular organization and biochemical function, and to elucidate interactions and pathways (Kitano, 2002). Studies of systems biology in non-mammalian models have recently been published (Ideker *et al.*, 2001; Raamsdonk *et al.*, 2001), and indeed the application of systems biology to a mammalian system has now been reported (Davidov *et al.*, 2002a, 2000b). Within the context of a fundamental understanding and treatment of disease, metabolomics and systems biology have the potential to impact pharmaceutical research and development across the drug development spectrum from target to clinic. For example, investigations at the metabolic level of perturbations of model systems with investigative pharmacological molecules have been exploited for drug response and efficacy studies and mechanism of action research. Furthermore the profiling of body fluids and tissues has proven to be key to discovering specific and sensitive biomarkers for safety and toxicity applications, interspecies response evaluation, clinical study applications and disease pathogenesis or diagnostic work.

In this chapter we discuss both metabolomic strategies as well as advances in enabling technologies in this field. Several specific metabolomics examples and applications are presented. We also explore the integration of metabolomics with the other "omics" approaches in a systems biology framework. As a point of clarity regarding nomenclature, the phrase 'metabolomics' as used in this chapter is meant to encompass terms used elsewhere in the literature such as 'metabonomics' (Nicholson *et al.*, 1999), 'metanomics' (Watkins, 2000) and other related terminology.

2. ANALYTICAL APPROACHES IN METABOLOMICS

2.1 History of Metabolic Profiling

While the various "omics" nomenclatures are relatively new, including 'metabolomics', bioanalytical research into metabolites and small molecules encompasses many decades of active research. A myriad of efforts can be cited in the field of mass spectrometry (MS) dating back to the 1970s and earlier. From that period, gas chromatography-MS (GC-MS) methods has been developed by many groups for the profiling of complex matrices such as blood and urine for clinical or nutritional studies. Many analyses related to studies of perturbations of biological systems, such as inborn errors of metabolism (*e.g.* Jellum 1977; Stanbury *et al.*, 1983). In these early works the importance of metabolic profiling for the detection of disease was clearly demonstrated. Studies on the influence of stress and physical exertion on human biochemistry and biochronological rhythms were also recorded (Spiteller 1985; Dietel and Spiteller, 1986; Heindl *et al.*, 1986). Important methods and developments in multivariate statistics were applied to metabolic profiling data in the 1980s, particularly in the field of pyrolysis MS. Successful profiling of complex biological material even after "destructive" pyrolysis techniques demonstrated that due to high platform reproducibility, informative profile information could be generated for diagnostic purposes, and the interpretation of observed chemical changes was possible (Droge *et al.*, 1987; Tas *et al.*, 1985, 1987, 1989a; Tas and van der Greef, 1993).

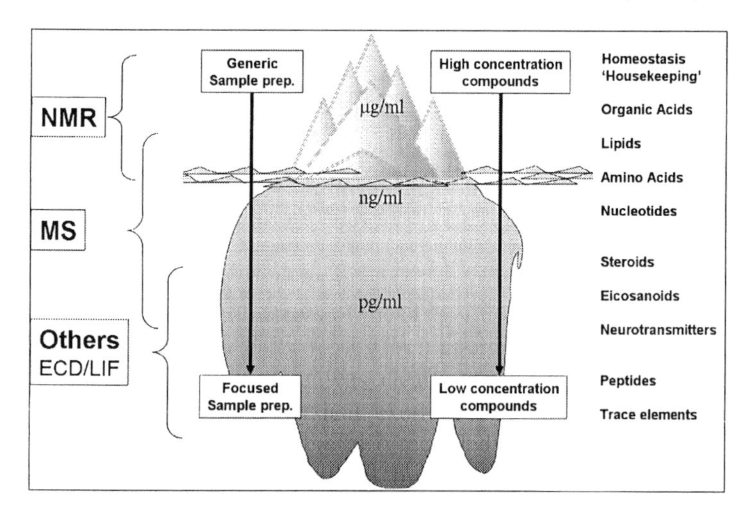

Figure 2. 'Iceberg'cartoon illustrating that comprehensive coverage of the metabolome requires multiple techniques and different sample preparation strategies to encompass wide concentration ranges. Listed are various techniques used in metabolomics (left column) and representative metabolite classes (right column). All metabolites span a range of concentrations, and the delineations in the central column are meant to be qualitative only.

In the early 1980s the use of soft-ionization MS coupled with pattern recognition strategies for metabolic profiling was explored by one of the authors (van der Greef *et al.*, 1983). The basis behind this development was that Curie-point pyrolysis was limited in detecting minor differences in intact molecules due to volatility and concentration limitations. A number of ionization approaches were employed and evaluated, such as direct-chemical ionization, field desorption and fast atom bombardment (van der Greef, 1986; Tas and van der Greef, 1993, 1995). Striking results were obtained in cases of urine metabolic profiling identifying significant gender differences (van der Greef and Leegwater, 1983), markers influenced by drugs (Tas *et al.*, 1989b), bacteria, fungi and algae characterization (Tas *et al.*, 1987, 1989b; van der Greef *et al.*, 1988a; 1988b) rapid drug metabolite tracing (van der Greef and Leegwater, 1983; van der Greef *et al.*, 1983b, 1984), disease biomarkers (Tas *et al.*, 1989c) and virus infected cells (Tas *et al.*, 1989d). It should be noted however that in all cases at the time, available soft-ionization technologies had inherent reproducibility limitations. As a result, during that period in mass spectrometric metabolomics, the importance of appropriate data preprocessing was recognized and subsequently described in detail (van der Greef *et al.*, 1986). In particular proper data normalization and other data preprocessing techniques were deemed crucial for the field of metabolic profiling and pattern recognition, particularly when comparing profiles from diverse sources and complex mixtures. Some of these tools are discussed in the following sections.

As noted above, the molecular structural and abundance diversity in the metabolome is significant. This is schematically represented in Fig. 2, which is meant to illustrate that the metabolome consists of compounds that span a considerable concentration and structural range. Such concentration ranges can even be found within a single class of compound, such as in the case of lipids that can range in concentration from milligram/mL to picogram/mL. Within certain molecular classes typical concentrations can be as low as femtogram/mL or even attogram/mL. In addition, the metabolome comprises molecules that span an enormous range of polarities including lipids to sugars and peptides. Due to such heterogeneity and diversity it is clear that a single method or technology cannot provide the performance necessary for comprehensive metabolomic study. In the late 1980s 'metabonomics' by nuclear magnetic resonance spectroscopy (NMR) was introduced by Nicholson (see Lindon *et al.*, 1999, 2000, 2001) to profile small molecules in biofluids. This approach provides a powerful way of analyzing the most abundant metabolite components of a biofluid mixture, and in concert with MS platforms affords a means to monitoring the metabolome. However, the limitation of the profiling methodologies available in the 1980s lay not only in the coverage of the metabolome but also in the data preprocessing techniques. In the latter case we have focused on four different issues over the last two decades, 1) correction of spectra due to instrumental shifts, 2) spectral normalization procedures, 3) development of robust peak detection and selection routines for multidimensional analytical techniques such as GC-MS and liquid-chromatography (LC)-MS, and 4) advanced bioinformatics tools for data interpretation. An understanding of these approaches was driven in large part by the objective of achieving the most comprehensive coverage possible of the metabolome.

2.2 Practical Metabolomics

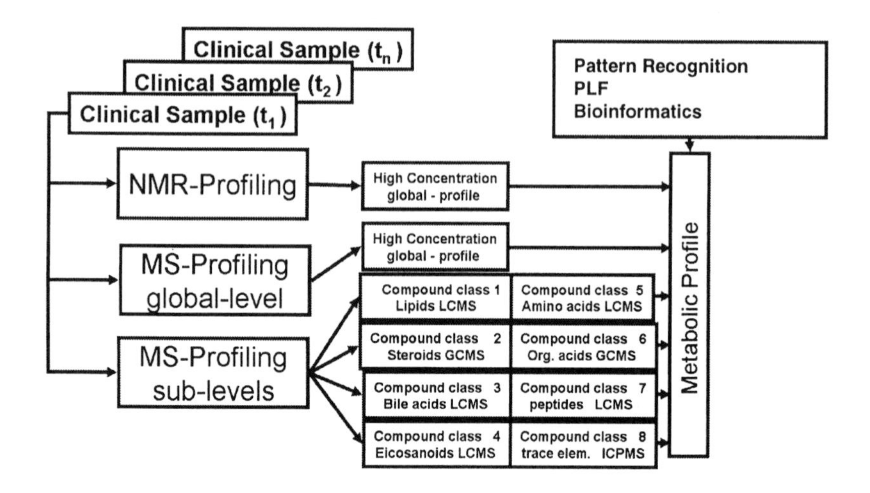

Figure 3. One possible arrangement of analytical techniques for systematic coverage of the metabolome.

The analytical challenges in metabolomics are different from those in traditional bioanalytical approaches where a selected number of compounds (often exogenous) needs to be accurately quantified in complex matrices. In traditional approaches, methods are developed in order to be as selective as possible to achieve the low picogram/mL concentrations necessary to allow analysis on the small amounts of biofluids typically available (*e.g.* 1 mL for plasma, and even less for samples such as cerebrospinal or synovial fluids). This selectivity is required to avoid contaminant interference and to achieve high recoveries and robustness for analyzing hundreds to thousands of samples. Several approaches to attaining requisite analytical selectivity have involved high resolution MS and tandem MS. However, in metabolomics it is desirable to measure every component present and thus sample preparation is often minimized in order to ensure no loss of individual components. Unfortunately, multiple tandem sample preparation steps are necessary in the analysis of very low abundant compounds because of the limitations of dynamic range for any single analytical platform. There is therefore a balance between minimizing sample preparation (and therefore minimizing analyte loss) on one hand, and removing larger abundant components to increase dynamic range and thereby maximizing the ability to detect low abundant compounds on the other. The authors are among those who have developed strategies that select platforms affording maximum coverage of the metabolome. In such a profiling platform, NMR is used in combination with global MS methods as a survey step. NMR is extremely useful in profiling components at the higher concentration range of the metabolome, in measuring compounds of different biosynthetic and structural natures in a single profile and is also unique in its quantitation performance. For further investigations, MS is used because of its unique ability to separate and quantify individual components in very complex mixtures. Quantification, however, is quite a challenge with MS when used in a broad profiling mode (as will be discussed later). Combinations of MS techniques such as GC-MS, LC-MS and multidimensional GC or LC separation techniques with MS or MS/MS open up the lower concentration range for study. In addition, other detection systems for GC or LC are attractive to explore in terms of selectivity or quantitation options. For instance quantitation by multi GC-flame ionization detection (FID) of lipids (Watkins, 2000) has been demonstrated to yield excellent profiles, and LC combined with electrochemical detection (ECD) is successfully able to profile neurotransmitters and other electrochemical active components (Tsai, 2000).

In Fig. 2 we depicted the challenge of metabolomics with an analogy to an iceberg, the peak of which represents the higher concentration components readily amenable to detection and analysis with NMR. However to detect and determine the presence of the vast majority (>90%) of metabolome constituents requires a suite of techniques including MS. As might be expected, the least explored region of the metabolome is at the low concentration range. An example of a basic platform is illustrated in Fig. 3. In optimizing the platform it should be emphasized that the number of sub-platforms to be selected depends on the available sample volume and on the sensitivity of the methods. An exception can be made for NMR, as this is generally a non-destructive technique, although it should be noted that even simple sample preparation procedures such as those used in NMR have an impact on the ability to perform subsequent additional analyses at lower concentration levels. Global profiles are obtained using both NMR and MS and additional platforms are generated by using GC-MS and LC-MS (or in combination with MS/MS) techniques or inductively coupled plasma-MS in the case of interest in non-organic components. Sample preparation for NMR can generally be very straightforward using freeze-drying and reconstitution in D_2O because the focus is on the higher concentration components, *i.e.* of typical concentration > 100 nanogram/mL. For MS a variety of sample preparation approaches are used ranging from solid phase extraction and liquid/liquid extraction to more specific methods using, for instance, affinity-based methods or derivitization procedures both for GC-MS as well as LC-MS. In the schematic diagram shown in Fig. 3, the choice of platform is based on certain desired targeted metabolite classes, by which selectivity in sample preparation is reduced to allow as many components to be measured as possible. At present it is still difficult to estimate the coverage of the human metabolome (Beecher, Chapter 17), as estimates of the number of components are vague and range from 10- to 30,000 components (including peptides for the higher range limit). For lipids, for instance, estimates around 600 components have been made. However one should bear in mind that, taking lipids as an example, oxidation or degradation products can also be present in low concentrations only and could be missed easily. Such results were demonstrated in published biomarker studies on chronic diseases such as rheumatoid arthritis, asthma, Alzheimer's disease, diabetes, atherosclerosis, Parkinson's disease and others (see Spiteller 1988, 1999).

3. DATA PRE-PROCESSING, PATTERN RECOGNITION AND BIOINFORMATICS

3.1 Data Pre-Processing

There are several important steps involved in handling raw data profiles from analytical instrumentation for metabolomic analysis purposes. These vary depending on the particular analysis technique used and the type of profile data obtained. For instance if a spectral profile is obtained, as with standard spectroscopic methods, a primary necessary step is to adjust for minor shifts in the spectra both in the intensity dimension as well as in the spectral or chromatographic dimension. Shifts can be due to instrumental factors, environmental conditions, or to varying concentrations of components (as is often the case in urine analysis). As an example, variation in NMR chemical shifts often occurs and needs to be accounted for, but the reproducibility and standardization in the intensity (or peak area) of a single profile (quantitation dimension) is typically very satisfactory. This is in contrast to MS where the peak intensity (ion abundance) dimension needs to be carefully adjusted or standardized due to lack of calibrants for each component present in the profile. In hyphenated techniques the reproducibility of the separation method (GC, LC or electromigration driven techniques such as capillary electrophoresis (CE)) needs to be carefully evaluated as well. In this respect near-infrared spectral profiles are impressive and correction in either dimension is hardly required.

In general, small instrumental shifts in the spectral (variable) dimension will be falsely interpreted as representing different components when a collection of data profiles is subjected to pattern recognition analysis. A straightforward way to cope with this problem is by using binning techniques in which the spectrum is reduced in resolution to a sufficient degree to ensure that a given peak remains its bin despite small spectral shifts between analyses. For example, in NMR the chemical shift axis may be discretized and coarsely binned, and in MS the spectral accuracies may be rounded to integer atomic mass unit values. Of course this is performed based on the knowledge of the instrumentation and within the constraints of the experiment.

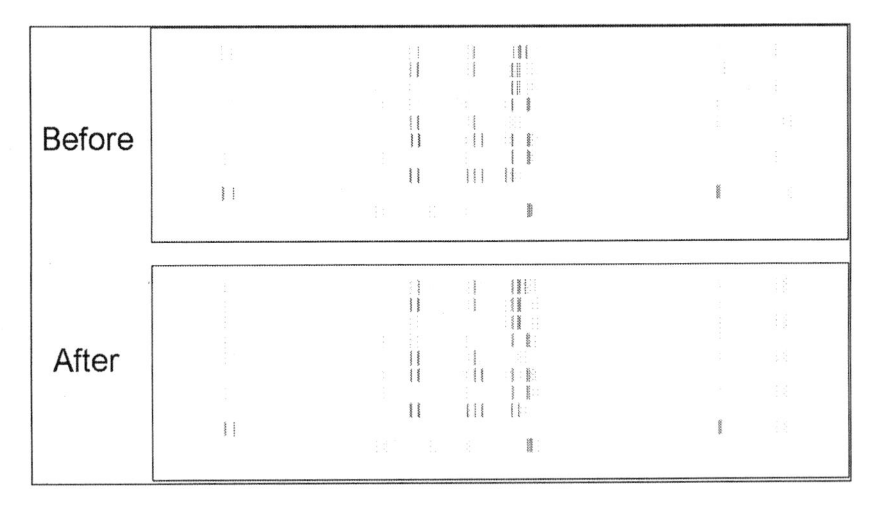

Figure 4. A grey-scale intensity plot of a set of ten NMR urine spectra before and after application of the PLF routine. A number of peaks have been selected to present a clear view. The resulting set shown in the bottom half of the figure has been aligned without resolution loss.

However, this intentional reduction of resolution is extremely undesirable, as it involves deliberate loss of information from spectral data. In the case of NMR, we recognized the drawback in data quality introduced by loss of resolution with conventional approaches and so developed instead a partial linear fit (PLF) algorithm (Vogels *et al.*, 1996a). This adjusts and aligns peaks in different spectra across a data set without resorting to binning or resolution reduction. The approach involves identifying major component peaks common to all data sets (and arising from common components), and segmenting the spectra followed by proper registration such that the major peaks are in alignment and a maximum calculated correlation is achieved across all constituent spectra of the data set. This is preformed prior to the application of pattern recognition techniques. Greatly improved subsequent statistical analyses are enabled, and the ability to detect minor peaks even on shoulders of larger peaks has been demonstrated (Vogels *et al.*, 1994, 1996a, 1996b, 1996c). The PLF process is schematically illustrated in Fig. 4.

In the case of hyphenated MS methods, binning is also often used. However, the challenges of alignment are also compounded by the addition of an extra dimension of separation corresponding to, for example, analyte retention times. For a long time this hampered the application of pattern recognition approaches to hyphenated MS data, even though many such MS methods were available and appropriate for the detection and measurement of components in complex mixtures. There have been a number of approaches developed to overcome these data shift issues. Most such methods involve some form of peak detection algorithm to address the problem of retention shifts and the selection of relevant peaks for subsequent pattern recognition analysis (Windig *et al.*, 1996). A particularly robust approach is the IMPRESS method (Gaspari *et al.*, 2001), discussed here. IMPRESS uses a data entropy-based peak and mass trace detection algorithm and assigns a significance score (IQ value) for each mass trace in a hyphenated MS data set. Once major peaks are selected and noted, alignment procedures may be applied.

The IMPRESS algorithm used for peak and mass trace detection calculates the data entropy E_i of each individual mass trace as follows:

$$Ei = -\sum_{t=0}^{n} p_t \ln(p_t).$$

where t is the scan time and p is the intensity of signal. The entropy values may then be transformed into significance scores by scaling the entropy by the maximum entropy (E_{max}) and then subtracting the resulting quotient from unity:

$$\mathrm{IQ} = 1 - E_i / E_{max}.$$

Traces with high IQ values (low entropy) contain relevant peaks and little noise, while the reverse is true for mass traces having a low IQ value. Ordering mass traces with respect to their IQ values provides the opportunity to differentiate mass traces according to their true information content. Peaks with signal-to-noise ratios as low as 3 can easily be selected. In the past several years a variety of enhancements have been made to the algorithm, including spurious spike removal, baseline drift correction and improved peak recognition.

The IMPRESS approach is illustrated in Fig. 5A, which shows an IQ-plot for an LC-MS lipid analysis of mouse plasma. The x-axis is related to the LC-MS traces and the y-axis is the IQ value. A selection based on IQ-value is used and the resulting profile, after conversion back to original intensities as shown in Fig. 5B, is used for pattern recognition. This approach has now been applied to a variety of separation protocols and metabolite analytes, including peptides.

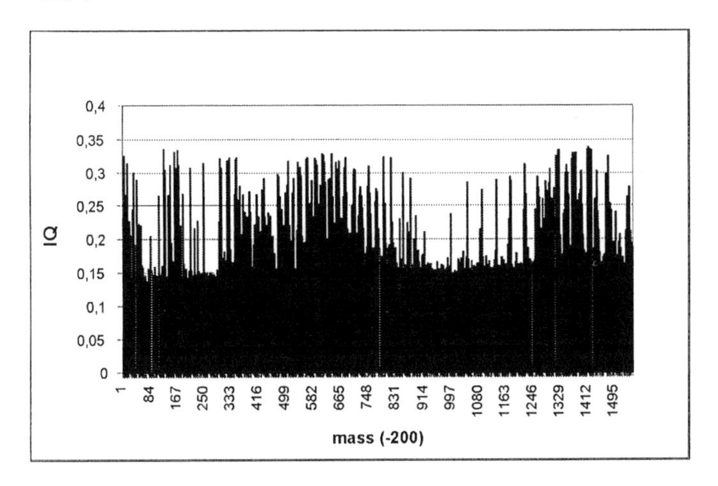

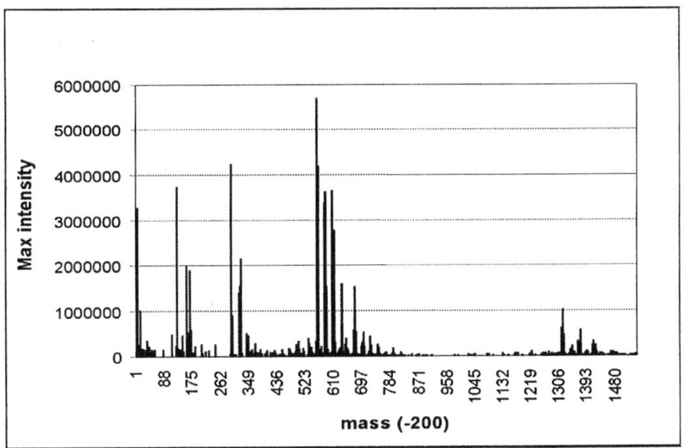

Figure 5. A. Example of IMPRESS.Quality (IQ) scored m/z traces from a complex LC-MS lipid profile. B. IQ-selected traces from Fig. 5A are converted into intensity based fingerprints that can be used directly for pattern recognition.

3.2 Pattern Recognition

After initial data pre-processing the spectral profiles are set for pattern recognition (multivariate analysis). The most frequently used and straightforward approach for pattern recognition is principal component analysis (PCA) or factor spectrum based strategies (Windig *et al.*, 1983). PCA was developed over seventy years ago for the analysis of high dimensionality data sets (Hotelling, 1933), and has since been applied to metabolite profiling data (van der Greef *et al.*, 1986b; Holmes *et al.*, 1998). More recently our group and others have adapted this approach for use with both metabolomic as well as proteomic data. The developments in statistical analysis for high dimensional data (small number of objects and large number of variables) have been greatly enhanced by genomics and microarray driven research. For practical reasons, PCA coupled with discriminant analysis (PCDA) approaches are extremely relevant in this context. It is typically based on triplicate analytical measurements on each sample, after which each sample is defined by a category and, after selection of a number of principal components (typically number of objects divided by 4), a discriminant analysis is performed (Hoogerbrugge *et al.*, 1983).

Any observed clustering or segregation of samples resulting from PCA or PCDA provides strong evidence for statistically significant differences between or among sample data sets. Numerous examples can be given for PCDA clustering as applied to body fluid profiling using NMR and MS. PCA or PCDA clustering is often the first step in pattern recognition, followed by interpretation that is performed by mining the multivariate data to determine those components of a data set that contribute to an observed pattern. Such interpretation reveals all variables that contribute to a given segregation of clusters and can directly lead to identification of the relevant biological factors. However in most cases the variables are not known, but are only identified by their chemical shift values (in the case of NMR) or their mass-to-charge (*m/z*) ratios (in the case of MS). Identification of relevant components thus becomes the next challenge in metabolomics. In-house proprietary databases and public domain databases are of great help, but especially at lower concentrations the number of unknowns increases rapidly. Depending on the technique various options exist to perform additional experiments for identification. With NMR, more advanced 2-D experiments can be performed. In the case of MS, high resolution or MS/MS experiments can be performed. The challenge of identifying unknowns remains high and spectrum interpretation expertise is of great importance at this point. Other options include the use of canonical correlation statistics based on comparison of two analytical approaches, such as NMR and MS, or MS and GC (Vogels *et al.*, 1994; Tas *et al.*, 1989). In appropriate cases, other tools can be used such as reflected discriminant analysis (Nierop *et al.*, 1994) or classifiers such as soft independent modeling of class analogy (SIMCA: Wold, 1976; Droge *et al.*, 1987; Nicholson *et al.*, 1999).

Often however, biological data sets, certainly in disease studies, are not amenable to distinct cluster assumptions but rather exhibit more continuous trends, particularly in studies related to early markers of disease or nutritional studies. Partial least squares (PLS) (Geladi and Kowalski, 1986) is an attractive approach for performing more elaborate studies.

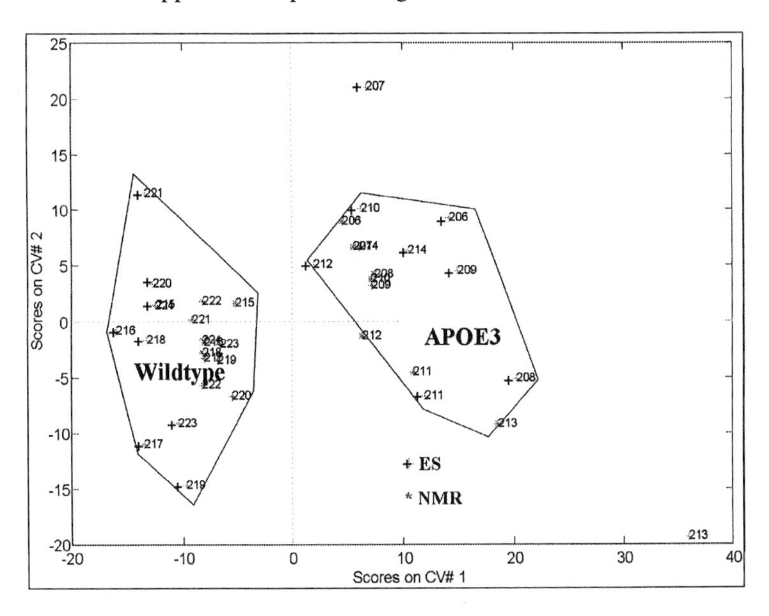

Figure 6. Canonical correlation of NMR and LC-MS (ES) data in a study on transgenic ApoE3-Leiden and wild-type mice. Each data point on the plot represents a separate measurement of plasma metabolites with either NMR (*) or LC-MS (+) techniques. Segregation of metabolic genotypes is observed in both approaches, with LC-MS seen to provide a higher degree of separation. The correlation of the contributing LC-MS peaks and their corresponding NMR peaks increases the success of the peak identification process.

An example of canonical correlation applied to NMR and MS is given in Fig. 6. The observed correlation between the components measured by the two techniques is very helpful in the identification process.

3.3 Bioinformatics

To extract maximum value from the data, multivariate analysis tools, as outlined above, may be used in concert with additional statistical and informatics strategies. Once statistically significant differences in metabolite abundances are determined and quantified among groups of samples, the objective is to understand the underlying biological reasons for, and the contexts of, the results. A first step is to identify metabolic components observed in data spectra and revealed by multivariate analysis as significantly different among samples. Such identification typically involves querying various databases of known metabolite component spectra and structures. A next step is to mine existing knowledge about molecular interactions through searches of public and private databases. This can go some way in explaining associations and behaviors observed in metabolomic profile results (see Section 4). However, because most of the metabolic, genomic, proteomic, and interaction databases depict biochemical events in a static state, progressively sophisticated analytical and mathematical tools are needed to integrate disjointed biological clues into dynamic models that are better suited to explain, for example, pathological processes. Indeed, both linear and non-linear multivariate analyses may uncover statistically significant associations between biomolecular components that will not be explained through the mining of existing databases or literature. Furthermore, it is ultimately desirable to progress from simple associations between components to proposing and verifying causal relationships.

Fig. 7 shows an example of an association network resulting from a non-linear analysis of data from metabolic profiles of components measured in a transgenic mouse plasma study. Each connection, indicated by a solid line, denotes a statistically significant correlation between two measured components. Many of these correlations correspond to known biochemical pathways, and serve as the building blocks upon which unanticipated correlations may be explained, or new pathways proposed.

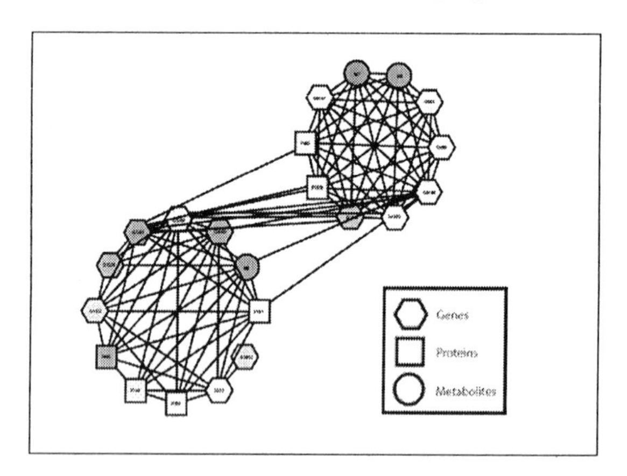

Figure 7. Association network produced by non-linear analysis of transcript, protein and metabolite data, as discussed in text. Hexagon = gene; square = protein; circle = metabolite.

In particular, a systems biology approach allows the integration of proteomic, genomic and other data with metabolomic results. As such, an integrative method provides additional evidence either for elucidating an affected biochemical pathway or for proposing causal explanations for unexpected associations observed in experimental results. For example, a correlation may be determined between a small molecule as measured by metabolomics and the enzyme which produces it as measured by proteomics, but a non-linearity in the correlation may indicate to the researcher additional biomolecular mechanisms, alternate pathways, dynamic effects, enzyme modifications and the like which may be involved. As such, metabolomic information is central to the ultimate objective of understanding the biochemical bases of experimental observations.

4. SELECTED EXAMPLES WITHIN PHARMACEUTICAL APPLICATIONS

Metabolomic applications for pharmaceutical purposes are increasingly establishing importance across a variety of drug development functions. Sustained demands on pharmaceutical productivity have prompted the industry to innovate research and development efforts, which to date have generally consisted of linear, serial processes based on screening chemical libraries against small numbers of pharmacologically relevant, but in some instances poorly defined, biological structures. Metabolomics research, as will be illustrated in the examples below, promises to have a impact on a number of discovery and development stages from identifying and validating targets to characterizing compounds in the clinic. Moreover, metabolomics within a systems biology structure promises further advances over current approaches.

4.1 Toxicity/Safety Evaluation

In pre-clinical development or during the late discovery phase, ranking lead compounds using a variety of clinical variables, particularly toxicity and safety, is important for further clinical development. The correlation of metabolite patterns recorded by NMR with toxicity endpoints has been extensively reported (see Lindon *et al.*, 1999, 2000, 2001).

In a toxicological study (Vogels, unpublished), a drug was evaluated in a product safety evaluation using male and female rats in 3 dose groups and a control group. Standard toxicological assays and immunohistochemistry were also performed. The metabolic profiling of 24 hr urine using NMR, with data pre-processed and analyzed using the techniques discussed earlier (*e.g.* PLF, PCDA), resulted in the cluster effects seen in Fig. 8. Gender differences in patterns are immediately apparent. More subtle but significant differences in the high dosage trends for male and female rats are also captured by the data analysis.

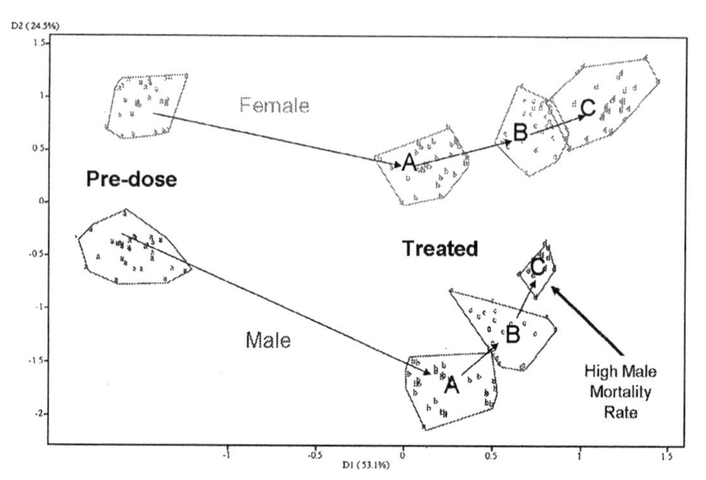

Figure 8. Example of a dose response toxicity study in rat. Clear differences can be detected between male/female patterns and responses at 24 hr.

The high dose treatment resulted in high mortality among males, while no mortality in the corresponding female group was observed. This is reflected in differences observed in the analysis results. At lower doses, the segregation observed in the metabolomic data was much larger than any effect detected by routine histology and immunohistochemistry. This illustrating that the sensitivity of metabolite profiling can surpass that of conventional methods. While in this study a subset of relevant peaks could be identified, the use of NMR metabolite screening alone often does not afford conclusive interpretation of the observed profiles and effects. The use of systems biology is key in understanding the mechanism of action, or in this case the mechanism of toxicity. However, in early screening, databases populated with chemical entities of known toxicity may enable classification and ranking through comparative profiling techniques.

4.2 'Reverse-Pharmacology' or 'Reverse-Discovery'

Several drugs or drug candidates are known to have beneficial effects, but their mechanism of action may be unknown. Discovery and characterization of the specific target that is perturbed by such agents is therefore of great interest to the pharmaceutical industry. One process followed can be referred to as 'reverse-discovery' or 'reverse-pharmacology', the terminology being analogous to reverse engineering in other fields. Working from metabolite measurements "backwards" toward the target is a unique application of metabolomics research, made even more powerful by the integration of other "omics" approaches in systems biology.

An interesting example of a reverse-discovery approach using MS based metabolomics involves an investigation of dimorphism-related differences in the fungus *Candida albicans* and the effect of sterol biosynthesis inhibitors (Tas *et al.*, 1989a). The reverse-discovery aspect of this study lies in the screening of imidazole derivatives followed by evaluation of their effects on *Candida* metabolic profiles generated by (pyrolysis) direct chemical ionization MS.

In Fig. 9, results are shown for treatment of both the hyphal form and the less pathogenic, yeast-like, form of *C. albicans* as analyzed by PCDA. Metabolic differences between the two forms could be assigned. The first significant observation found by metabolic profiling was the convergence of the two forms after treatment with imidazole inhibitor in the culture medium.

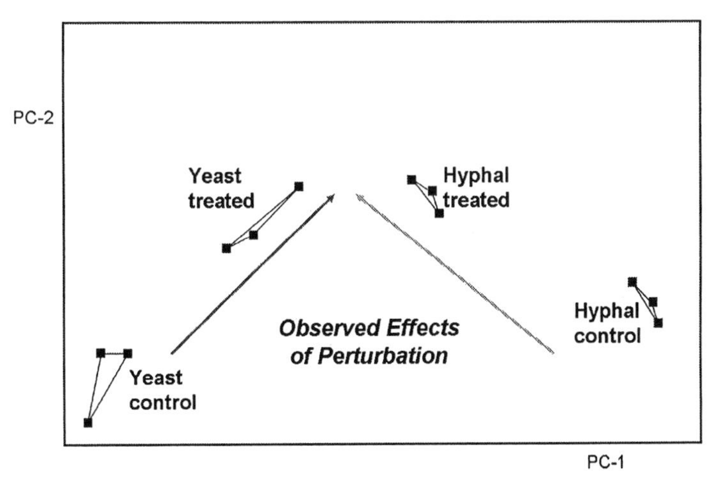

Figure 9. Pattern recognition of control and treated yeast cells, revealing changes in pathways due to perturbation with a drug, as highlighted by direction of arrows.

Differences between the two forms decreased (morphological changes were observed by electron and phase contrast microscopy) after treatment and more clustered yeast-like cells were noted. Both the hyphal and yeast metabolic profiles shifted as determined by PCDA. Inspection of the factor spectra revealed an increase in peaks that could be attributed to lombazole and its metabolites. However, a decrease in a non-drug related peak at *m/z* 379 and an increase in peaks at *m/z* 407, 409 and 423, among others, was also recorded. This is highlighted in Fig. 10, in which the relevant sections of the mass spectra profiles are shown to highlight the aforementioned effect. Peak intensities in this part of the spectrum are typically less than 1% of that of the major peaks (not shown)in the spectra.

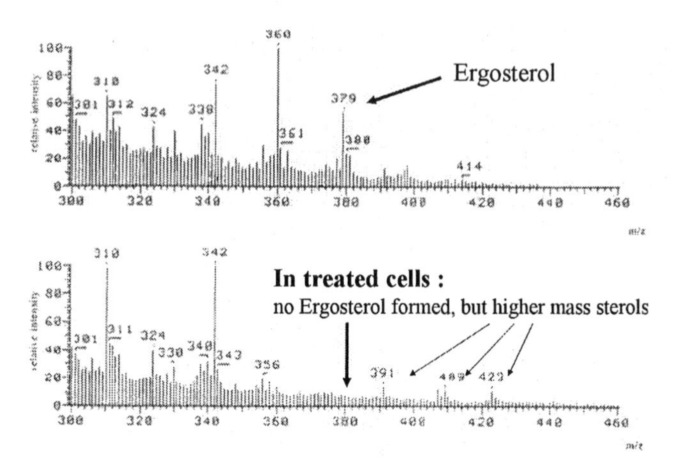

Figure 10. Inspection of the perturbed signals via pattern recognition point to perturbation of the C-14 demethylation process in sterol biosynthesis.

These results pointed to a blockage of the C-14 demethylation step in the ergosterol biosynthesis. Identification of the higher mass sterols (dehydrolanosterol, lanosterol and 24-demethylene-24,25-dihydrolanosterol) was performed by high resolution and MS/MS measurements. This type of experiment can now be performed even more efficiently by searching metabolic databases, such as the Kyoto Encyclopedia of Genes and Genomes (KEGG). Directly linking the experimental observations to known metabolic pathways can help identify the pathway perturbation points. The results of the above study revealed that lombazole selectively inhibits the cytochrome P450-dependent C-14 lanosterol α-demethylase, a key enzyme in ergosterol biosynthesis.

Metabolomics and proteomics in combination with pattern recognition before and after drug treatment can be an effective way to discover modes of action of drugs or discover minor differences in a response profile, which can ultimately be used in prioritization of lead compounds.

4.3 Drug Response Profiling

Drug response profiling using metabolomics often comprises the analysis of body fluids such as plasma or urine before and after treatment. From an analytical perspective the approach is similar to that discussed above for reverse-discovery. However, the main objective in this case is to correlate profiles with an effect such as therapeutic response. The sensitivity and applicability of metabolic profiling for such studies has been demonstrated by urine analysis by NMR of calves treated with different beta-agonists and a corticosteroid (Vogels *et al.*, 1996b). In this study the effect of treatment with prohibited growth promoters was studied, and even 24 days after treatment, effects on the endogenous metabolic profiles persisted. Differences between beta-agonists were observed in addition to an orthogonal effect between the anabolic (beta-agonists) and catabolic (corticosteroid) drug. In this study, a simple classification approach based on three nearest neighbors (k-Nearest Neighbor method) was used.

Interestingly, an important feature of these studies was the environmental effects on metabolic profiles. This is also observed in metabolomic analyses in human clinical studies. In the present case, for example, it was observed that if urine samples from calves were taken from a variety of different farms, and if this feature was not taken into account, differences in feeding conditions became dominant discriminators. The corollary for humans is typically 'lifestyle' differences, which are major discriminators between individuals and an important effect that must be considered. For the calves, prior clustering into different feed categories before classification provided satisfactory results. Such simple pre-classification is not likely to succeed with human subjects due to the lifestyle diversity typically represented in such studies. For that reason, alternative methods have been explored to compensate for lifestyle effects on metabolic profiles. One such technique is trend analysis. This approach is based on the early findings in urine profiling of rats before and after drug treatment (van der Greef and Leegwater, 1983; van der Greef *et al.*, 1984). Using a quotient spectrum from the same rat and evaluating the quotient spectra strongly increased the ability to detect drug metabolites. In effect this self-normalization routine results in metabolic profiles that are less dependent on individual biological background variations. This approach has been evaluated in a study on pre-menstrual syndrome (Tas *et al.*, 1989d) and can be further refined using more sophisticated multivariate approaches.

4.4 Disease Biomarkers

An important area in which metabolomics has great potential is in the discovery of biomarkers related to disease. In many cases useful biomarkers for efficacy or safety are lacking and hamper clinical studies.

It is becoming increasingly apparent that a single biomarker cannot provide sufficient information and specificity and that more complex biomarker profiles are often necessary. The concept of finding biomarkers can be refined by recognizing that early markers of disease will differ from markers of later stages of disease progression. This represents an opportunity for applying metabolomics to the discovery and investigation of biomarkers both from the prognostic and diagnostic perspectives, as well as new perspectives.

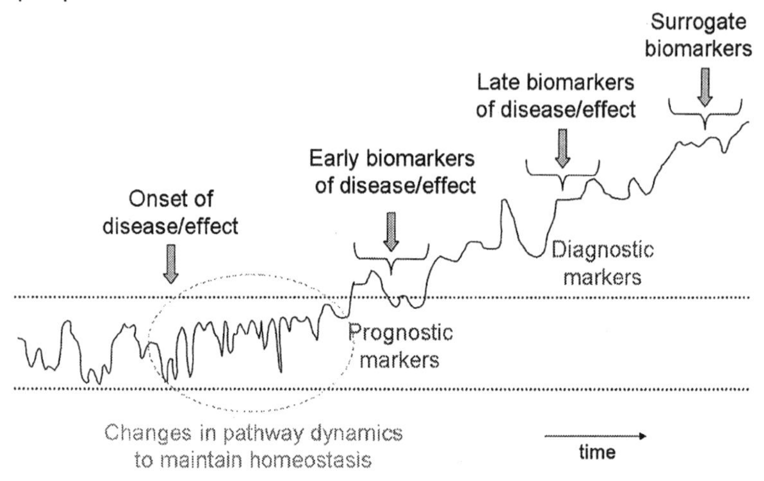

Figure 11. A schematic illustration of the behavior of a hypothetical biomarker parameter (solid line) as a function of disease progression.

Fig. 11 is a schematic representation of these concepts. Here, horizontal lines are meant to reflect the boundaries of some property of the marker under homeostasis. Initially the system resides within these limits. But after a perturbation (such as disease) the system necessarily changes its internal regulation dynamics in order to maintain homeostasis and these dynamics will be evidenced in the biomarker behavior. This will continue until the system becomes unstable and the disease further develops, as depicted by the upward trend in the figure. However, when the system loses homeostasis, other, more indirect, effects may also occur. This implies that if one starts measuring when the disease is clearly present, the chance of finding fewer specific biomarkers is diminished. Working nearer the onset of disease when the dynamics have changed but the system is still within the boundaries of homeostasis is an area where important information can be obtained not only for biomarker selection but also for target prioritization.

Biomarkers at an early stage of disease can have a high prognostic value and, although not discussed in this chapter, are key for nutritional studies with a functional food or nutraceutical focus (Watkins *et al.*, 2001). A first exploratory longitudinal study on osteoarthritis biomarkers and the effect of

Vitamin C has been finalized (Lamers *et al.*, 2002). At present research is directed on study these dynamic changes from a systems biology perspective and is closely related to the elegant work (Glass and Mackey, 1988) referred to in studies of dynamic modeling in dynamical diseases.

As an example of disease biomarker research, Fig. 12 presents results from a study of multiple sclerosis ('t Hart *et al.*, 2002). In chronic cases of multiple sclerosis a common pathological hallmark is the occurrence of demyelinated lesions of various size and stage in the CNS white matter. Direct non-invasive visualization of pathological alterations in the CNS white matter of multiple sclerosis patients can be measured by magnetic resonance imaging to observe structural alterations in the brain and spinal cord but unfortunately are not sufficiently specific to provide the requisite insight into the extent of demyelination or axonal pathology within lesions. Therefore a metabolomics study was initiated based on urine analysis of a multiple sclerosis animal model, the marmoset monkey experimental autoimmune encephalomyelitis (EAE) model, which is a suitable experimental system for efficacy and safety testing of new therapies.

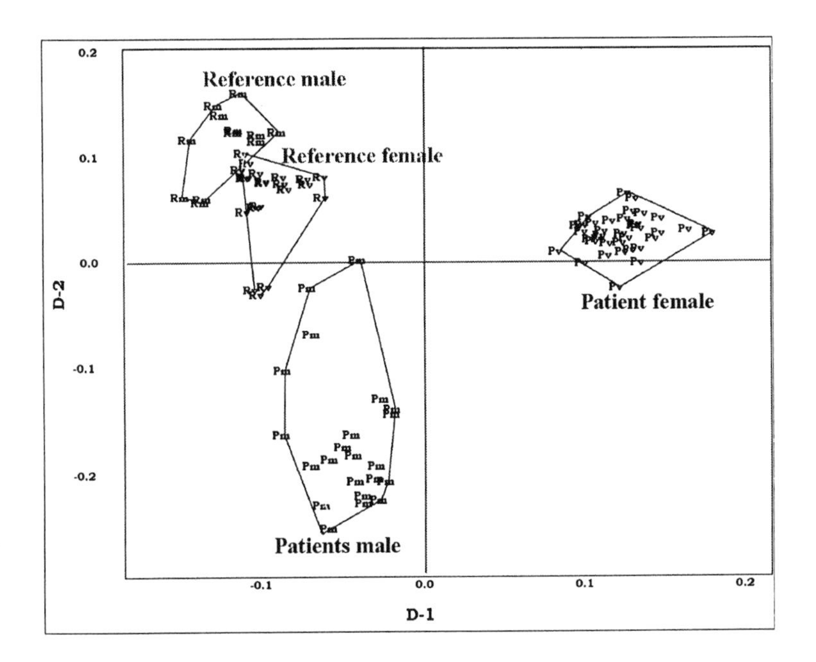

Figure 12. Gender and disease difference in multiple sclerosis biomarkers.

A corresponding study of human multiple sclerosis subjects and controls was also undertaken. The objective was to detect biomarkers for multiple sclerosis development. The introduction of new therapies for multiple sclerosis would benefit enormously from the availability of easily accessible bodyfluid biomarkers that correlate directly with the extent of white matter destruction. The biomarker profiles found were investigated and trends related to disease progression as well as other interesting biomarker patterns were examined in the pharmacological model and in patients. A striking example of an unexpected finding was the observed gender difference in multiple sclerosis biomarkers, as shown in Fig. 12. In order to begin validating observed effects it is necessary to explore whether differences are specific to multiple sclerosis or are more common or indirectly linked to neurological disease in general. For that reason 20 additional non-multiple sclerosis neurological diseases were investigated and satisfactory specificity and differentiation criteria for the biomarkers discovered were established. Interestingly, the discriminatory profiles often contained components that were common to many neurological diseases, however a pattern or set of components, carefully selected, proved sufficient to be specific for any one disease. It is important to note that while the use of metabolomics in detecting differences in disease versus control seems to offer promise, the validation of such results entails extensive follow-up research. The understanding of the pathogenesis of disease in an early stage is a primary objective of such studies.

5. METABOLOMICS AND SYSTEMS BIOLOGY

With new insights into molecular hallmarks of disease, researchers are investigating system-level approaches to reconcile the relationship between the genotype-to-phenotype gap and disease. Systems biology is a relatively new discipline that aims to understand phenotypic variation and build comprehensive models of cellular organization and function as well as elucidate the interaction and functions of cellular, organ, and even organism-wide systems. Recently, metabolomics has emerged as a unifying "functional glue" for translation of data from genomics, expression and proteomic analyses into underlying maps of metabolic networks. Functional genomics and proteomics approaches in conjunction with metabolic control analysis (Cascante *et al.*, 2002) are increasingly used to study the metabolic status of cells in an effort to understand the metabolic effects of specific perturbations at the gene and protein levels (Sanford *et al.*, 2002; Fiehn, 2000).

Similar strategies have been applied, for example, to reveal functional differences of silent phenotypes in the sugar phosphate conversion pathway in *Saccharomyces cerevisiae* (Raamsdonk *et al.*, 2001). The primary result of such measurements is a metabolic profile that reflects the system's genetic regulatory status and can be instrumental in identifying key enzymatic steps in biochemical pathways. The physicochemical parameters such as rate limiting steps, stoichiometric relationships, kinetic and control coefficients can be obtained from metabolic flux measurements and used in modeling of metabolic networks. The information gained from these procedures augments and unifies data provided by gene and protein analysis. The implication of having detailed metabolic maps of therapeutically important tissues can have a dramatic impact on the drug development process.

Recent studies have shown interesting results from partial parallel analyses of genetically perturbed organisms by correlating gene expression with protein function or physiological profiles (Ideker *et al.*, 2001; Stoll *et al.*, 2001). However, to truly take advantage of multi-tiered quantitative analysis, a platform requires an accounting of metabolic components and sophisticated statistical tools to integrate data. Beyond Genomics Inc. is one organization where an applied systems biology approach interfaces data generating technology with specialized statistical analysis and normalization tools that allow for high resolution modeling and assembly of integrated biochemical data. To demonstrate the utility of this platform, a recent study reports a targeted systems biology approach to elucidate lipoprotein metabolism disregulation in a well-established model for atherosclerosis and hyperlipidemia (Davidov *et al.,* 2002a, 2002b).

The intrinsic value of systems biology is its ability to unite individual fields of research devoted to structural, functional and dynamic aspects of biology into one powerful discipline. The deliverables for systems biology will span the spectrum from drug development and clinical trials to personalized medicine. As it matures, this nascent discipline promises to become a dominant approach in drug discovery and development by overcoming the limitations of the individual "omics" technologies.

6. FUTURE PERSPECTIVES

The metabolomics field has enormous potential in a wide area of applications in the pharmaceutical sciences. To fully explore and utilize this potential, interesting challenges remain and current technology and methodology must undergo many refinements and innovations. Analytical techniques will benefit from novel sample preparation methodologies to achieve better coverage of the metabolome. Sample volume, dynamic range, robustness, selectivity and sensitivity will be key variables. Although these are amongst the normal bioanalytical variables in experimental design, the focus and perspective is different for metabolic profiling approaches

(Harrigan, 2002). Electromigration techniques such as isotachophoretic based sample focusing or hyphenated approaches using capillary electrochromatography may contribute to these objectives in terms of concentration capabilities and different approaches to achieving selectivity (Mazereeuw *et al.*, 2000; Gucek *et al.*, 2000). The requirements for large analytical dynamic range, combined with inherent sample complexity, can be addressed with multi-dimensional techniques based on GC-GC or LC-LC with orthogonal separation qualities in each dimension. MS will become an increasingly important technique, and using methods with high quantification capabilities will become especially attractive as separation capacities are increased. For instance traditional FID or nitrogen-phosphorus detection in GC-based techniques, or ECD or laser-induced fluorescence detection in HPLC could become more widely applicable in metabolomics using multi-dimensional separation approaches for both quantitation and selectivity. Important contributions for reducing required sample volumes might result from these approaches or applying microfabrication-based electromigration driven techniques. In certain cases profiling at the single-cell level by MALDI-TOF-MS has been demonstrated for peptides in neurons and melanotropic cells (Van Veelen *et al.*, 1993, Jiménez *et al.*, 1994, Van Strien *et al.*, 1996) demonstrating that cell processes can be followed using such approaches, albeit with a limited dynamic range. In certain cases individual cell profiles provide information not obtainable by measuring the average of large cell populations.

Developments in instrumentation at the detector end will continue and next-generation NMR and mass spectrometers will provide improvements in throughput, sensitivity and identification. Such improvements include novel ion mobility MS technologies currently being developed for small molecule and peptide analysis (Barnes and Clemmer, 2001). Metabolite identification needs to be further improved at low concentrations and novel high sensitivity tools for MS/MS based on, for instance, linear traps or high resolution capabilities in analytical FT-MS or TOF-based hybrid instrumentation.

The data pre-processing domain will remain key in further optimization of metabolomic techniques, with development continuing on normalization, spectral adjustment, 2-D image analysis and peak detection and integration algorithms. Furthermore, multivariate approaches continue to be applied to metabolomic data. New algorithms are needed to correlate and integrate bioimaging results with physiological metabolomics data. In particular, an emphasis on non-linear clustering methods will increase the level of understanding of biological systems and will be important for integrating metabolomics into a systems biology framework (Davidov *et al.*, 2002a, 2002b).

In addition, the study of temporal aspects of systems together with non-linear techniques will put high demands on molecular quantification, but will also enable the study of systems in their most relevant context. The analysis of temporal data will be a major focus, and the use of time-series, time warping, multi-block, multi-way analysis and other informatics approaches will become significant.

However biology remains an integrated puzzle—a "biopuzzle"—each piece equally important, unique and necessary to create the whole. From that understanding, systems biology is anticipated to have an enormous impact on our knowledge of human biology and the life sciences, with metabolomics as an essential discipline within this framework.

REFERENCES

Barnes CAS, Clemmer DE. Assessment of purity and screening of peptide libraries by nested ion mobility-TOFMS: identification of RNase S-protein binders *Anal Chem* 73: 424-433 (2001).

Cascante M, Boros LG, Comin-Anduix B *et al.* Metabolic control analysis in drug discovery and disease. *Nature Biotechnol* 20: 243-249 (2002).

Davidov E, Clish CB, Meyes M *et al.* Systems biology approach: parallel analysis of the ApoE3-Leiden transgenic mouse model. *Nature Biotechnol* submitted (2002a).

Davidov E, Marple EW, Naylor S. Advancing drug discovery and development through systems biology. *Drug Discov Today* submitted (2002b).

Dietel P, Spiteller G. Changes in the excretion of organic acids in human urine after physical exertion. *J Chromatogr* 378: 1-8 (1986).

Droge JBM, Rinsma WJ, van 't Klooster HA *et al.* An evaluation of SIMCA: Part 2-Classification of pyrolysis mass spectra of *Pseudomonas* and *Serratia* bacteria by pattern recognition using the SIMCA classifier. *J Chemometrics* 1: 231-241 (1987).

Fiehn O, Kopka J, Dormann P *et al.* Metabolite profiling for plant functional genomics. *Nature Biotechnol* 18: 1157-1161 (2000).

Gaspari M, Vogels J, Wulfert F *et al.* Novel strategies in mass spectrometric data handling. In *Advances in Mass Spectrometry.* Gelpi E (Ed) pp. 283-296, John Wiley and Sons, Chichester (2001).

Geladi P, Kowalski BR. An example of 2-block predictive partial least-squares regression with simulated data. *Anal Chim Acta* 185: 1-17 (1986).

Geng M, Ji J, Regnier, FE. Signature-peptide approach to detecting proteins in complex mixtures. *J Chromatogr* 870: 295-313 (2000).

Glass L, Mackey MC. *From Clocks to Chaos: The Rhythms of Life.* Princeton University Press, New Jersey (1988).

Gucek M, Gaspari M, Walhagen K *et al.* Capillary electrochromatography/nanoelectrospray mass spectrometry for attomole characterization of peptides. *Rapid Comm Mass Spectrom* 14: 1448-1454 (2000).

Gygi SP, Rist B, Gerber SA *et al.* Quantitative analysis of complex protein mixtures using isotope-coded affinity tags. *Nature Biotechnol* 17: 994-999 (1999).

Harrigan GG. Metabolic profiling: pathways in drug discovery. *Drug Discov Today* 7: 351-352 (2002).

Hoogerbrugge R, Willig SJ, Kistemaker PG. Discriminant analysis by double stage principal component analysis. *Anal. Chem* 55: 1710-1712 (1983).
Heindl P, Dietel P, Spiteller G. Distinction between urinary acids originating from nutrition and those produced in the human body. *J Chromatogr* 377: 3-14 (1986).
Holmes E, Nicholson JK, Nicholls AW *et al.* The identification of novel biomarkers of renal toxicity using automatic data reduction techniques and PCA of proton NMR spectra of urine. *Chemom Intel Lab Sys* 44: 245-255 (1998).
Hotelling, H. Analysis of a complex of statistical variables into principal components. *J Ed Psychol* 24: 417-441, 498-520 (1933).
Ideker T, Thorsson V, Ranish JA *et al.* Integrated genomic and proteomic analyses of a systematically perturbed metabolic network. *Science* 292: 929-934 (2001).
International Human Genome Sequencing Consortium. Initial sequencing and analysis of the human genome. *Nature* 409: 860-921 (2001).
Jellum E. Profiling of human body fluids in healthy and disease states using gas chromatography and mass spectrometry, with special reference to organic acids. *J Chromatogr* 143: 427-462 (1977).
Jiménez CR, van Veelen PA, Li KW *et al.* Neuropeptide expression and processing as revealed by direct matrix-assisted laser desorption ionization mass spectrometry of single neurons. *J Neurochem* 62: 404-407 (1994).
Kitano H. Systems biology: a brief overview. *Science* 295: 1662-1664 (2002).
Lamers RAN, Faber EJ, Jellema RH *et al.* Metabolic fingerprinting: identification of disease related biomarkers: a pilot study of osteoarthritis and vitamin C. *J Nutr* submitted (2002)
Lindon JC, Nicholson JK, Everett JR. NMR spectroscopy of biofluids. *Ann Rep NMR Spectr* 38:1-88 (1999).
Lindon JC, Nicholson JK, Holmes E, Everett JR. Metabonomics: metabolic processes studied by NMR spectroscopy of biofluids. *Concepts Magn Reson* 12:289-320 (2000).
Lindon JC, Holmes E, Nicholson JK. Pattern recognition methods and applications in biomedical magnetic resonance. *Prog Nuc Magn Reson Spectr* 39:1-40 (2001).
Mazereeuw M, Spikmans MV, Tjaden UR, van der Greef J. On-line isotachophoretic sample focusing for loadability enhancement in capillary electrochromatography-mass spectrometry. *J Chromatogr* 879: 219-233 (2000).
McLuckey SA, Wells JM. Mass analysis at the advent of the 21st century. *Chem Rev* 101: 571-606 (2001).
Nicholson JK, Lindon JC, Holmes E. 'Metabonomics': understanding the metabolic responses of living systems to pathophysiological stimuli via multivariate statistical analysis of biological NMR spectroscopic data. *Xenobiotica* 29: 1181-1189 (1999).
Nierop AFM, Tas AC, van der Greef J. Reflected discriminant analysis. *Chemom Intel Lab Sys* 25: 249-263 (1994).
Raamsdonk LM, Teusink B, Broadhurst D *et al.* A functional genomics strategy that uses metabolome data to reveal the phenotype of silent mutations. *Nature Biotechnol* 19: 45-50 (2001).
Regnier FE, Riggs L, Zhang R *et al.* Comparative proteomics based on stable isotope labeling and affinity selection. *J Mass Spectrom* 37: 133-45 (2002).
Sanford K, Soucaille P, Whited G, Chotani G. Genomics to fluxomics and physiomics - pathway engineering. *Curr Opin Microbiol* 5: 318-322 (2002).
Spiteller G. Kombination chromatographisher trennmethoden mit der massaspectrometrie – ein moderned verfahren zur stoffwechseeluntersuchung. *Angew Chem* 97: 461-476 (1985).
Spiteller G. Linoleic acid peroxidation – the dominant peroxidation process in low density lipoprotein- and its relationship to chronic diseases. *Chem Phys Lipids* 95: 105-162 (1988).

Spiteller G. Investigation of aldehyde lipid peroxidation products by gas chromatography – mass spectrometry. *J Chromatogr* 843: 29-98 (1999).

Stanbury JB, Wyngaarden JB, Frederickson DS *et al. The Metabolic Basis of Inherited Disease.* 5th Edn. McGraw-Hill, New York (1983).

Stoll M, Cowley AW, Tonellato PJ *et al.* A genomic-systems biology map for cardiovascular function. *Science* 294: 1723-1726 (2001).

't Hart, BA, Vogels JTWE, Gerwin Spijksma G *et al.* ^{1}H-NMR spectroscopy combined with pattern recognition analysis reveals characteristic chemical patterns in urines of MS patients and non-human primates with MS-like disease. *J Neurol Sci* submitted (2002).

Tas AC, van der Greef J, de Waart J *et al.* Comparison of direct chemical ionization and direct probe electron impact/chemical ionization pyrolysis for characterization of *Pseudomonas* and *Serratia* bacteria. *J Anal Appl Pyrolysis* 7: 249-255 (1985).

Tas AC, de Waart J, Bouwman J *et al.* Rapid characterization of *Salmonella* strains with direct chemical ionization pyrolysis. *J Anal Appl Pyrol* 11: 329-340 (1987).

Tas AC, Bastiaanse HB, van der Greef J, Kerkenaar A. Pyrolysis-direct chemical ionization mass spectrometry of the dimorphic fungus *Candida albicans* and the pleomorphic fungus *Ophiostoma ulmi. J Anal Appl Pyrol* 14: 309-321 (1989a).

Tas AC, ten Noever de Brauw MC, van der Greef J, Wieten G. Multivariate relations between data sets: canonical correlation of mass spectral and chromatographic results. In *Advances in Mass Spectrometry*. Vol. 11. Longevialle P (Ed) pp. 1146-1147, Heyden and Son, London (1989b).

Tas AC, Odink J, van der Greef J *et al.* Characterization of virus infected cell cultures by pyrolysis/direct chemical ionization mass spectrometry. *Biomed Environ Mass Spectrom* 18: 757-760 (1989c).

Tas AC, van den Berg H, Odink J *et al.* Direct chemical ionization - mass spectrometric profiling in premenstrual syndrome. *J Pharm Biomed Anal* 7: 1239-1247 (1989d).

Tas AC, van der Greef J. Pyrolysis: mass spectrometry under soft ionization conditions. *Trends Anal Chem* 12: 60-66 (1993).

Tas AC, van der Greef J. Mass spectrometric profiling and pattern recognition. *Mass Spectrom Rev* 13: 155-181 (1995).

Tsai H. Separation methods used in the determination of choline and acetylcholine. *J Chromatogr* 747: 111-122 (2000).

Valentine SJ, Kulchania M, Barnes CAS, Clemmer DE. Multidimensional separations of complex peptide mixtures: a combined high-performance liquid chromatography/ion mobility/time-of-flight mass spectrometry approach. *Int J Mass Spectrom* 212: 97-109 (2001).

van der Greef J, Tas AC, Bouwman J *et al.* Evaluation of field desorption and fast atom bombardment mass spectrometric profiles by pattern recognition techniques. *Anal Chim Acta* 150: 45-52 (1983).

van der Greef J, Leegwater D. Urine profile analysis by field desorption mass spectrometry, a technique for detecting metabolites of xenobiotics. *Biomed Mass Spectrom* 10: 1-14 (1983).

van der Greef J, Bouwman J, Odink J *et al.* Evaluation of field desorption mass spectrometric profiles by quotient weighting. *Biomed Mass Spectrom* 11: 535-538 (1984).

van der Greef J. Field desorption mass spectrometry in bioanalysis. *Trends Anal Chem* 5: 241- 246 (1986).

van der Greef J, Tas AC, Bouwman J, ten Noever de Brauw MC. Pattern recognition of complex matrix profiles generated by soft ionization methods. *Adv Mass Spectrom* 10: 1227-1228 (1986).

van der Greef J, Tas AC, ten Noever de Brauw MC. Direct chemical ionization-pattern recognition: characterization of bacteria and body fluid profiling. *Biomed Environ Mass Spectrom* 16: 45-50 (1988a).

van der Greef J, de Waart J, Tas AC. Characterization of algae by pyrolysis-direct chemical ionization mass spectrometry. In *COST 48: Aquatic Primary Biomass- Marine Macroalgae. Proc 2nd Workshop of the COST 48 Subgroup 3: Biomass Conversion Removal and Use of Nutrients*. de Waart J, Nienhuis PH (Ed) pp. 34-49, TNO-CIVO Zeist, DIHO Yerseke (1988b).

van Strien FJC, Jespersen S, van der Greef J *et al.* Identification of POMC processing products in single melanotrope cells by matrix-assisted laser desorption/ionization mass spectrometry. *FEBS Lett* 379: 165-170 (1996).

van Veelen PA, Jiménez CR, Li KW *et al.* Direct peptide profiling of single neurons by matrix-assisted laser desorption-ionization mass spectrometry. *Org Mass Spectrom* 28: 1542-1546 (1993).

Venter JC, Adams MD, Myers EW *et al.* The sequence of the human genome. *Science* 291: 1304-1351 (2001).

Verner, J. Large-scale prediction of phenotype: concept. *Biotech Bioeng* 69: 664-678 (2000).

Vogels JTWE, Tas AC, van der Greef J. Canonical correlation of proton nuclear magnetic resonance and pyrolyis-direct chemical ionization mass spectroscopic data used in the authentication of wines. In *Trends in Flavour Research.* Maarse H, van der Heij DG (Ed) pp. 99-106, Elsevier, Amsterdam (1994).

Vogels JTWE, Tas AC, Venekamp J, van der Greef J. Partial linear fit: a new NMR spectroscopy preprocessing tool for pattern recognition applications. *J Chemometrics* 10: 425-438 (1996a).

Vogels JTWE, Arts CJM, Tas AC *et al.* Application of proton nuclear magnetic resonance spectroscopy and multivariate analysis as an indirect screening method for monitoring the illegal use of growth promoters. In *EuroResidue III: Conference on Residues of Veterinary Drugs in Food.* Haagsma N, Ruiter A (Ed) pp 968-972, University of Utrecht, Veldhoven (1996b).

Vogels JTWE, Terwel L, Tas AC *et al.* Detection of adulteration in orange juices by a new screening method using proton NMR spectroscopy in combination with pattern recognition techniques. *J Agric Food Chem* 44: 175-180 (1996c).

Watkins SM. Comprehensive lipid analysis: a powerful metanomic tool for predictive and diagnostic medicine. *Israel Med Assoc J* 2: 722-724 (2000).

Watkins SM, Hammock BD, Newman JW, German JB. Individual metabolism should guide agriculture toward foods for improved health and nutrition. *Am J Clin Nutr* 74: 283-286 (2001).

Winding W, Haverkamp J, Kistemaker PG. Interpretation of sets of pyrolysis mass spectra by discriminant analysis and graphical rotation. *Anal Chem* 55: 81-88 (1983).

Windig W, Phalp JM, Payne AW. A noise and background reduction method for component detection in liquid chromatography mass spectrometry. *Anal Chem* 68: 3602-3606 (1996).

Wold, SJ. Pattern recognition by means of disjoint principal components models. *J Pattern Recogn* 8: 127-139 (1976).

Chapter 11

USE OF METABOLOMICS TO DISCOVER METABOLIC PATTERNS ASSOCIATED WITH HUMAN DISEASES

Oliver Fiehn[1] and Joachim Spranger[2]
[1]Max-Planck-Institute of Molecular Plant Physiology, 14424 Potsdam/Golm, Germany [2]German Institute of Human Nutrition (DifE), 14558 Bergholz-Rehbrücke, Germany

1. INTRODUCTION

Functional genomics has been a major emphasis in recent years as researchers attempt to unravel the structure and function of complex biological mechanisms. These functional genomic strategies aim to monitor all gene products simultaneously in order to establish a more comprehensive overview of disease mechanisms and to find suitable therapeutic targets. The analysis of primary gene products has also been considered as a diagnostic and screening tool for disease recognition. There are several test cases demonstrating the validity of such approaches. However, there are also severe practical and theoretical constraints in applying mRNA or protein profiling as universal tools for an improved understanding and diagnosis of disease patterns. The paradigm of linear control from gene expression to transcription and translation to metabolic phenotypes has been challenged by multiple experimental observations. These include low or missing correlations between mRNA and protein abundances, fluctuations in RNA or protein turnover rates, and the complexity of protein interaction networks. Furthermore both transcript arrays and proteomics are associated with high sample costs, limiting the number of analyzed biological replicates to the point that valid statistical evaluations cannot be carried out. It has been estimated that even for isogenic mouse lines, biological variability requires at least 15 replicates per genotype tested. Only in rare cases can research budgets allow such high replicate numbers. Researchers try to circumvent the likelihood of false positive findings by setting high thresholds for differential gene expression to greater than 5- or even 10-fold. By doing so, the balance reverts to an increased probability of false negative findings *i.e.* genes that were, in fact, differentially expressed but only at values less than threshold settings.

At this point, low cost metabolomic assessments may be considered. Metabolism may be regarded as the result of all regulatory steps that respond to external stimuli and perturbations including the onset or progression of disease. Small differences in metabolic flux rates may result in amplification of metabolite contents through the pathway network. Although disease mechanisms and gene functions are unlikely to be pinned down by metabolite profiles alone, this amplification effect can be used for screening and diagnostic purposes. Furthermore, metabolomics including the related metabolic fingerprinting approaches has costs two to three orders of magnitude lower than transcriptomics or proteomics. This chapter will therefore focus on potential implications of metabolomics as a tool to identify novel metabolic patterns or markers associated with disease status. We will illustrate the potential of this approach using the association between metabolic profiles and the development of type 2 diabetes.

2. GROWING IMPACT OF TYPE 2 DIABETES MELLITUS ON HUMAN HEALTH

Worldwide, type 2 diabetes mellitus (T2DM) is among the major diseases caused by human malnutrition (Warram *et al.*, 1997). High socio-economic burdens arise from T2DM and other nutritional diseases, yet disease manifestation can be prevented by timely therapeutic intervention. Unfortunately, risk assessments in healthy humans are rare, mostly due to a lack of adequate assays with sufficient predictive value. Lifestyle factors, as well as genetic factors, are also major contributors to the development of diabetes (Hu *et al.*, 2001). If diet or alterations in lifestyle such as increased physical activity are included in intervention therapies, over half of all T2DM manifestations can be avoided (Tuomilehto *et al.*, 2001, Knowler *et al.*, 2002). Cross-sectional as well as therapeutic studies have demonstrated the efficacy of such therapies to prevent the onset of T2DM in individuals with impaired glucose tolerance.

It is therefore crucial to develop accurate tools to predict the risk of individuals to T2DM. Commonly used predictors of T2DM include biochemical markers (cholesterol, triglycerides, LDL and HDL, and lipoproteins) and anthropometric parameters (body-mass index, waist-hip ratio). However, the performance of theses risk factors is rather poor. Only 10% of individuals that later develop T2DM can be identified when these parameters are taken together. Correspondingly, 90% of the people who are classified at risk by these known biomarkers will not develop diabetes (Report of the Expert Committee, 1997). Similar results have been found for other metabolism-related diseases and it must be concluded that classical biomarkers routinely fail to predict risks. These biomarkers are too prone to yield high numbers of both false positive and false negative results. Therefore, additional diagnostic strategies are of the utmost clinical importance. It makes perfect sense to utilize metabolomics in diagnoses of metabolic diseases and, despite the concerns illustrated above, metabolites still remain among the best risk predictors.

The pathogenesis of many complex diseases is considerably influenced by environmental factors as has been shown for cancer, coronary heart disease and T2DM. The prevalence of T2DM is rapidly increasing with the adoption of industrial (Western) lifestyles. The lowest levels of T2DM are still found in rural areas where indigenous people still follow their traditional lifestyles (Amos *et al.*, 1997). However, indigenous communities show a large increase in the number of T2DM manifestations on changing to typical Western diets as illustrated by the Pima Indians in Arizona, Micronesians in Nauru and Aborigines in Australia (Bennett, 1999). In the year 2025, 300 million adults will suffer from diabetes worldwide, with a prevalence of T2DM of 5.4%. At this time, the majority of diabetes cases will be observed in developing countries, with India and China having more cases than any other country (King *et al.*, 1995). This dramatic rise in T2DM and accompanying complications such as end-stage renal impairment and heart disease will impart severe socio-economic consequences in these countries

Diet and nutrition are broadly accepted as major factors in the onset and pathogenesis of T2DM. However, specific dietary factors have never been delineated. Although numerous epidemiological studies have shown that a diet rich in vegetables and fruits has significant protective effects (Hu *et al.*, 2001), the exact role of fat and carbohydrates with respect to diabetic risk remains hotly debated. Little is known about the potential benefits of other phytochemicals such as phenolic antioxidants.

Low-fat and high-carbohydrate diets are often recommended for the prevention of major nutritional diseases such as diabetes, coronary heart disease and other chronic diseases. However, 'fat' and 'carbohydrate' are poorly defined and very heterogenous compound classes. It is increasingly evident that different types of fats and different types of carbohydrates have different effects on insulin sensitivity, insulin secretion, and glucose homeostasis. Free fatty acids have long been associated with the development of insulin resistance by competing with glucose oxidation (Randle hypothesis, Randle *et al.*, 1965), an observation extensively researched over the past few decades (Boden, 1997). For saturated as well as monounsaturated and polyunsaturated fatty acids (with the exception of n-3 fats) a causal relationship to insulin resistance has been shown in animal models (Storlien *et al.*, 1996). More specifically, high saturated fat intakes are associated with increased risk of glucose intolerance, elevated fasting glucose and a boost in insulin concentrations in epidemiological studies (Hu *et al.*, 2001b). Moreover, increases in fasting insulin levels, lowered insulin sensitivity and higher risk of T2DM manifestation have also been shown to be associated with higher proportions of saturated fatty acids in serum lipids and muscle phospholipids (Vessby *et al.*, 1994, Folsom *et al.*, 1996). In humans, higher proportions of long-chain polyunsaturated fatty acids in skeletal muscle phospholipids have been related to better insulin sensitivity in humans (Borkmann *et al.*, 1993). The role of monounsaturated fatty acids is still unclear, although some reports suggest high levels of monounsaturated fatty acids may be harmful (Feskens *et al.*, 1995). Thus, insulin sensitivity in healthy individuals was shown to increase during a three-month dietary intervention in which monounsaturated fatty acids were substituted. Furthermore, intake and composition of fats both contributed to this beneficial effect: improved insulin sensitivity by substitution of monounsaturated fatty acids was found only for subjects who were consuming less than 37% of energy as fat (Vessby *et al.*, 2001). However, there is still insufficient data to give accurate recommendations on healthy fat compositions, since the amount and composition of carbohydrates, polyphenolic antioxidants, micronutrients and physical activity also affect the pathogenesis of complex diseases such as T2DM. Accordingly, nutritional research should focus on the contribution and relative importance of different dietary and lifestyle factors simultaneously, in order to avoid over-emphasis on the role of any single nutrient. It may well turn out that the onset of T2DM is more likely to follow a pattern of dietary components in relation to other factors such as health status, physical activity, age and sex. For such broad overviews of metabolic health status, profiling techniques may prove very helpful for screening metabolic changes in response to specific diets.

It is obvious that intervention is highly advantageous and much more successful when started prior to diabetes manifestation. But how can we accurately predict the risk of healthy persons to T2DM? As mentioned earlier, screening programs relying on known risk factors result in a high number of individuals falsely identified to be at high risk. Correspondingly, such programs are highly expensive with cost estimations ranging between $4000-8000 per new case of diabetes identified. If screening programs were more accurate, the high socio-economic burdens caused by T2DM and its related metabolic diseases could be minimized. The American Diabetes Association calculated that, in 1997, diabetes cost the US economy $27 billion in direct medical expenses and $32 billion in lost productivity. Several reports have been published demonstrating that aggressive anti-diabetic therapies in high-risk cohorts can delay or avoid complications and that such pre-manifestation interventions can reduce indirect health-related costs by up to $9000 per person per annum.

Rapid and unbiased techniques screening for metabolic changes could thus improve the correct identification of high-risk persons and thereby lower T2DM related costs. Moreover, once metabolic patterns can efficiently characterize T2DM risk, costs and efficiencies can be compared to target analyses based on only a few metabolite biomarkers. Finally, comprehensive and unbiased analysis of the metabolome may provide insights to molecular and pathogenetic mechanisms in T2DM.

3. PROFILING TECHNIQUES

What are the most suitable techniques for unraveling complex associations between disease phenotypes and the underlying high biological variation in typical cohorts? RNA microarrays are increasingly used to find differences in complex phenotypes by clustering methods. Specific RNA clusters have been associated with the progression of breast cancer (van't Veer *et al.*, 2002) thus enabling risk predictions of individuals for specific disease endpoints. However, transcriptomics and proteomics are still too costly for large-scale screening of healthy populations. For complex disorders, metabolomic analyses could easily become the preferred tool for diagnostics and prediction. In metabolic diseases, including obesity, metabolites are central to understanding pathogenesis. Recently, a new anorexia-causing lipid, whose activity is dependent on feeding patterns, has been reported (de Fonseca *et al.*, 2001).

Profiling of metabolic diseases is, of course, not a novel technique to clinicians (Tanaka *et al.*, 1980a, 1980b). For decades, inborn errors of metabolism have been efficiently assayed in neonatal screening and such approaches have now extended to screening for many other types of metabolic disease (Jellum *et al.*, 1988). However, despite successful applications in acidurias (Kimura *et al.*, 1999; Halket, 1999), cervical cancer (Kim *et al.*, 1998), or mitochondrial myopathy (Ning *et al.*, 1996), comprehensive analysis of metabolite patterns is still not a routine diagnostic tool. Instead, analytical chemists have expanded their technologies into such diverse areas of functional genomics, plant studies and bacterial characterizations. Four different directions for metabolite analysis can thus be identified and broadly classified as follows:

Classical hypothesis-driven research still uses *metabolite target analysis*. If isotope labeled reference compounds are applied, quantification and identification of biomarkers can be unrivalled. However, interpretations tend to over-simplify research issues since causal effects cannot readily be distinguished from simple associations or side-effect correlations.

This concern led to the idea of assessing biochemical reactions in a broader way. By *metabolite profiling*, select pathways or compound classes are studied simultaneously by defining dozens to hundreds of target molecules. Analyte identification is usually assured by a combination of compound spectra and chromatographic retention indices. However accurate quantification is compromised by the large number of target compounds.

Both target analysis and metabolite profiling focus on pre-defined compounds. Yet the complexity and flexibility of biochemical networks ensures that unpredicted metabolic events will ensue upon disease development. Therefore, *metabolomic analysis* tries to detect and quantitate all metabolic signals in a truly unbiased way, irrespective of the size, volatility, or other physicochemical parameters of the analytes. It is thus an extension of metabolite profiling in the sense that known and unknown metabolites are considered equally important. Metabolomics is challenged by huge differences in molecule structures and abundances, and it poses high demands on software capabilities of the analytical instruments as well as on database structure and implementation. Many recorded signals can remain undetected in preliminary evaluations, but can be later identified once that signal proves to be consistently important in biological or clinical studies.

If the aim is classification of samples rather than comparing detailed biochemical pathways, *metabolic fingerprinting* may be advantageous. Fingerprinting refuses to distinguish all metabolites. Instead, physical spectra of crude mixtures are acquired without use of chromatography. This allows the analysis of multiple samples in short time. Afterwards, all spectra are investigated by supervised or unsupervised learning methods in order to cluster samples according to origin. In some instances, signal abundance or spectra characteristics may be sufficient to point to particular metabolites. However, precise quantitation and identification of single metabolite components is not a primary objective in fingerprinting approaches.

Metabolite fingerprinting has utilized various physical techniques and has been applied in a variety of different research fields. Most commonly (at least in toxicological applications), proton nuclear magnetic resonance (^{1}H NMR) is used (Gavaghan *et al.*, 2000; Raamsdonk *et al.*, 2001). However even Fourier transform infrared spectroscopy (FT-IR, Goodacre *et al.*, 2000) has provided sufficient information to *e.g.* distinguish salt stressed tomato fruits from unstressed ones and information derived from that spectra to identify candidate biomarkers (Johnson *et al.*, 2000). More promising might be direct infusion MS. Efficiency of compound ionization and adduct formation in liquid MS (and other techniques such as matrix assisted laser desorption/ionization (MALDI)) inevitably depends on matrix effects within the crude mixture itself. Subtle differences in total composition might directly lead to improved or reduced ionization of sample components that therefore give altered intensities in the corresponding mass spectra. Whereas any effort to derive information upon changes in biochemical pathways would remain of dubious value, even if high resolution Fourier transform MS were to be used, this amplification of matrix differences is advantageous for classification purposes. Recently Vaidyanathan *et al.* (2001) could show that direct infusion-MS spectra were sufficient to rapidly detect and discriminate between bacteria strains at competitive costs and high speed. Whereas this application aimed at consumer safety, it can easily be envisioned that other areas could benefit from this approach, including the screening of large cohorts of healthy individuals for primary assessments of metabolic distances and risk stratifications. Individuals identified at risk could then be subjected to further analysis with more detailed metabolomic techniques to validate the classification of the initial screen and to attempt to pin down changes in particular biochemical pathways. As defined above, metabolomics aims at the simultaneous detection and quantitation of all individual metabolites in a particular biological tissue, as further explained in recent reviews (Fell, 2001; Fiehn, 2002). To date, metabolomic approaches predominantly use MS techniques due to its inherent selectivity, universality, and sensitivity. The major problem to date is in the unambiguous deconvolution and annotation of severely overlapping peaks in multiple runs. Therefore, even gas chromatography-mass spectrometry (GC-MS) still remains primarily a metabolite profiling technique (Fiehn *et al.*, 2000a). Improved deconvolution algorithms and faster spectral acquisition by time-of-flight (TOF) measurements (Shellie *et al.*, 2001) have, however, resulted in the detection of over 1,000 components from plant leaf extracts at a throughput of over 1,000 samples per month (Weckwerth *et al.*, 2001). In a test case, over 400 metabolites analyzed by GC-TOF showed average analytical errors of less than 10% relative standard deviation even when low abundance compounds were included (which typically account for the largest deviations, Fiehn 2002, unpublished results). This data proves that GC-TOF is the 'gold standard' for high throughput metabolomics. However, any GC-based analytical method will suffer from inherent bias against large and thermolabile compounds. If the aim is to elucidate diabetic biomarkers and mechanisms in an unbiased way, methods based on liquid chromatography are required for quantifying neutral and polar lipids, glycosylated or glucuronidated compounds, antioxidants or co-factors. Numerous methods, based on reversed-phase chromatography, do exist that couple LC to MS and tandem MS, particularly for metabolites such as phenolics (Justesen *et al.*, 1998). LC-tandem MS may also be extended to highly polar and large metabolites by applying a variant of normal phase, hydrophilic interaction chromatography (Tolstikov and Fiehn, 2002). However, even recent software releases from well-established LC-MS manufacturers lack the capability to reliably find and quantitate metabolites in multiple chromatographic runs by deconvolution, which today is standard in GC-based methods (Stein, 1999). This hampers direct application of LC-MS methods for metabolomic screening: it is not sufficient to detect 1,000 peaks. The ability to detect and distinguish these 1,000 peaks from other potentially novel peaks in multiple runs must be assured with high precision.

As alternative to LC-MS, LC-coulometry has been applied to disease recognition (Vigneau-Callahan *et al.*, 2001). Coulometry enables detection of all compounds that contain oxidizable or reducable moieties. Instead of sweeping the applied voltage across a redox potential, an array of electrodes with fixed potentials can be used. A 16-channel LC-coulometry array run could detect over 1,000 peaks in human plasma, demonstrating that, in principle, this technology is sufficiently advanced to be used as complement to MS metabolomic approaches (Vigneau-Callahan *et al.*, 2001). However, as the authors admit, peak finding routines were barely adequate for routine operation since only 25% of all peaks were found in more than 5 of 8 comparisons of chromatographic runs from sample pools. To date, no further metabolomic strategy has been developed that aims at recognition of disease progression, effect of therapies or drug target discovery. However, such approaches have been discussed in a few recent conferences, and the number of publications and patents is expected to increase in the next few years.

With any MS, UV, fluorescence, or coulometry-based metabolomic technique, a large proportion of detected signals will not be assigned to an exact chemical structure. In part, this is because no comprehensive metabolite spectral libraries exist in the public domain. Furthermore, the complexity of plant and animal biochemistry is still underestimated due to decades of hypothesis-driven approaches. However there is a considerable literature that clearly demonstrates that unidentified metabolites play a role in classifying samples and are often central in biochemical correlation networks. Therefore, *de novo* identification of unknown metabolites is a necessity in truly metabolomic approaches. In GC-MS, identification is often hampered by the lack of abundant molecular ions. This may be circumvented by using semi-exact masses and isotope ratios after modified derivatization schemes in order to calculate elemental compositions (Fiehn *et al.*, 2000b). In LC-MS, identification is less error prone and easier to achieve. Starting from chromatographically separated molecular ions, ion trap MS can be used to elucidate fragmentation pathways (Drexler *et al.*, 1998). This can be aided by quadrupole-TOF hybrid instruments that deliver exact masses (Blom, 2001) for spectral interpretation software. Although in many cases suitable hits can be produced if appropriate chemical and biochemical databases are interrogated, for many metabolites even this information may prove sufficient. Metabolites have large degrees of freedom not only by the position and sequence of atoms, but also to form isobaric and isomeric enantiomers and diastereomers. This level of detail eventually requires use of one or two-dimensional NMR for structure elucidation as has been successfully demonstrated for phytochemicals (Pauli, 2000).

4. METHOD VALIDATION FOR DISEASE RECOGNITION

Metabolomic as well as fingerprinting methods have great potential in health-related research fields to serve as rapid, reliable, sensitive, and cost-effective methods for disease risk stratifications. However, if applied to large cohorts or as a routine clinical step, the need of method validation cannot be underestimated. Specifically, all steps from sampling human or animal tissues or plasma samples to final data acquisition need to be carefully investigated in order to reduce sources of error. Such errors would contribute to the large variability in mammalian samples resulting in a decreased likelihood of detecting dietary, genetic, developmental or behavioral disease factors. Biological variability requires data from a large collection of metabolic snapshots, as has been pointed out elaborately (Kell and Mendes, 2000). Consequently, the large number of analyte samples again calls for highly robust and automated throughput technologies that include automated peak annotation, data export and database setup prior to statistical evaluations. Such rigorous validations must be carried out even in academic laboratories (Krull and Swartz, 1999) if results are expected to reach beyond low-impact research journals. For disease recognition in population screens, robustness and repeatability have clear priority over other aspects of validation, *i.e.* universality, sensitivity, selectivity or comprehensiveness. Robustness needs to be tested by as many alterations to existing protocols as possible, since it cannot be assumed that clinical or laboratory co-workers would strictly follow protocol details in all cases. Steps that could lead to gross errors should be identified and, if possible, automated. For example, solvent composition as well as extraction times and temperatures have huge impact on peak intensities in metabolite profiles. Therefore, validation studies focusing on method robustness must include both small and large variation of protocols, especially if protocols are adopted from other biological disciplines such as plant biology or cell cultures. For example, within-run and between-run repeatability must be tested by comparing aliquots of pooled master samples, including multiple injections and precision tests under varying instrument conditions or variation of derivatization conditions. Methods should be critically assessed whether or not they enable distinguishing isomers, *e.g.* unsaturated fatty acids or monosaccharides. Even if method validation is only aiming at relative quantitation or fingerprinting, sensitivity and selectivity should be assessed (and documented) at least once by spiking internal references, *i.e.* stable isotope labeled standard compounds. Finally, it is of the utmost importance for high throughput operations that rigid quality control routines are implemented, including the establishment of quality control charts. Intervention and abort criteria must be developed at all steps of method control, specifically at the level of tolerable and intolerable result output of standard samples and instrumental conditions.

Once these basic steps have been implemented, clinical proof-of-concept studies with known outcome may be carried out, for example, on diseased-healthy comparisons. If successful, the developed routine operation may then be used for defining biological variability under a variety of nutritional regimes in order to accurately distinguish metabolite patterns that are associated with pathogenesis or disease endpoints from metabolic outliers of healthy subjects found at special circumstances. A project like the one outlined above may then generate large databases that allow the application of appropriate multivariate statistics tools or classification algorithms, aiming at defining certain patterns or single biomarker molecules associated with the disease or a clinical treatment.

It is not yet obvious whether fingerprinting or metabolomics is more appropriate for such an approach. It can be assumed that hyphenated metabolomic techniques have larger problems in instrument quality control and in chromatographic retention time shifts. On the other hand, data variation is also known for metabolic fingerprinting, *e.g.* shimming effects in NMR analysis. Moreover, problems associated with instrument inlet parameters (like injection inlets in GC, or electrospray interfaces and skimmers in LC-MS) can be easily found and eliminated if quality control is based on individual metabolites rather than overall spectra of crude mixtures. In a test case demonstrating the ability to adopt sample preparation protocols derived from plant biology, the authors performed extraction, sample preparation and analysis of different tissues of knock out and control lines of syngenic mice. Each tissue gave a very specific metabolic profile, indicating the ability to provide an unbiased look at metabolism (Fig. 1A).

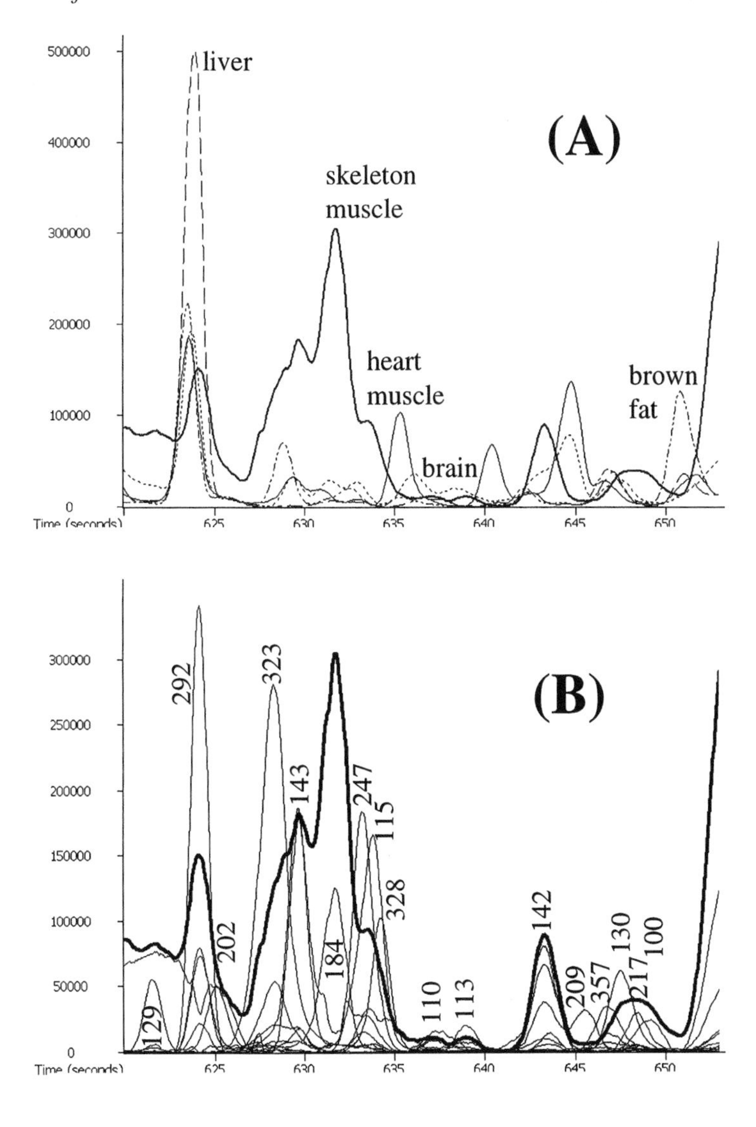

Figure 1. Comparison of tissue extracts of isogenic mice lines by GC-TOF analysis (unpublished results). **A**: Metabolic profiles of liver, brain, fat, and muscle extracts shown for a 35 s window. **B**: For the example of skeletal muscle, metabolites that were only partly resolved by chromatography could clearly be distinguished after *mass* spectral deconvolution and automatic quantitation using optimal ion traces (given in mass-to-charge numbers).

Even for regions in which too many metabolites were co-eluting to obtain clear chromatographic separation, mass spectral deconvolution resulted in automatically purified mass spectra and annotation of so-called unique ions that are best suited for relative quantification. An example is given for the complex mixture of the skeleton muscle tissue extract (Fig. 1B). Comparing the metabolite levels from multiple samples is facilitated by this approach. In Fig. 2, an example is given for the automatic peak finding in liver tissue extracts and the relative quantification by determining peak areas for all peaks using the model ion traces. In this example, the knock out mouse line showed a clear up-regulation of oleic acid after normalization to internal standards, whereas other fatty acids such as palmitic acid remained unaffected. In high throughput applications, all analytical results are then exported to databases for subsequent statistics and data mining.

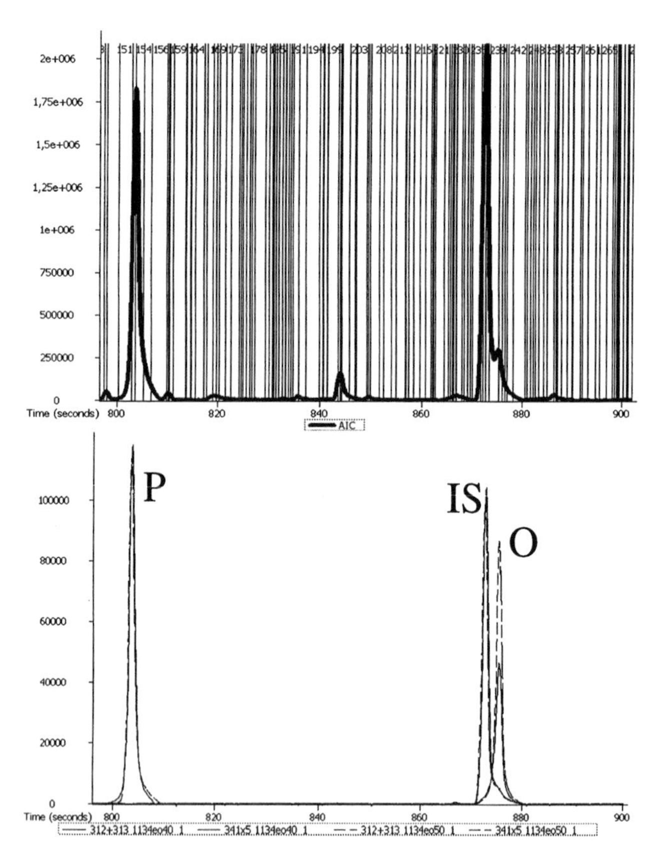

Figure 2. GC-TOF chromatograms of liver extracts of control and knock out mice lines. Upper panel: automatic peak detection of hundreds of metabolites as indicated by vertical lines, here shown for a 100 s retention time window. Lower panel: Quantification for known metabolites using unique ion traces. Dotted line: knock out mouse line, solid line: control mouse line. P = palmitic acid, O = oleic acid, IS = internal standard.

5. DATA MINING IN METABOLIC DATABASES

Any rigorous assessment of disease-related metabolic patterns will need support from well documented and curated databases. Metabolomics and fingerprinting methods are prone to data over-fitting, especially if inappropriate classification tools are used, or if success rates and background knowledge cannot be adequately compared with test samples. Although the number of variables from a single metabolomic experiment may be an order of magnitude smaller than that of transcript arrays, cost effectiveness allows twice as many samples to be analyzed for a given experiment. Correspondingly, it can be anticipated that large volumes of clinical and analytical data will populate metabolic databases. The analytical result files (raw data) must be accompanied by standardized biological, clinical, and method-related information (metadata) in order to ensure the ability to mine and extract health-related information. The data model must refer to agreed nomenclatures and ontology, and it must allow flexibility in experimental designs as well preserving a rigid and searchable database structure. It needs to support general concepts such as organ specimen, sex, nutritional regime *etc.*, and it must distinguish raw data from data derived from, for example, different types of data normalization. Logical correctness and minimization of duplicate entries have to be ensured as well as system usability, *e.g.* by web interfaces for cooperating users. In this respect, a typical expression database does not necessarily need to include data mining tools. Rather it should aim at ease of data downloads and re-formatting, taking into account the spreadsheet formats most commonly used by biologists (typically Microsoft office packages) and even differences in the numerical use of dots and commas between European and American users. Such seemingly trivial problems often hamper existing collaborations and can easily be avoided thus enabling clinicians and biologists to select data according to their needs *e.g.* by disease type, sex, biological matrix or sampling date. Finally, and most importantly in the data storage management of human studies, privacy concerns must be addressed at all times.

It might be asked which data mining tools should be used? Although the obvious answer is that there is simply no best algorithm, some guidelines should be followed. First, underlying assumptions have to be clarified. This relates most importantly to the independence of samples, but also to frequency distributions or data structures such as canonical or linear data, connected or independent data. It is highly advisable to have biostatisticians onboard when entering large-scale health related projects, since poor experimental designs may nullify valid conclusions. Researchers should also follow a general truism in statistics: more replicates and repeats are better than relying on the absolute minimum number of independent samples. This is particularly true if the minimum number of samples were calculated from (assumed) biological variability and the limits of analytical precision that were determined in the method validation steps. Next, it is generally accepted that results from data mining are more reliable if they are established by the use of different and complementary clustering methods. Any commercial package such as Pirouette, MatLab, SAS, or SPSS can be used but great care should be devoted not to violate any assumptions underlying different statistical tools. If only classification is asked for, supervised learning methods such as discriminant function analysis, support vector machines, artificial neural networks, decision trees, or evolutionary computing might be most appropriate (Gilbert *et al.*, 2000; Johnson *et al.*, 2000). Several of these techniques should be applied to a given data set to ensure classification results, taking into consideration that some of these algorithms will allow one to follow the path that led to classification (such as evolutionary computing) whereas others will only focus on optimal separation (such as neural networks). Although the use of cross-validation steps is often unavoidable in classification schemes, it is generally advisable to keep the training data set completely separate from the test data set in order to minimize the risk of biased results. With the large number of replicates possible in metabolomic or fingerprinting approaches, this recommendation should not pose a major obstacle to practical work.

In certain cases, unsupervised learning methods such as hierarchical cluster analysis or principal component analysis may be appropriate, for example, if distance measures are to be applied in various classes, and specifically, if the number of expected classes is not known *a priori*. Supervised and unsupervised methods may also be combined, as has been shown for discrimination of silent mutations from wild type yeast strains (Raamsdonk *et al.*, 2001). Once classes or sub-populations have been defined unambiguously, more sophisticated methods may be applied such as comprehensive analysis of co-variance matrices in large metabolic networks (Kose *et al.*, 2001), or time-dependent patterns (Lukashin *et al.,* 2001) that might be most appropriate for analyzing the onset and progression of diseases, and responses to successful therapeutic treatments.

6. CONCLUSIONS

Unbiased detection of unexpected events in biological or clinical studies is within the reach of well-equipped clinical research laboratories. A combination of metabolomic and metabolic fingerprinting methods may be appropriate if classification and subsequent detailed biochemical analysis is required. For large scale disease-related projects such as diagnosis and prediction of T2DM, rigorous method validation followed by assessments of 'metabolic baselines' of healthy cohorts is a prerequisite that needs to be complemented by large-scale efforts at the level of experimental design, database setup, and appropriate data mining and interpretation schemes. A current bottleneck in the development of metabolomics is the vast diversity and complexity of human and animal metabolomes; complexity that extends far beyond the classical knowledge found in textbooks or databases. However, the benefits of systematic evaluation of nutritional diseases may also impact other disciplines, such as defining the concept of 'healthy food' and in evaluating the composition of putatively beneficial food ingredients such as antioxidants, polyphenolics, carotenoids, and vitamins, particularly in relation to the impact of dietary fats, fibers, and carbohydrates.

REFERENCES

Amos AF, McCarty DJ, Zimmet P. The rising global burden of diabetes and its complications, estimates and projections to the year 2010. *Diabet Med* 14: 1-85 (1997).

Bennett PH. Type 2 diabetes among the Pima Indians of Arizona, an epidemic attributable to environmental change? *Nutr Rev* 57, 51-54 (1999).

Blom KF. Estimating the precision of exact mass measurements on an orthogonal time-of-flight mass spectrometer. *Anal Chem* 73: 715-719 (2001).

Boden G. Role of fatty acids in the pathogenesis of insulin resistance and NIDDM. *Diabetes* 46: 3-10 (1997).

Borkman M, Storlien LH, Pan DA *et al.*The relation between insulin sensitivity and the fatty-acid composition of skeletal-muscle phospholipids. *N Engl J Med* 328: 238-244 (1993).

de Fonseca FR, Navarro M, Gomez R *et al.* An anorexic lipid mediator regulated by feeding. *Nature* 414: 209-212 (2001).

Drexler DM, Tiller PR, Wilbert SM *et al.* Automated identification of isotopically labeled pesticides and metabolites by intelligent 'real time' LC-tandem MS using a bench-top ion trap mass spectrometer. *Rapid Comm Mass Spectrom* 12: 1501-1507 (1998).

Fell DA. Beyond genomics. *Trends Genet* 17: 680-682 (2001).

Feskens EJ, Virtanen SM, Rasanen L *et al.* Dietary factors determining diabetes and impaired glucose tolerance. A 20-year follow-up of the Finnish and Dutch cohorts of the Seven Countries study. *Diabetes Care* 18: 1104-1112 (1995).

Fiehn O. Combining genomics metabolome analysis and biochemical modeling to understand metabolic networks. *Compar Funct Genom* 2: 155-168 (2001).

Fiehn O. Metabolomics- the link between genotypes and phenotypes. *Plant Mol Biol* 48: 155-171 (2002).

Fiehn O, Kopka J, Dörmann P *et al.* Metabolite profiling for plant functional genomics. *Nature Biotechnol* 18: 1157-1161 (2000a).

Fiehn O, Kopka J, Trethewey RN, Willmitzer L. Identification of uncommon plant metabolites based on calculation of elemental compositions using gas chromatography and quadrupole mass spectrometry. *Anal Chem* 72: 3573-3580 (2000b).

Folsom AR, Ma J, McGovern PG, Eckfeldt H. Relation between plasma phospholipid saturated fatty acids and hyperinsulinemia. *Metabolism* 45: 223-228 (1996).

Gavaghan CL, Holmes E, Lenz E *et al.* An NMR-based metabonomic approach to investigate the biochemical consequences of genetic strain differences: application to the C57BL10J and Alpk:ApfCD mouse. *FEBS Lett* 484: 169-174 (2000).

Gilbert RJ, Rowland JJ, Kell DB. Genomic computing, explanatory modelling for functional genomics. In *Proc Genetic and Evolutionary Computation Conference*. Whitley D, Goldberg D, Cantú-Paz E (Ed) pp. 551-557, Morgan Kaufman, San Francisco (2000).

Goodacre R, Shann B, Gilbert RJ *et al.* Detection of the dipicolinic acid biomarker in *Bacillus* spores using curie-point pyrolysis mass spectrometry and Fourier transform infrared spectroscopy. *Anal Chem* 72: 119-127 (2000).

Halket JM, Przyborowska A, Stein S *et al.* Deconvolution gas chromatography mass spectrometry of urinary organic acids - Potential for pattern recognition and automated identification of metabolic disorders. *Rapid Comm Mass Spectrom* 13: 279-284 (1999).

Hu FB, Manson JE, Stampfer MJ *et al.* Diet, lifestyle, and the risk of type 2 diabetes mellitus in women. *N Engl J Med* 345: 790-797 (2001).

Hu FB, van Dam RM, Liu S. Diet and risk of Type II diabetes, the role of types of fat and carbohydrate. *Diabetologia* 44: 805-817 (2001).

Jellum E, Kvittingen EA, Stokke O. Mass spectrometry in diagnosis of metabolic disorders. *Biomed Environ Mass Spectrom* 16: 57-62 (1988).

Johnson HE, Gilbert RJ, Winson MK *et al.* Explanatory analysis of the metabolome using genetic programming of simple interpretable rules. *Genet Program Evolv Mach* 1: 243-258 (2000).

Justesen K, Knuthsen P, Leth T. Quantitative analysis of flavonols, flavone and flavanones in fruits, vegetables and beverages by high-performance liquid chromatography with photo-diode array and mass spectrometric detection. *J Chromatogr* 799: 101-110 (1998).

Kell DB, Mendes P. Snapshots of systems. In *Technological and Medical Implications of Metabolic Control Analysis*. Cornish-Bowden AJ, Cárdenas ML (Ed) pp. 3-25, Kluwer Academic Publishers, Dordrecht (2000).

Kim K-R, Park H-G, Paik M-J *et al.* Gas chromatographic profiling of urinary organic acids from uterine myoma patients and cervical cancer patients. *J Chromatogr* 712: 11-22 (1998).

Kimura H, Yamamoto T, Seiji Y. Automated metabolic profiling and interpretation of GC/MS data for organic aciduria screening, a personal computer-based system. *Tohuku J Exp Med* 188: 317-344 (1999).

King H, Aubert RE, Herman WH. Global burden of diabetes, 1995-2025, prevalence, numerical estimates, and projections. *Diabetes Care* 21: 1414-1431 (1998).

Knowler WC, Barrett-Connor E *et al.*Reduction in the incidence of type 2 diabetes with lifestyle intervention or metformin. *N Engl J Med* 346: 393-403 (2002).

Kose F, Weckwerth W, Linke T, Fiehn O. Visualising plant metabolomic correlation networks using clique-metabolite matrices. *Bioinformatics* 17: 1198-1208 (2001).

Krull IS, Swartz M. Analytical method development and validation for the academic researcher. *Anal Lett* 32: 1067-1080 (1999).

Lukashin AV, Fuchs R. Analysis of temporal gene expression profiles, clustering by simulated annealing and determining the optimal number of clusters. *Bioinformatics* 17: 405-414 (2001).

Ning C, Kuhara T, Inoue Y *et al.* Gas chromatographic mass spectrometric metabolic profiling of patients with fatal infantile mitochondrial myopathy with de Toni-Fanconi-Debre syndrome. *Acta Paed Japon* 38: 661-666 (1996).

Pauli GF. Higher order and substituent chemical shift effects in the proton NMR of glycosides. *J Nat Prod* 63: 834-838 (2000).

Raamsdonk LM, Teusink B, Broadhurst D *et al.* A functional genomics strategy that uses metabolome data to reveal the phenotype of silent mutations. *Nature Biotechnol* 19: 45-50 (2001).

Randle PJ, Garland PB, Newsholme EA, Hales CN. The glucose fatty acid cycle in obesity and maturity onset diabetes mellitus. *Ann NY Acad Sci* 131: 324-333 (1965).

Report of the Expert Committee on the Diagnosis and Classification of Diabetes Mellitus. *Diabetes Care* 20:1183-1197 (1997).

Shellie R, Marriot P, Morrison P. Concepts and preliminary observations on the triple dimensional analysis of complex volatile samples by using GC x GC – TOF MS. *Anal Chem* 73: 1336-1344 (2001).

Stein SE. An integrated method for spectrum extraction and compound identification from gas chromatography/mass spectrometry data. *J Am Soc Mass Spectrom* 10: 770-781 (1999).

Storlien LH, Baur LA, Kriketos AD *et al.* Dietary fats and insulin action. *Diabetologia* 39: 621-631 (1996).

Tanaka K, Hine DG, West-Dull A, Lynn TB. Gas-chromatographic method of analysis of urinary organic acids I Retention indices of 155 metabolically important compounds. *Clin Chem* 26: 1839-1846 (1980a).

Tanaka K, West-Dull A *et al.* Gas-chromatographic method of analysis of urinary organic acids II Description of the procedure and its application to diagnosis of patients with organic acidurias. *Clin Chem* 26: 1847-1853 (1980b).

Tolstikov VV, Fiehn O. Analysis of highly polar compounds of plant origin, combination of hydrophilic interaction chromatography and electrospray ion trap mass spectrometry. *Anal Biochem* 301: 298-307 (2002).

Tuomilehto J, Lindstrom J, Eriksson JG *et al.* Prevention of type 2 diabetes mellitus by changes in lifestyle among subjects with impaired glucose tolerance. *N Engl J Med* 344: 1343-1350 (2001).

Vaidyanathan S, Rowland JJ, Kell DB, Goodacre R. Discrimination of aerobic endospore-forming bacteria via electrospray-ionization mass spectrometry of whole cell suspensions. *Anal Chem* 73: 4134-4144 (2001).

van't Veer LJ, Dai HY, van de Vijver MJ *et al.* Gene expression profiling predicts clinical outcome of breast cancer. *Nature* 415: 530-536 (2002).

Vessby B, Aro A, Skarfors E *et al.* The risk to develop NIDDM is related to the fatty acid composition of the serum cholesterol esters. *Diabetes* 43: 1353-1357 (1994).

Vessby B, Tengblad S, Lithell H. Insulin sensitivity is related to the fatty acid composition of serum lipids and skeletal muscle phospholipids in 70-year-old men. *Diabetologia* 37: 1044-1050 (1994).

Vessby B, Unsitupa M, Hermansen K *et al.* Substituting dietary saturated for monounsaturated fat impairs insulin sensitivity in healthy men and women, The KANWU Study. *Diabetologia* 44: 312-319 (2001).

Vigneau-Callahan KE, Shestopalov AI *et al.* Characterization of diet-dependent metabolic serotypes: Analytical and biological variability issues in rats. *J Nutr* 131, 924-932 (2001).

Warram JH, Kopczynski J, Janka HU, Krolewski AS. Epidemiology of non-insulin-dependent diabetes mellitus and its macrovascular complications. A basis for the development of cost- effective programs. *Endocrinol Metab Clin North Am* 26: 165-188 (1997).

Weckwerth W, Tolstikov VV, Fiehn O. Metabolomic characterization of transgenic potato plants using GC/TOF and LC/MS analysis reveals silent metabolic phenotypes. *Proc 49th ASMS Conference on Mass Spectrometry and Allied Topics*. Chicago (2001).

Chapter 12

BIOACTIVE LIPIDS IN REPRODUCTIVE DISEASES

Opportunities for New Diagnostics

Jaideep Chaudhary, Tim Compton and Jeff A. Parrott
Atairgin Technologies, 101 Theory, Irvine CA 92612, USA

1. INTRODUCTION

Phospholipids, predominantly phosphoglycerides, have long been recognized as a major component of cell and organellar membranes where they form the phospholipid bilayer. The phospholipid bilayer creates the permeability barrier of all cells and serves as a matrix for a large number of proteins involved in diverse cellular functions including cellular communication, replication, trafficking *etc*. Although devoid of any catalytic activity, the phospholipids are in a constant state of flux within the membrane and play a significant role in regulating biochemical signaling and the activity of the proteins associated with the membranes (Dowhan, 1997).

Highly defined and controlled enzymatic processes are involved in the metabolism of phospholipids. An example of such a periodic biosynthesis and degradation is the doubling of membrane phospholipids during cell cycle progression (Lykidis and Jackowski, 2001). Membrane homeostasis, maintained primarily by phospholipases such as phospholipase A2, D and C (Kobayashi *et al*., 1998; Wang and Dennis, 1999) is therefore expected to be a highly regulated process synchronized with other components of the cell cycle (Lykidis and Jackowski, 2001). Therefore any cellular process involving re-organization of cellular membranes may generate changes in phospholipid metabolism.

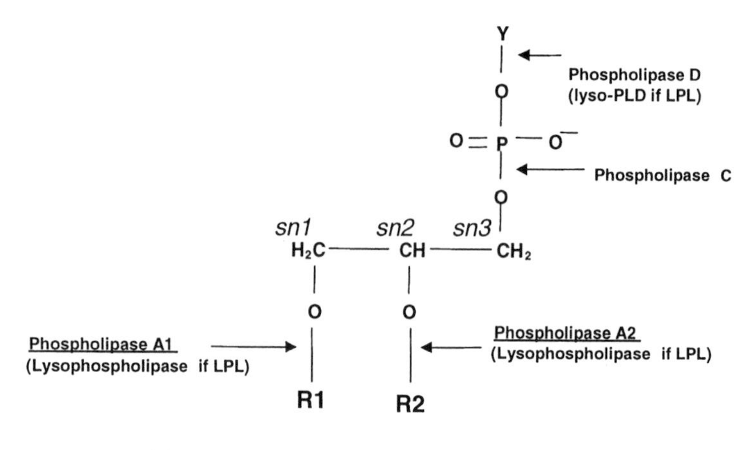

R1/R2 - Molecular Species	
16:0	Palmitoyl
18:0	Stearoyl
18:1	Oleoyl
18:2	Linoleoyl
20:4	Arachidonyl
22:6	Docosahexanoyl
*H	at R1 or R2 if LPL
Y - Classes	
H	PA/LPA
Choline	PC/LPC
Inositol	PI/LPI
Serine	PS/LPS
Ethanolamine	PE/LPE
Glycerol	PG/LPG
Functionally Related Molecules	
Sphingoid base derived bioactive lipids	S1P, SPC
Acetyl derivative at sn-2	PAF

Figure 1. Structural elements of phospholipids (PL) and lysophospholipids (LPL). **Y:** Classes of lipids (*e.g.* PA, LPA, PS, LPS, etc.) are structurally defined by the nature of the head group esterified to the phosphoryl moiety at sn-3. **R1/R2:** Two hydrocarbon chains are bonded to the glycerol backbone in phospholipids at positions **R1** and **R2**. Variability in the nature of the linkages (acyl, alkyl, or alkenyl) between glycerol and the hydrocarbon chains is possible. Common molecular species that are observed in PL and LPL are tabulated, but other molecular species are also present. Lysophospholipids have a single hydrocarbon chain (R1 or R2) bonded to sn-1 or sn-2. A hydroxyl residue is present at the sn-1 or sn-2 position that is not bonded to the hydrocarbon. Lyso-derivatives of sphingolipids have analogous structural components (not shown) and can be functionally related to lysophospholipids (*e.g.* S1P and LPA), but have different synthetic routes. Also shown are the sites of hydrolysis of several types of enzymes. (Abbreviations: LPI: Lysophosphatidylinositol, LPS: Lysophosphatidylserine, LPE: Lysophosphatidylethanolamine, LPG: Lysophosphatidylglycerol, PC: Phosphatidylcholine, PI: Phosphatidylinositol, PS: Phosphatidylserine, PE: Phosphatidylethanolamine, PG: Phosphatidylglycerol, Lyso-PLD: Lysophospholipase D; S1P: Sphingosine 1 phosphate; for others see text).

The action of phospholipases on membrane phospholipids results in the generation of metabolites including lysophospholipids with different head groups and hydrocarbon chains. The lysophospholipids, which include lyso-derivatives of all known phosphoglycerides, are important functional mediators and precursors to diverse cellular components.

The simplest of all lysophospholipids, the lysophosphatidic acid (LPA; *e.g.* 1-acyl-2-hydroxy-sn-glycero-3-phosphate, Fig. 1) subfamily (Pages *et al.*, 2001), has gained particular attention because these molecules have been shown to have mitogenic properties (Jalink *et al.*, 1994; Goetzl, 2001; Tigyi, 2001). The mitogenic effects of LPA are indistinguishable from those activated by polypeptide growth factors which supports its inclusion into the growing family of phospholipid growth factors. In addition to LPA, other lysophospholipids and related molecules such as platelet activating factor (PFA; 1-hexadecyl-2-acetyl-glycerophosphocholine) (Bussolino and Camussi, 1995), phosphatidic acid (PA; 1-acyl, 2-acyl–sn-glycero-3-phosphate) (Sliva *et al.*, 2000), sphingosine-1-phosphate (Pyne and Pyne, 2000) and sphingosylphosphorylcholine (SPC) (Chin and Chueh, 1998) are also candidate phospholipid growth factors. The cellular responses elicited by LPA and other phospholipid growth factors are a result of specific binding to their cognate G-protein coupled receptors. Experimental evidence and homology searching of the human genome have led to the identification of three LPA receptors (LPA_1, LPA_2 and LPA_3), at least 5 sphingosine-1-phosphate receptors ($S1P_1$, $S1P_2$, $S1P_3$, $S1P_4$ and $S1P_5$), one PAF receptor, at least two lysophosphatidylycholine (LPC) and SPC receptors (OGR1 and GPR4) and a PA receptor (Moolenaar, 1999; Contos *et al.*, 2000; Fukushima *et al.*, 2001; Chun *et al.*, 2002).

The functional significance and structural diversity of lysophospholipids strongly suggest a role of this lipid class in eliciting diverse biological responses. The list of diseases in which the role of lysophospholipids is implicated is diverse and includes cancer (Mills *et al.*, 2002), neurological disorders (Farooqui *et al.*, 2000; Klein, 2000; Yoshida and Ueda, 2001), gynecological conditions (Murphy *et al.*, 1998; Xu *et al.*, 2001; Budnik and Mukhopadhyay, 2002;), cardiovascular diseases (Sato, 2000; Levade *et al.*, 2001) and immune responses (Sakata-Kaneko *et al.*, 1998). Of particular interest is the role of lysophospholipids in the pathophysiology of ovarian cancer. High concentrations of LPA have been observed in ascitic fluid and blood in patients with widespread ovarian cancer (Erickson *et al.*, 2001; Mills *et al.*, 2002). In addition to ovarian cancer, elevated levels of LPA have also been reported in serum from multiple myeloma patients (Sasagawa *et al.*, 1999) and increased expression of LPA receptors have been observed in thyroid tissue from thyroid cancer patients (Schulte *et al.*, 2001).

2. BIOSYNTHESIS OF LYSOHOSPHOLIPIDS

The diversity of lysophospholipid molecular species and cognate G protein coupled receptors is consistent with the wide array of bioactivities attributed to the lysophospholipids. Molecular species of lysophospholipids can be generated by intracellular *de novo* synthesis and intra- or extra-cellular metabolism of phospholipids (Spiegel and Merrill, 1996; Gaits *et al.*, 1997; Goetzl and An, 1998; Hannun *et al.*, 2001; Pages *et al.*, 2001). For example, the actions of phospholipases and lecithin-cholesterol acyltransferase can generate lysophospholipids from phospholipids that may then be converted to LPA by lysophospholipase D (Figs. 1 and 2) (le Balle *et al.*, 1999; Subramanian *et al.*, 1999; Six and Dennis, 2000; Tokumura *et al.*, 2000).

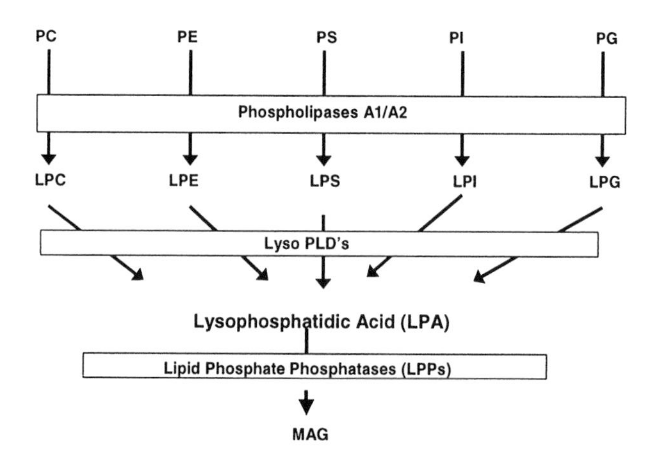

Figure 2. Summary of biochemical pathways for LPA metabolism. LPA is a common structural element of all glycerophospholipids and can be generated by enzymatic hydrolysis of phospholipid and/or lysophospholipid precursors. *De novo* routes of synthesis are not shown. (Abbreviations: MAG: Monoacylglycerol; for others see Fig 1. and text).

Phospholipases generate lysophospholipids by hydrolyzing the ester linkage of one of the fatty-acyl moieties in phospholipids. The altered expression of phospholipases in disease states ranging from inflammation to cancer may be linked to the observed changes in lysophospholipid levels (Balsinde *et al.*, 1999). However, a multitude of enzymes may likewise be involved. Lecithin-cholesterol acyltransferase transfers fatty-acyl moieties from phosphatidylcholine to cholesterol thereby producing LPC and cholesterol esters. Lysophospholipase D activity generates LPAs from other lysophospholipids by hydrolyzing the phosphoester linkage between the phosphoryl moiety and the head group (Figs. 1 and 2). *De novo* synthesis of lysophospholipids generally involves the production of monoacylglycerol as an intermediate and monoacylglycerol kinase to make LPAs, the simplest type of lysophospholipid.

Alternatively, the actions of lysophospholipases and lipid phosphate phosphatases (LPP) result in the degradation of lysophospholipids (Fig. 2) (Roberts *et al.*, 1998; Wang and Dennis, 1999; Imai *et al.*, 2000). Considering the various classes and molecular species of lysophospholipids and the partial list of enzymes and associated metabolic pathways indicated above, a complex interaction of many molecules is implicated in the normal and pathological regulation of lysophospholipids. This complexity has made the investigation of lysophospholipid biology challenging.

3. LIPOGENOMICS

Altered patterns of bioactive lipids may result from mutations and/or altered expression of genes in the biosynthetic pathways leading to phospholipid metabolism and responses. A comprehensive understanding of these so-called lipogenomic pathways, especially through the use of microarrays, may help create a better understanding of the initial and ongoing stages of disease progression. Recent advances in parallel processing techniques such as microarrays of genes (Khan *et al.*, 1999; Alizadeh *et al.*, 2001; Oliver *et al.*, 2002) and metabolic profiling of proteins (Albala, 2001; Templin *et al.*, 2002) has provided valuable information on the molecular aspects of disease processes.

Single nucleotide polymorphisms and protein mutations that can be strong prognostic indicators of a particular disease such as cancer may not be applicable to lipids. Since many lipids are metabolic products with significant biological activity, it is evident that profiling genes or proteins involved in lipid metabolism and function is an attractive complement to global lipid profiling. A number of observations on individual genes or gene products support this hypothesis especially in cancer:

a. Sphingosine kinase (SphK): SphK is the rate-limiting enzyme in the synthesis of the phospholipid growth factor sphingosine-1-phosphate from sphingosine and its expression regulates multiple cellular functions including apoptosis (Edsall *et al.*, 2001). In addition, SphK has also been shown to act as an oncogene (Xia *et al.*, 2000). Up-regulation of SphK is therefore expected to transform cells and increase the production of sphingosine-1-phosphate.

b. Phospholipase D (PLD): PLD activity is significantly elevated in many tumors and transformed cells, suggesting that PLD might be involved in tumorigenesis (Eder *et al.*, 2000). Increased PLD levels may regulate multiple lipid metabolic pathways including LPA production.

c. Phospholipase A2 (PLA2): The levels of membrane-bound PLA2 (mPLA2) are increased in breast cancer cells and may result in production of higher amounts of lysophosphospholipids. Higher levels of mPLA2 in breast cancer biopsies are associated with poorer prognosis (Yamashita *et al.*, 1993, 1994a, 1994b). Increased levels of PLA2 can influence phospholipid metabolism at multiple levels including intra- and extra-cellular sites.

d. Lipid phosphate phosphatase or phosphohydrolase (LPP): The plasma membrane-bound lipid phosphate phosphohydrolases hydrolyze PA, LPA and sphingosine-1-phosphate. Levels of LPP may therefore regulate the bio-availability of phospholipid growth factors (Jasinska *et al.*, 1999). Altered levels of LPP can be expected in tumors dependent on these bioactive lipids for growth and survival.

e. Lysophospholipid receptors: Malignant transformation of some types of cells results in the appearance/predominance of one or more LPA receptors not expressed by the equivalent non-malignant cell. Numerous human ovarian cancer cell lines express high levels of LPA_2 receptor not detectable in either primary cultures or immortalized lines of normal ovarian surface epithelium (Goetzl *et al.*, 1999a). Several lines of human breast cancer cells have higher levels of $S1P_2$ and $S1P_3$ and lower levels of LPA_1 than normal breast epithelial cells in primary cultures (Goetzl *et al.*, 1999b). Altered levels of these receptors may help regulate the cellular responses to lysophospholipids.

f. Phosphatidylinositol-3-kinase, alpha unit (PI3KCA): The majority of the receptor-mediated effects of lysophosphatidic acid involve activation of intracellular PI3KCA. The amplification of the of chromosomal region 3q26 harboring the PI3KCA gene is the most frequent amplification in cervical (Ma *et al.*, 2000), breast and ovarian cancers (Mills *et al.*, 2001). Amplification of PI3K3CA gene is expected to amplify the responses mediated by phospholipid growth factors.

Taken together, microarrays that can measure global changes in lipid-associated genes or "lipogenomics" will allow a better understanding of the molecular mechanisms involved in the metabolism and actions of lipids and may help identify novel diagnostics and therapeutic targets. Lipogenomic profiling may also provide valuable insight into the observed diversity of lipids in diseases (*e.g.* molecular species related to disease states) and the metabolic and functional networks involving and/or regulated by lipids.

4. CELLULAR EFFECTS OF LYSOPHOSPHOLIPIDS

Apoptosis is a major pathway targeted by lysophospholipids such as LPA (Fang *et al.*, 2000). Other cellular responses affected by LPA are cell proliferation, cell migration and matrix assembly (Goetzl, 2001; Panetti *et al.*, 2001). Studies have concluded that LPA alone acts as a potent survival factor for many cells. The G-protein receptor-mediated growth factor-like activities of lysophospholipids are primarily mediated by the activation of pertussis toxin-sensitive and insensitive $G\alpha/\beta/\gamma$ heterotrimers. The downstream biochemical events linking lysophospholipids to their pleiomorphic effects are complex and are dictated by the local expression of receptor sub-types and subsequent G-protein coupling pathways (Goetzl, 2001; Kranenburg and Moolenaar, 2001). Most of these effects are mediated intracellularly by the activation of phosphatidylinositol-3-kinase (Koh *et al.*, 1998) and MAPK/ERK pathways. The pleiomorphic effects of LPA may also be mediated by ligand-independent transactivation of EGFR/Gab1 (Laffargue *et al.*, 1999; Cunnick *et al.*, 2000) and PDGFR signaling pathway (Herrlich *et al.*, 1998; Rui *et al.*, 2000). Over the past several years, studies have elucidated the mechanisms of action of LPA, LPC, sphingosine-1-phosphate, sphingosylphosphorylcholine and PAF. These studies form the basis of our current understanding of the cellular effects of lysophospholipids.

Oxidized low-density lipoprotein (LDL) is a key factor in the pathogenesis of atherosclerosis and its thrombotic complications. Oxidation of LDL results in the generation of LPA, which in turn leads to activation of platelets and endothelial cells (Siess *et al.*, 1999; Siess, 2002). LPA has also been shown to play an important role in several aspects of the inflammatory process including platelet aggregation and proliferation of fibroblasts and smooth muscle cells (Tigyi, 2001). Given the diversity in these bioactivities, it is apparent that in order to understand their role in the disease process, highly sensitive and specific quantitative profiling of lysophospholipid molecular species in biological fluids and tissues is needed.

5. STRUCTURE ACTIVITY RELATIONSHIPS OF LYSOPHOSPHOLIPIDS

The various classes of lysophospholipids are defined by the presence of a head group containing a phosphate at the sn-3 position (Fig. 1). Structural diversity within a class of lysophospholipids is derived from the length of the hydrocarbon chain, the nature of the covalent linkage between the hydrocarbon chain and the glyceroyl moiety (*e.g.* acyl, alkyl or alkenyl), the position of the hydrocarbon on the glyceroyl moiety (sn-1 or sn-2), and the number and position of *cis* double bonds in the hydrocarbon chain. Precise structure-activity relationships (SAR) for bioactive lipids and their cognate receptors have not been fully characterized (Lynch *et al.*, 1997; Lynch and Macdonald, 2001). However, several studies have shown that particular molecular species of LPA are more potent in cell-based assays designed to examine specific receptors (Jalink *et al.*, 1995; Hopper *et al.*, 1999; Bandoh *et al.*, 2000; Erickson *et al.*, 2000). While other structural determinants are evidently required for reactivity with receptors (*e.g.* head group, nature of covalent bond between glyceroyl moiety and the hydrocarbon chain, *etc.*), specificity within a class of lysophospholipids (*e.g.* acyl-linked LPA) is apparently conferred by the nature of the hydrocarbon chain (*i.e.* degree of saturation and chain length) (Bandoh *et al.*, 2000).

Evidence for molecular species-specific differences in bioactivity comes from many types of studies including receptor activation/desensitization studies. In one study of some LPA molecular species, the order of potency for stimulating calcium flux in insect Sf9 cells transfected with LPA_3 was $\Delta 9$-unsaturated fatty-acyl chains (*e.g.* 18:1, 18:2 and 18:3) at sn-2 followed by 16:2 and 20:4 at sn-2 (Bandoh *et al.*, 2000). In contrast, the 16:0 but not 18:1 LPC has been shown to be bioactive (Zhu *et al.*, 2001).The observations from many of the SAR studies need to be placed in context with the cell-based assay system used, expression levels of receptors, subsequent coupling/signaling mechanisms and the experimental presentation of lysophospholipids to cells. However, it is important to note that differential responses to molecular species in controlled experiments have been observed.

While molecular species-specific bioactivities have been discovered, these discoveries have not necessarily correlated to clinical applications. For example, many molecular species of LPA that are considered to be poor ligands for lysophospholipid receptors (*e.g.* 16:0 LPA) are apparently generated by ovarian tumors and are present in ascites fluid of malignant effusions (Baker *et al.*, 2002; Xu *et al.*, 1995b). Such LPAs are often capable of stimulating responses in lysophospholipid receptor assays when present at high concentrations, like those observed in ascitic fluid from patients with ovarian cancer. More complete characterization of the relationships of lysophospholipid molecular species and the presence of diseases remains to be elucidated.

Given the structural diversity of lysophospholipids and their physiological and pathophysiological effects, an analysis scheme that measures the concentrations of lysophospholipid molecular species in conjunction with monitoring changes associated with genetic information involved in lipid synthesis, lipid degradation, and cell signaling in lipid-controlled pathways would provide a unique view of the role of these analytes in disease.

6. LIPOMICS AND BIOLIPOMICS: LIPID PROFILING TO DIAGNOSE COMPLEX DISEASES

Lipomics or profiling specific classes of lipid biomarkers also provides a unique opportunity to identify diseases other than cancer. Biolipomics describes profiling classes of lipomic analytes that are bioactive and may be a more direct analysis of functional lipomics. Unlike Cystic Fibrosis that is caused by mutations in a single gene on chromosome 7 that encodes the Cystic Fibrosis Transmembrane Conductance Regulator protein (Riordan *et al.*, 1989), the causes of most diseases are not so simple. Many diseases require several "hits" or alterations in order to show clinical manifestations. In addition some diseases may be caused or influenced by complex metabolic pathways in cells, organs, and endocrine systems. As a result, it can be difficult to understand and diagnose some of these diseases. In recent years, there has been an attempt to profile genes (genomics) and proteins (proteomics) to better understand these diseases.

Genomics is a powerful tool to examine the profile of specific genes that have been inherited, possibly mutated, and expressed in an individual but provides little or no information on the metabolic profile associated with the disease. Proteomics, which can profile various proteomes that result from an individual's genomic profile and physiological environment, may reveal metabolic products of various proteins. Lipid biomarkers are another class of molecules that provide valuable information on the metabolic alterations associated with diseases. In addition levels of bioactive lipids are highly regulated in cells and bodily fluids. These analytes are both indicators of various complex metabolic pathways as well as powerful bioactive regulators controlling processes such as proliferation, survival, angiogenesis, immune functions, and cardiovascular functions. Therefore lipomics and biolipomics offer unique opportunities to examine snapshots of both metabolism and functional biology. This information will be useful to expand our knowledge of, and to develop, commercially viable diagnostic assays for more complex diseases. Discussed below are few examples of complex diseases that may soon benefit from the lipomics and biolipomics approaches:

6.1 Ovarian Cancer

Ovarian cancer causes more deaths than any other reproductive organ cancer. The American Cancer Society estimates that there will be about 23,300 new cases of ovarian cancer in this country in 2002 and an estimated 13,900 women will die of the disease. Unfortunately, there is no cost effective screening technology currently available that may help in early stage diagnosis resulting in most cases of ovarian cancer being diagnosed at advanced stage of the disease. However ovarian cancer is one of the most treatable cancers if diagnosed early (ACS, 2002).

One reported unique biomarker for ovarian cancer is LPA. Elevated levels of LPA in the plasma from patients with stage I ovarian cancer suggests that LPA may represent a useful early marker for the disease (Xu *et al.*, 1998, 2001; Mills *et al.*, 2002). LPA is released by transformed ovarian surface epithelial cells and may also be involved, with other bioactive lysophospholipids, in the pathogenesis of ovarian cancer (Mills *et al.*, 2002). Increased levels of LPA are also found in malignant effusions from ovarian cancer patients (Baker *et al.*, 2002).

Increased levels of lysophospholipids and their molecular species have been reported in malignant ascitic fluid including total acyl-LPA, total alkyl- and alkenyl-LPA, total lysophosphatidylinositol (LPI), sphingosylphosphorylcholine (SPC) and total LPC (Shen *et al.*, 2001; Lu *et al.*, 2002). Total LPI and total alkyl- and alkenyl-LPA have also been used to distinguish malignant from non-malignant ascitic fluid (Xiao *et al.*, 2000). It is possible that the LPA pools in plasma and malignant effusions may be different (Baker *et al.*, 2002) suggesting a more complex mechanism involved in the metabolism of lysophospholipids in these biological fluids.

These published studies and other unpublished work have supported the development of blood-based diagnostic tests for ovarian cancer based on patterns of individual bioactive lipids such as molecular species of LPA and LPI. The possibility that these lipid-based tests may help diagnose early stage ovarian cancer emphasizes the importance of such new tests.

6.2 Breast Cancer

According to the World Health Organization, more than 1.2 million women will be diagnosed with breast cancer in the year 2001 (ACS, 2002). In the US, more than 203,500 cases of breast cancer will be diagnosed in the year 2002 resulting in about 40,000 mortalities (ACS, 2002). Early detection of breast cancer remains a challenge despite the widespread use of X-ray mammography. Health care professionals and patients will benefit from additional reliable diagnostic information to more accurately predict probability or presence of malignancy. A unique and reliable serum/plasma tumor marker for breast cancer would most certainly improve the diagnostic sensitivity and specificity of the approaches currently available to practitioners. In addition, an accurate breast cancer test would also likely lead to decreased mortality from the disease because of increased surveillance accompanying early detection.

Changes in blood lipid profiles or patterns provide a novel opportunity to develop breast cancer diagnostic tests. Initial studies have suggested the presence of several molecular species of bioactive derivatives of PAF (platelet-activating factor) including C_{16}-alkyl PAF, C_{18}-LPC, C_{16}-LPC, Lyso-PAF, and C_{16}-acylPAF in PAF-lipid extracts from breast carcinoma (Montrucchio *et al.*, 1998). Elevated levels of PAF-bioactivity observed in the lipid extracts of breast carcinoma may be associated with angiogenesis and may correlate with decreased lymph node metastasis (Montrucchio *et al.*, 1998).

The enzymes involved in the breakdown of phospholipids such as PLA2 and PLD are expressed in breast cancer cells. Breast cancer cells have been shown to respond to LPA, lysophosphatidylserine, and SPC *in vitro* (Xu *et al.*, 1995a). Altered levels of LPC have also been reported in breast cancer tissue (Merchant *et al.*, 1991). Lipomics and biolipomics are expected to expand the list of potential diagnostic candidates for breast cancer.

The new diagnostic test based on lipid profiling may overcome the barriers to widespread x-ray mammography, including low resolution, providing the ability to detect small tumors several years earlier. This may allow detection of tumors before they can metastasize with no exposure to X-rays and no discomfort during the procedure.

6.3 Endometriosis

Currently, surgical confirmation is necessary to diagnose endometriosis, hence the true prevalence is not known (Brosens, 1997; Eskenazi *et al.*, 2001; Murphy, 2002). Endometriosis has been found in approximately 4 percent of asymptomatic women undergoing laparoscopy for sterilization. However evidence of disease is present in almost 20 percent of women undergoing laparoscopic investigation for infertility (Eskenazi and Warner, 1997). Approximately 24 percent of women who complain of pelvic pain are subsequently found to have endometriosis (Eskenazi and Warner, 1997). The overall prevalence, including symptomatic and asymptomatic women, has been conservatively estimated to be 5 to 10 percent (Lu and Ory, 1995) and the incidence is rising. Historically identification of patients that have the disease has been problematic. For example, there are currently approximately 5.5 million women identified with endometriosis, but before 1921 there were only 20 reports on it in the world literature (Older, 1984; Ballweg, 1995). Clearly a sensitive and specific means to diagnose the disease is needed as an alternative to surgery. Biolipomics provides a unique opportunity to develop such a tool.

Similar, but not identical, cellular mechanisms are involved in the establishment and progression of many ovarian cancers and lesions associated with endometriosis. In women with endometriosis, these lesions may originate from retrograde menstruation, alterations of peritoneal cells through "coelomic metaplasia," or differentiation of müllarian remnants (Gerbie and Merrill, 1988; Wellbery, 1999; Vinatier *et al.*, 2001). In any case, endometrial tissue must survive, attach, invade, form new blood vessels, and sometimes proliferate in order to develop into a lesion. These cellular mechanisms are similar to those in ovarian cancer cells and both diseases occur in the peritoneal environment. There are differences between the diseases highlighted by the observations that endometriosis is a benign disease and ovarian cancer usually involves a rapidly growing tumor. The similarities and differences support the hypothesis that these diseases may produce and respond to similar bioactive lipids, and that each disease may create unique biolipomic profiles that can be identified and used to detect the presence of each disease.

Several empirical observations support the use of biolipomics to identify and diagnose endometriosis. Endometriosis occurs in the peritoneal cavity containing peritoneal fluid (PF) rich in cells and soluble factors including cytokines, growth factors, and lipids. These cellular and biochemical components of the PF are known to play important roles in the pathogenesis of the disease (Ramey and Archer, 1993; Bedaiwy *et al.*, 2002; Cheong *et al.*, 2002). Specific alterations in lipid-metabolizing enzymes and lipids have been described in endometriosis and may be the result of the highly oxidative environment associated with the disease. PF or PF cells from patients with endometriosis contains higher levels of platelet-activating factor acetylhydrolase (Hemmings *et al.*, 1993) and phospholipase A2 (Sano *et al.*, 1994). In addition, higher levels of lysophosphatidylcholine (LPC) have been described in patients with endometriosis (Murphy *et al.*, 1998). To date no systematic approach such as biolipomics has been used to identify patients with endometriosis. This suggests that there are ample opportunities to develop novel and effective diagnostic tests for this broad-reaching disease.

6.4 Pregnancy Complications (*e.g.* Preeclampsia)

Utilizing biolipomics during pregnancy provides an opportunity to diagnose complications of pregnancy such as preeclampsia (PE). Not all cases of PE are correctly diagnosed so the true incidence is not known in all countries. Conservative estimates of incidence are 6% to 8% of all pregnancies and the overall incidence is rising (ACOG, 1996; WHO, 1988). PE is the second most common cause of maternal mortality in the United States accounting for 12-18% of all pregnancy-related maternal deaths (Koonin *et al.*, 1988; Rochat *et al.*, 1988 ACOG, 1996; Berg *et al.*, 1996). If the condition is not accurately diagnosed in a timely manner and treated, it can also lead to high morbidity and mortality for the fetus. Current diagnosis of PE is only possible later in pregnancy after the disease has progressed systemically and has obvious symptoms. In some cases, pregnant women present with borderline symptoms that may or may not be actual PE. False diagnosis of PE in patients that simply show elevated blood pressure leads to unnecessary stress, care, and costs. Negative diagnosis (*i.e.* normal) in patients that actually have PE leads to dramatically increased risk of morbidity and mortality. Some of the complications associated with these untreated abnormal pregnancies may increase the risk of adult-onset diseases even after an apparently healthy delivery. Earlier diagnosis of the condition and/or more accurate diagnosis in borderline cases would dramatically improve the management of these pregnancies. Biolipomics provides a unique opportunity to develop such diagnostic capabilities.

During normal pregnancies, several changes occur in the mother that result in changes in blood volume, blood cell profiles, biochemical alterations, hormonal changes, and changes in metabolism. Some or all of these changes may be accompanied by changes in the biolipomic profile in the mother's fluids such as plasma and serum. Preeclampsia is a disorder during pregnancy that arises from abnormalities in the placenta but leads to a variety of systemic complications in the mother including newly onset hypertension, newly onset proteinuria, and newly onset nondependent edema (ACOG, 1996; Genbacev *et al.*, 1999; Salas, 1999; Norwitz and Repke, 2000; Walker, 2000, Lopez-Jaramillo *et al.*, 2001; Stemmer, 2001). There is a generalized inflammatory state where several plasma factors that regulate endothelial cell functions are altered (Roberts, 1998; Gratacos, 2000). Several of these factors suggest that the biolipomic profile in the mother's blood is altered in preeclampsia.

Some changes in lipids and lipid-modifying enzymes are well documented, but systematic biolipomics has not yet been performed. Abnormal lipid metabolism may also be directly involved in the pathogenesis of PE, not merely a manifestation of the condition. Platelet activating factor (Rowland *et al.*, 2000), triglycerides and free fatty acids are elevated in PE and lipoproteins show predominantly the atherogenic low density lipoproteins (LDL). A lysophospholipase activity that acts on lysophosphatidylcholine (LPC) (Endresen *et al.*, 1993) and levels of phospholipase A2 (Lim *et al.*, 1995; Pulkkinen *et al.*, 1996) are elevated in PE. Many of these lipid alterations appear to be involved in endothelial dysfunction, lipoprotein changes, oxidative stress, and immune functions and some may be involved directly in the placental abnormalities (Jendryczko and Drozdz, 1990; Jendryczko *et al.*, 1989). Many of these lipid alterations in pregnant women with PE suggest important roles of bioactive lipids during normal and abnormal pregnancies, but no systematic approach such as biolipomics has been used to identify these conditions. This provides a unique opportunity to develop novel and effective diagnostic tests for PE and other complications of pregnancy.

7. MEASURING LYSOPHOSPHOLIPIDS

In order to examine biolipomic profiles, a large number of quantitative measurements of small molecular weight analytes must be performed in complex biological samples such as plasma and serum. A variety of methods exist that may be capable of analyzing simple or "clean" sample matrices, but extensive sample preparations would be time- and cost-prohibitive in a commercial assay and would likely increase the variability of the methods. Many of the analytes in a biolipomic profile have relatively similar sizes, structures, and chemical properties making them difficult to separate. This poses technical considerations that are critical to successful examination of lipomic and biolipomic profiles.

The masses of most of the biolipomic analytes under consideration are relatively small at approximately 400 – 700 atomic mass units (amu). In addition, there may be a large number of analytes being analyzed within a much smaller mass range such as 400 – 450 amu. Each of these analytes may have the same glycerol backbone (*e.g.* the lyso-derivatives of glycerophospholipids), a single hydrocarbon chain, and a phosphate or modified phosphate head group. Historically these analytes have been examined using analytical methods such as organic extraction followed by thin layer chromatography (TLC). This approach is time-intensive, is prone to large inter-assay variability and has relatively low sensitivity. TLC-gas chromatography and TLC-mass spectrometry (MS) approaches have been successfully used in research laboratories to quantitate LPA in blood and ascitic fluid of ovarian cancer patients (Xu et al., 1995b, 1998; Xiao *et al.*, 2000; Shen *et al.*, 2001). However, this approach is still limited by the time, cost, and assay variability of TLC, and the sensitivity of GC may not be sufficient for higher throughput commercial biolipomics of these analytes. Utilization of ^{1}H- and ^{31}P-NMR has probably been the closest attempt to lipid profiling, has generated some functional and metabolic profiles of lipids associated with disease progression, and has implicated lipid biochemistry in the pathogenesis of disease (Podo, 1999; Klein, 2000; Katz-Brull *et al.*, 2002; Sullentrop et al., 2002). However high cost and limited availability of NMR technology has limited its use as a routine commercial analytical tool.

An alternative method for examining lipomic and biolipomic profiles is liquid chromatography-MS (LC-MS). High-performance LC (HPLC) can be used to separate many phospholipids and some lysophospholipids as shown in several published (Creer and Gross, 1985; Brouwers *et al.*, 1999; Lesnefsky *et al.*, 2000) and unpublished studies. However, the resolution and sensitivity of these methods are not sufficient to examine detailed lipomic and biolipomic profiles. Detection of these analytes using MS greatly improves resolution and sensitivity. Therefore coupled LC-MS approaches allow the rapid quantitation of both classes and molecular species of the lipids of interest.

Complex biological samples can be easily prepared for LC-MS analysis using automated liquid handlers and typical reagents. Samples can be separated by HPLC and specific analytes can be quantitated using small, highly sensitive mass spectrometers. Of course, several aspects of sample preparation, separation by HPLC, and detection by MS must be thoroughly developed and validated for high accuracy and precision. Several types of mass spectrometers may be used including single quadrupole (*i.e.* MS), triple quadrupole (*i.e.* MS/MS), ion trap, and the new combination instruments depending on the required analytical sensitivity and specificity. These types of instruments have become cost-effective, can be used commercially and are becoming common tools for major reference laboratories and hospitals. However, specific commercial methods on these instruments for generating biolipomic profiles are not currently available. These methods must be robust and may require significant time and cost for commercial development. Once established, these methods may be routinely used on commercially available instruments in a variety of laboratory and hospital settings. Commercial development of such methods is ongoing.

8. CONCLUSIONS

New efforts in lipid profiling offer a unique opportunity to examine normal biology and disease states beyond genomics and proteomics.

Focused efforts in biolipomics (*i.e.* particular bioactive lipids) and lipogenomics (*i.e.* genes involved in lipid metabolism and actions) will provide unique information about particular diseases. Utilizing particular patterns or profiles of lipids may in fact provide better diagnostic and prognostic indicators of disease since these analyses combine metabolic and functional information. Contrary to other profiling efforts such as proteomics, many of these analytes are already recognized and will not require extensive characterization for commercial and clinical use. These lipomic and lipogenomic patterns for a particular disease, may also be combined with various environmental, clinical (*e.g.* family history, risk factors, and diet), and diagnostic (*e.g.* traditional markers, proteomics, and others) inputs to provide sensitive and specific population-based diagnostic and prognostic disease indicators (Fig 3).

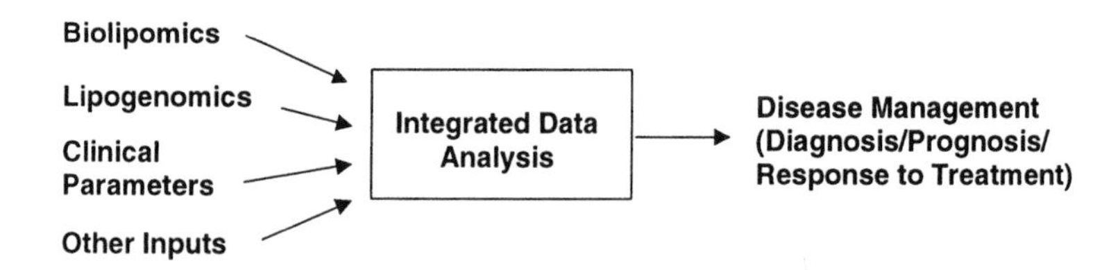

Figure 3. Integration of information from biolipomics, lipogenomics and other sources. Changes in lipid profiles and associated clinical parameters will aid in the understanding of and ability to diagnose complex diseases.

ACKNOWLEDGEMENTS

We would like to thank Dr. Walter Tribley for assistance in the writing and reviewing of this chapter. We would also like to thank all of the employees at Atairgin Technologies for their dedicated efforts to help develop methodologies to diagnose ovarian cancer and other diseases.

REFERENCES

ACOG. Hypertension in pregnancy. In American College of Obstetricians and Gynecologists, Washington DC, Technical Bulletin No. 219 (1996).

ACS. Cancer Facts and Figures 2002 and Imaging Breast Health Information. American Cancer Society (2002).

Albala JS. Array-based proteomics: the latest chip challenge. *Expert Rev Mol Diagn* 1: 145-152 (2001).

Alizadeh AA, Ross DT, Perou CM, van de Rijn M. Towards a novel classification of human malignancies based on gene expression patterns. *J Pathol* 195: 41-52 (2001).
Baker DL, Morrison P, Miller B *et al.* Plasma lysophosphatidic acid concentration and ovarian cancer. *J Am Med Assoc* 287: 3081-3082 (2002).
Ballweg ML. *The Endometriosis Sourcebook.* Contemporary Books, Lincolnwood (1995).
Balsinde J, Balboa MA, Insel PA, Dennis EA. Regulation and inhibition of phospholipase A2. *Ann Rev Pharmacol Toxicol* 39: 175-189 (1999).
Bandoh K, Aoki J, Taira A *et al.* Lysophosphatidic acid (LPA) receptors of the EDG family are differentially activated by LPA species. Structure-activity relationship of cloned LPA receptors. *FEBS Lett* 478: 159-165 (2000).
Bedaiwy MA, Falcone T, Sharma RK *et al.* Prediction of endometriosis with serum and peritoneal fluid markers: a prospective controlled trial. *Hum Reprod* 17: 426-431 (2002).
Berg CJ, Atrash HK, Koonin LM, Tucker M. Pregnancy-related mortality in the United States, 1987-1990. *Obstet Gynecol* 88:161-167 (1996).
Brosens I. Diagnosis of endometriosis. *Semin Reprod Endocrinol* 15: 229-233 (1997).
Brouwers JF, Vernooij EA, Tielens AG, van Golde LM. Rapid separation and identification of phosphatidylethanolamine molecular species. *J Lipid Res* 40: 164-169 (1999).
Budnik LT, Mukhopadhyay AK. Lysophosphatidic acid and its role in reproduction. *Biol Reprod* 66: 859-865 (2002).
Bussolino F, Camussi G. Platelet-activating factor produced by endothelial cells. A molecule with autocrine and paracrine properties. *Eur J Biochem* 229: 327-337 (1995).
Cheong YC, Shelton JB, Laird M *et al.* IL-1, IL-6 and TNF-alpha concentrations in the peritoneal fluid of women with pelvic adhesions. *Hum Reprod* 17: 69-75 (2002).
Chin TY, Chueh SH. Sphingosylphosphorylcholine stimulates mitogen-activated protein kinase via a Ca^{2+}-dependent pathway. *Am J Physiol* 275: C1255-1263 (1998).
Chun J, Goetzl EJ, Hla T *et al.* International Union of Pharmacology. XXXIV. Lysophospholipid receptor nomenclature. *Pharmacol Rev* 54: 265-269 (2002).
Contos JJ, Ishii I, Chun J. Lysophosphatidic acid receptors. *Mol Pharmacol* 58: 1188-1196 (2000).
Creer MH, Gross RW. Separation of isomeric lysophospholipids by reverse phase HPLC. *Lipids* 20: 922-928 (1985).
Cunnick JM, Dorsey JF, Munoz-Antonia T *et al.* Requirement of SHP2 binding to Grb2-associated binder-1 for mitogen-activated protein kinase activation in response to lysophosphatidic acid and epidermal growth factor. *J Biol Chem* 275: 13842-13848 (2000).
Dowhan W. Molecular basis for membrane phospholipid diversity: why are there so many lipids? *Ann Rev Biochem* 66: 199-232 (1997).
Eder AM, Sasagawa T, Mao M *et al.* Constitutive and lysophosphatidic acid (LPA)-induced LPA production: role of phospholipase D and phospholipase A2. *Clin Cancer Res* 6: 2482-2491 (2000).
Edsall LC, Cuvillier O, Twitty S *et al.* Sphingosine kinase expression regulates apoptosis and caspase activation in PC12 cells. *J Neurochem* 76: 1573-1584 (2001).
Endresen MJ, Lorentzen B, Henriksen T. Increased lipolytic activity of sera from pre-eclamptic women due to the presence of a lysophospholipase. *Scand J Clin Lab Invest* 53: 733-739 (1993).
Erickson JR, Espinal G, Mills GB. Analysis of the EDG2 receptor based on the structure/activity relationship of LPA. *Ann NY Acad Sci* 905: 279-281 (2000).

Erickson JR, Hasegawa Y, Fang X *et al.* Lysophosphatidic acid and ovarian cancer: a paradigm for tumorogenesis and patient management. *Prostaglandins Other Lipid Mediat* 64: 63-81 (2001).

Eskenazi B, Warner M, Bonsignore L *et al.* Validation study of nonsurgical diagnosis of endometriosis. *Fertil Steril* 76: 929-935 (2001).

Eskenazi B, Warner ML. Epidemiology of endometriosis. *Obstet Gynecol Clin North Am* 24: 235-258 (1997).

Fang X, Yu S, LaPushin R *et al.* Lysophosphatidic acid prevents apoptosis in fibroblasts via G(i)-protein-mediated activation of mitogen-activated protein kinase. *Biochem J* 352: 135-143 (2000).

Farooqui AA, Horrocks LA, Farooqui T. Deacylation and reacylation of neural membrane glycerophospholipids. *J Mol Neurosci* 14: 123-135 (2000).

Fukushima N, Ishii I, Contos JJ *et al.* Lysophospholipid receptors. *Ann Rev Pharmacol Toxicol* 41: 507-534 (2001).

Gaits F, Fourcade O, Le Balle F *et al.* Lysophosphatidic acid as a phospholipid mediator: pathways of synthesis. *FEBS Lett* 410: 54-58 (1997).

Genbacev O, DiFederico E, McMaster M, Fisher SJ. Invasive cytotrophoblast apoptosis in pre-eclampsia. *Hum Reprod* 14: 59-66 (1999).

Gerbie AB, Merrill JA. Pathology of endometriosis. *Clin Obstet Gynecol* 31: 779-786 (1988).

Goetzl EJ. Pleiotypic mechanisms of cellular responses to biologically active lysophospholipids. *Prostaglandins Other Lipid Mediat* 64: 11-20 (2001).

Goetzl EJ, An S. Diversity of cellular receptors and functions for the lysophospholipid growth factors lysophosphatidic acid and sphingosine 1-phosphate. *FASEB J* 12: 1589-1598 (1998).

Goetzl EJ, Dolezalova H, Kong Y *et al.* Distinctive expression and functions of the type 4 endothelial differentiation gene-encoded G protein-coupled receptor for lysophosphatidic acid in ovarian cancer. *Cancer Res* 59: 5370-5375 (1999a).

Goetzl EJ, Dolezalova H, Kong Y, Zeng L. Dual mechanisms for lysophospholipid induction of proliferation of human breast carcinoma cells. *Cancer Res* 59: 4732-4737 (1999b).

Gratacos E. Lipid-mediated endothelial dysfunction: a common factor to preeclampsia and chronic vascular disease. *Eur J Obstet Gynecol Reprod Biol* 92: 63-66 (2000).

Hannun YA, Luberto C, Argraves KM. Enzymes of sphingolipid metabolism: from modular to integrative signaling. *Biochemistry* 40: 4893-4903 (2001).

Hemmings R, Miron P, Falcone T *et al.* Platelet-activating factor acetylhydrolase activity in peritoneal fluids of women with endometriosis. *Obstet Gynecol* 81: 276-279 (1993).

Herrlich A, Daub H, Knebel A *et al.* Ligand-independent activation of platelet-derived growth factor receptor is a necessary intermediate in lysophosphatidic, acid-stimulated mitogenic activity in L cells. *Proc Natl Acad Sci USA* 95: 8985-8990 (1998).

Hopper DW, Ragan SP, Hooks SB *et al.* Structure-activity relationships of lysophosphatidic acid: conformationally restricted backbone mimetics. *J Med Chem* 42: 963-970 (1999).

Imai A, Furui T, Tamaya T, Mills GB. A gonadotropin-releasing hormone-responsive phosphatase hydrolyses lysophosphatidic acid within the plasma membrane of ovarian cancer cells. *J Clin Endocrinol Metab* 85: 3370-3375 (2000).

Jalink K, Hengeveld T, Mulder S *et al.* Lysophosphatidic acid-induced Ca^{2+} mobilization in human A431 cells: structure-activity analysis. *Biochem J* 307: 609-616 (1995).

Jalink K, Hordijk PL, Moolenaar WH. Growth factor-like effects of lysophosphatidic acid, a novel lipid mediator. *Biochim Biophys Acta* 1198: 185-196 (1994).

Jasinska R, Zhang QX, Pilquil C *et al.* Lipid phosphate phosphohydrolase-1 degrades exogenous glycerolipid and sphingolipid phosphate esters. *Biochem J* 340: 677-686 (1999).

Jendryczko A, Drozdz M. Increased placental phospholipase A2 activities in pre-eclampsia. *Zentralbl Gynakol* 112: 889-891 (1990).

Jendryczko A, Tomala J, Drozdz M, Wloch S. Placental phospholipase A2 activities in pre-eclampsia. *Ginekol Pol* 60: 280-282 (1989).

Katz-Brull R, Seger D, Rivenson-Segal D *et al.* Metabolic markers of breast cancer: enhanced choline metabolism and reduced choline-ether-phospholipid synthesis. *Cancer Res* 62: 1966-1970 (2002).

Khan J, Bittner ML, Chen Y *et al.* DNA microarray technology: the anticipated impact on the study of human disease. *Biochim Biophys Acta* 1423: M17-28 (1999).

Klein J. Membrane breakdown in acute and chronic neurodegeneration: focus on choline-containing phospholipids. *J Neural Transm* 107: 1027-1063 (2000).

Kobayashi T, Gu F, Gruenberg J. Lipids, lipid domains and lipid-protein interactions in endocytic membrane traffic. *Semin Cell Dev Biol* 9: 517-526 (1998).

Koh JS, Lieberthal W, Heydrick S, Levine JS. Lysophosphatidic acid is a major serum noncytokine survival factor for murine macrophages which acts *via* the phosphatidylinositol 3-kinase signaling pathway. *J Clin Invest* 102: 716-727 (1998).

Koonin LM, Atrash HK, Rochat RW, Smith JC. Maternal mortality surveillance, United States, 1980-1985. *Mor Mortal Wkly Rep CDC Surveill Summ* 37: 19-29 (1988).

Kranenburg O, Moolenaar WH. Ras-MAP kinase signaling by lysophosphatidic acid and other G protein-coupled receptor agonists. *Oncogene* 20: 1540-1546 (2001).

Laffargue M, Raynal P, Yart A *et al.* An epidermal growth factor receptor/Gab1 signaling pathway is required for activation of phosphoinositide 3-kinase by lysophosphatidic acid. *J Biol Chem* 274: 32835-32841 (1999).

le Balle F, Simon MF, Meijer S *et al.* Membrane sidedness of biosynthetic pathways involved in the production of lysophosphatidic acid. *Adv Enzyme Reg* 39: 275-284 (1999).

Lesnefsky EJ, Stoll MS, Minkler PE, Hoppel CL. Separation and quantitation of phospholipids and lysophospholipids by high-performance liquid chromatography. *Anal Biochem* 285: 246-254 (2000).

Levade T, Auge N, Veldman RJ *et al.* Sphingolipid mediators in cardiovascular cell biology and pathology. *Circ Res* 89: 957-968 (2001).

Lim KH, Rice GE, de Groot CJ, Taylor RN. Plasma type II phospholipase A2 levels are elevated in severe preeclampsia. *Am J Obstet Gynecol* 172: 998-1002 (1995).

Lopez-Jaramillo P, Casas JP, Serrano N. Preeclampsia: from epidemiological observations to molecular mechanisms. *Braz J Med Biol Res* 34: 1227-1235 (2001).

Lu J, Xiao YJ, Baudhuin LM *et al.* Role of ether-linked lysophosphatidic acids in ovarian cancer cells. *J Lipid Res* 43: 463-476 (2002).

Lu PY, Ory SJ. Endometriosis: current management. *Mayo Clin Proc* 70: 453-463 (1995).

Lykidis A, Jackowski S. Regulation of mammalian cell membrane biosynthesis. *Prog Nucleic Acid Res Mol Biol* 65: 361-393 (2001).

Lynch KR, Hopper DW, Carlisle SJ *et al.* Structure/activity relationships in lysophosphatidic acid: the 2-hydroxyl moiety. *Mol Pharmacol* 52: 75-81 (1997).

Lynch KR, Macdonald TL. Structure activity relationships of lysophospholipid mediators. *Prostaglandins Other Lipid Mediat* 64: 33-45 (2001).

Ma YY, Wei SJ, Lin YC *et al.* PIK3CA as an oncogene in cervical cancer. *Oncogene* 19: 2739-2744 (2000).

Merchant TE, Meneses P, Gierke LW *et al.* ^{31}P magnetic resonance phospholipid profiles of neoplastic human breast tissues. *Br J Cancer* 63: 693-698 (1991).
Mills GB, Eder A, Fang X *et al.* Critical role of lysophospholipids in the pathophysiology, diagnosis, and management of ovarian cancer. *Cancer Treat Res* 107: 259-283 (2002).
Mills GB, Lu Y, Fang X *et al.* The role of genetic abnormalities of PTEN and the phosphatidylinositol 3-kinase pathway in breast and ovarian tumorigenesis, prognosis, and therapy. *Semin Oncol* 28: 125-141 (2001).
Montrucchio G, Sapino A, Bussolati B *et al.* Potential angiogenic role of platelet-activating factor in human breast cancer. *Am J Pathol* 153: 1589-1596 (1998).
Moolenaar WH. Bioactive lysophospholipids and their G protein-coupled receptors. *Exp Cell Res* 253: 230-238 (1999).
Murphy AA. Clinical aspects of endometriosis. *Ann NY Acad Sci* 955: 1-10: discussion 34-16, 396-406 (2002).
Murphy AA, Santanam N, Morales AJ, Parthasarathy S. Lysophosphatidylcholine, a chemotactic factor for monocytes/T-lymphocytes is elevated in endometriosis. *J Clin Endocrinol Metab* 83: 2110-2113 (1998).
Norwitz ER, Repke JT. Preeclampsia prevention and management. *J Soc Gynecol Investig* 7: 21-36 (2000).
Older J. Leeches and laudanum: Grandmother and you: Historical highlights. In *Endometriosis*. Scribner's, New York (1984).
Oliver DJ, Nikolau B, Wurtele ES. Functional genomics: high-throughput mRNA, protein, and metabolite analyses. *Metab Eng* 4: 98-106 (2002).
Pages C, Simon MF, Valet P, Saulnier-Blache JS. Lysophosphatidic acid synthesis and release. *Prostaglandins Other Lipid Mediat* 64: 1-10 (2001).
Panetti TS, Magnusson MK, Peyruchaud O *et al.* Modulation of cell interactions with extracellular matrix by lysophosphatidic acid and sphingosine 1-phosphate. *Prostaglandins Other Lipid Mediat* 64: 93-106 (2001).
Podo F. Tumour phospholipid metabolism. *NMR Biomed* 12: 413-439 (1999).
Pulkkinen MO, Poranen AK, Kivikoski AI, Nevalainen TJ. Elevated serum group II phospholipase A2 levels are associated with decreased blood flow velocity in the umbilical artery. *Gynecol Obstet Invest* 41: 93-95 (1996).
Pyne S, Pyne NJ. Sphingosine 1-phosphate signalling in mammalian cells. *Biochem J* 349: 385-402 (2000).
Ramey JW, Archer DF. Peritoneal fluid: its relevance to the development of endometriosis. *Fertil Steril* 60: 1-14 (1993).
Riordan JR, Rommens JM, Kerem B *et al.* Identification of the cystic fibrosis gene: Cloning and characterization of complementary DNA. *Science* 245: 1066-1073 (1989).
Roberts JM. Endothelial dysfunction in preeclampsia. *Semin Reprod Endocrinol* 16: 5-15 (1998).
Roberts R, Sciorra VA, Morris AJ. Human type 2 phosphatidic acid phosphohydrolases. Substrate specificity of the type 2a, 2b, and 2c enzymes and cell surface activity of the 2a isoform. *J Biol Chem* 273: 22059-22067 (1998).
Rochat RW, Koonin LM, Atrash HK, Jewett JF. Maternal mortality in the United States: report from the Maternal Mortality Collaborative. *Obstet Gynecol* 72: 91-97 (1988).
Rowland BL, Vermillion ST, Roudebush WE. Elevated circulating concentrations of platelet activating factor in preeclampsia. *Am J Obstet Gynecol* 183: 930-932 (2000).
Rui L, Archer SF, Argetsinger LS, Carter-Su C. Platelet-derived growth factor and lysophosphatidic acid inhibit growth hormone binding and signaling *via* a protein kinase C-dependent pathway. *J Biol Chem* 275: 2885-2892 (2000).

Sakata-Kaneko S, Wakatsuki Y, Usui T *et al.* Lysophosphatidylcholine upregulates CD40 ligand expression in newly activated human CD4$^+$ T cells. *FEBS Lett* 433: 161-165 (1998).
Salas SP. What causes pre-eclampsia? *Baillieres Best Pract Res Clin Obstet Gynaecol* 13: 41-57 (1999).
Sano M, Morishita T, Nozaki M *et al.* Elevation of the phospholipase A2 activity in peritoneal fluid cells from women with endometriosis. *Fertil Steril* 61: 657-662 (1994).
Sasagawa T, Okita M, Murakami J *et al.* Abnormal serum lysophospholipids in multiple myeloma patients. *Lipids* 34: 17-21 (1999).
Sato TN. A new role of lipid receptors in vascular and cardiac morphogenesis. *J Clin Invest* 106: 939-940 (2000).
Schulte KM, Beyer A, Kohrer K *et al.* Lysophosphatidic acid, a novel lipid growth factor for human thyroid cells: Over-expression of the high-affinity receptor edg4 in differentiated thyroid cancer. *Int J Cancer* 92: 249-256 (2001).
Shen Z, Wu M, Elson P *et al.* Fatty acid composition of lysophosphatidic acid and lysophosphatidylinositol in plasma from patients with ovarian cancer and other gynecological diseases. *Gynecol Oncol* 83: 25-30 (2001).
Siess W. Athero- and thrombogenic actions of lysophosphatidic acid and sphingosine-1-phosphate. *Biochim Biophys Acta* 1582: 204-215 (2002).
Siess W, Zangl KJ, Essler M *et al.* Lysophosphatidic acid mediates the rapid activation of platelets and endothelial cells by mildly oxidized low density lipoprotein and accumulates in human atherosclerotic lesions. *Proc Natl Acad Sci USA* 96: 6931-6936 (1999).
Six DA, Dennis EA. The expanding superfamily of phospholipase A(2) enzymes: Classification and characterization. *Biochim Biophys Acta* 1488: 1-19 (2000).
Sliva D, Mason R, Xiao H, English D. Enhancement of the migration of metastatic human breast cancer cells by phosphatidic acid. *Biochem Biophys Res Comm* 268: 471-479 (2000).
Spiegel S, Merrill AH. Sphingolipid metabolism and cell growth regulation. *FASEB J* 10: 1388-1397 (1996).
Stemmer SM. Current progress in early pregnancy investigation. *Early Pregnancy* 5: 4-8 (2001).
Subramanian VS, Goyal J, Miwa M *et al.* Role of lecithin-cholesterol acyltransferase in the metabolism of oxidized phospholipids in plasma: studies with platelet-activating factor-acetyl hydrolase-deficient plasma. *Biochim Biophys Acta* 1439: 95-109 (1999).
Sullentrop F, Moka D, Neubauer S *et al.* ^{31}P NMR spectroscopy of blood plasma: determination and quantification of phospholipid classes in patients with renal cell carcinoma. *NMR Biomed* 15: 60-68 (2002).
Templin MF, Stoll D, Schrenk M *et al.* Protein microarray technology. *Trends Biotechnol* 20: 160-166 (2002).
Tigyi G. Physiological responses to lysophosphatidic acid and related glycero-phospholipids. *Prostaglandins Other Lipid Mediat* 64: 47-62 (2001).
Tokumura A, Yamano S, Aono T, Fukuzawa K. Lysophosphatidic acids produced by lysophospholipase D in mammalian serum and body fluid. *Ann NY Acad Sci* 905: 347-350 (2000).
Vinatier D, Orazi G, Cosson M, Dufour P. Theories of endometriosis. *Eur J Obstet Gynecol Reprod Biol* 96: 21-34 (2001).
Walker JJ. Pre-eclampsia. *Lancet* 356: 1260-1265 (2000).
Wang A, Dennis EA. Mammalian lysophospholipases. *Biochim Biophys Acta* 1439: 1-16 (1999).

Wellbery C. Diagnosis and treatment of endometriosis. *Am Fam Physician* 60: 1753-1762, 1767-1758 (1999).

WHO. Geographic variation in the incidence of hypertension in pregnancy. World Health Organization international collaborative study of hypertensive disorders of pregnancy. *Am J Obstet Gynecol* 158: 80-83 (1988).

Xia P, Gamble JR, Wang L *et al.* An oncogenic role of sphingosine kinase. *Curr Biol* 10: 1527-1530 (2000).

Xiao Y, Chen Y, Kennedy AW *et al.* Evaluation of plasma lysophospholipids for diagnostic significance using electrospray ionization mass spectrometry (ESI-MS) analyses. *Ann NY Acad Sci* 905: 242-259 (2000).

Xu Y, Fang XJ, Casey G, Mills GB. Lysophospholipids activate ovarian and breast cancer cells. *Biochem J* 309: 933-940 (1995a).

Xu Y, Gaudette DC, Boynton JD *et al.* Characterization of an ovarian cancer activating factor in ascites from ovarian cancer patients. *Clin Cancer Res* 1: 1223-1232 (1995b).

Xu Y, Shen Z, Wiper DW *et al.* Lysophosphatidic acid as a potential biomarker for ovarian and other gynecologic cancers. *J Am Med Assoc* 280: 719-723 (1998).

Xu Y, Xiao YJ, Baudhuin LM, Schwartz BM. The role and clinical applications of bioactive lysolipids in ovarian cancer. *J Soc Gynecol Investig* 8: 1-13 (2001).

Yamashita S, Ogawa M, Sakamoto K *et al.* Elevation of serum group II phospholipase A2 levels in patients with advanced cancer. *Clin Chim Acta* 228: 91-99 (1994a).

Yamashita S, Yamashita J, Ogawa M. Overexpression of group II phospholipase A2 in human breast cancer tissues is closely associated with their malignant potency. *Br J Cancer* 69: 1166-1170 (1994b).

Yamashita S, Yamashita J, Sakamoto K *et al.* Increased expression of membrane-associated phospholipase A2 shows malignant potential of human breast cancer cells. *Cancer* 71: 3058-3064 (1993).

Yoshida A, Ueda H. Neurobiology of the Edg2 lysophosphatidic acid receptor. *Jpn J Pharmacol* 87: 104-109 (2001).

Zhu K, Baudhuin LM, Hong G, *et al.* Sphingosylphosphorylcholine and lysophosphatidylcholine are ligands for the G protein-coupled receptor GPR4. *J Biol Chem* 276: 41325-41335 (2001).

Chapter 13

EVOLUTIONARY COMPUTATION FOR THE INTERPRETATION OF METABOLOMIC DATA

Royston Goodacre[1,2] and Douglas B. Kell[2]
[1]Institute of Biological Sciences, University of Wales, Aberystwyth, SY23 3DD, UK [2]Department of Chemistry, University of Manchester Institute of Science and Technology, PO Box 88, Sackville St., Manchester, M60 1QD, UK

> "The fewer data needed, the better the information. And an overload of information, that is, anything much beyond what is truly needed, leads to information blackout. It does not enrich, but impoverishes."
> *Peter F. Drucker - Management: Tasks, Responsibilities, Practices*

1. INTRODUCTION

Post-genomic science is producing bounteous data floods, and as the above quotation indicates the extraction of the most meaningful parts of these data is key to the generation of useful new knowledge. Atypical metabolic fingerprint or metabolomics experiment is expected to generate thousands of data points (samples times variables) of which only a handful might be needed to describe the problem adequately. Evolutionary algorithms are ideal strategies for mining such data to generate useful relationships, rules and predictions. This chapter describes these techniques and highlights their exploitation in metabolomics.

In a recent study Lyman and Varian estimated that in 2000 the world produced between 1 and 2 exabytes ($1\text{-}2.10^{18}$ bytes) of unique information (www.sims.berkeley.edu/how-much-info). This data flood is roughly 250 megabytes for every man, woman and child on earth! IBM's (www.ibm.com) estimates are that information within the life sciences doubles every 6 months, and this data explosion comes from genomic sequencing, the 'omics' (transcriptome, proteomics, metabolomics), high-throughput screening as well as the more traditional pre-clinical and clinical trials.

Metabolomics is the third level of 'omics' analysis. The metabolome is the quantitative complement of all the low molecular weight molecules present in cells in a particular physiological or developmental state (Oliver *et al.*, 1998; Fiehn, 2002) and whilst complementary to transcriptomics and proteomics may be seen to have special advantages. In particular, we know from the theory underlying metabolic control analysis (MCA) (Kell and Westerhoff, 1986; Fell, 1996; Heinrich and Schuster, 1996; Mendes *et al.*, 1996; Kell and Mendes, 2000) as well as from experiment (Fiehn *et al.*, 2000a; Raamsdonk *et al.*, 2001), that while changes in the levels of individual enzymes may be expected to have little effect on metabolic fluxes, they can and do have significant effects on the concentrations of a variety of individual metabolites (Westerhoff and Kell, 1996). In addition, as the 'downstream' result of gene expression, changes in the metabolome are amplified relative to changes in the transcriptome and the proteome.

Currently the 'gold standard' for measuring the metabolome is gas chromatography-mass spectrometry (GC-MS) (Fiehn *et al.*, 2000a, 2000b), and whilst a single run generates the name of a metabolite (or unique designation) with its (relative) concentration, GC-MS suffers from being chemically biased because of the extraction solvents employed. It is also relatively slow both for the chromatography itself (typical run times are 30 min per sample) and for the subsequent deconvolution steps. By contrast, rather than attempting to measure every metabolite, metabolic fingerprinting methods are sufficiently rapid to enable the classification of samples according to the origin or their biological relevance (Fiehn, 2002). For high-throughput metabolic fingerprinting the methods typically employed include nuclear magnetic resonance spectroscopy (NMR) (Lindon *et al.*, 2000), direct infusion electrospray ionization-MS (Vaidyanathan *et al.*, 2001, 2002; Allen *et al.*, 2002), and Fourier transform infrared (FT-IR) spectroscopy (Winson *et al.*, 1997; Goodacre *et al.*, 1998; Oliver *et al.*, 1998). These profiling strategies generate large amounts of data, and it is obvious (Fiehn *et al.*, 2001; Mendes, 2002) that current informatic approaches need to adapt and grow in order to make the most of these data

2. MULTIVARIATE ANALYSIS

Multivariate data such as those from a metabolic fingerprint consist of the results of observations on a number of individuals (objects, or samples) of many different characters (variables) such as the spectral intensities at different mass-to-charge ratios, chemical shifts from NMR, or absorbance at different wavenumbers from FT-IR (Martens and Næs, 1989). Each variable may be regarded as constituting a different dimension, such that if there are *n* variables each object may be said to reside at a unique position in an abstract entity referred to as *n*-dimensional hyperspace (Goodacre *et al.*, 1996). This hyperspace is necessarily difficult to visualize (Wilkinson, 1999) and the underlying theme of multivariate analysis is thus *simplification* (Chatfield and Collins, 1980) or dimensionality reduction (Tukey, 1977). This usually means that one wants to summarize a large body of data by means of *relatively* few parameters, preferably the two or three which lend themselves to graphical display, with minimal loss of information.

Within chemometrics there are three varieties of algorithms that are used to analyze multivariate data.

2.1 The Clustering Variety

These algorithms are based on *unsupervised* learning (Duda *et al.*, 2001; Hastie *et al.*, 2001) and seek to answer the question 'How similar to one another are these samples based on the metabolite fingerprints I have collected?'

Conventionally the reduction of multivariate data has been carried out using principal components analysis (PCA; (Jolliffe, 1986; Everitt, 1993)) or hierarchical cluster analysis (HCA; (Manly, 1994)). PCA is a well-known technique for reducing the dimensionality of multivariate data whilst preserving most of the variance, and is used to identify *correlations* amongst a set of variables and to transform the original set of variables to a new set of *uncorrelated* variables called principal components (PCs). These PCs can then be plotted and clusters in the data visualized; moreover this technique can be used to detect outliers. In its more conventional form, HCA calculates distances (usually Euclidean, but often Mahanalobis or Manhattan) between the objects in either the original data or a derivative thereof (*e.g.* the PCs) and uses these to construct a similarity matrix using a suitable similarity coefficient. These distance measures are then processed by an agglomerative clustering algorithm (although divisive algorithms are also used) to construct a dendrogram. In post-genomics such methods are sometimes referred to as 'guilt-by-association' (Altshuler *et al.*, 2000; Oliver, 2000).

2.2 The Classification/Quantification Variety

These algorithms are based on *supervised* learning (e.g. (Mitchell, 1997; Beavis *et al.*, 2000; Kell and King, 2000; Hastie *et al.*, 2001)) and seek to give answers of biological interest which have much-lower dimensionality, such as "Based on the metabolite fingerprint of this new sample I have just collected, which class in my database does it (most likely) belong to?" and/or "what are the levels of these metabolites in my biological sample?"

The basic idea behind supervised learning is that there are some patterns (*e.g.* metabolic fingerprints) that have desired responses which are known (*i.e.* whether an animal has been challenged with a drug or placebo). These two types of data (the representation of the objects and their responses in the system) form pairs that are conventionally called inputs (*x*-data) and targets (*y*-data). The goal of supervised learning is to find a *model* or *mapping* that will correctly associate the inputs with the targets (Fig. 1).

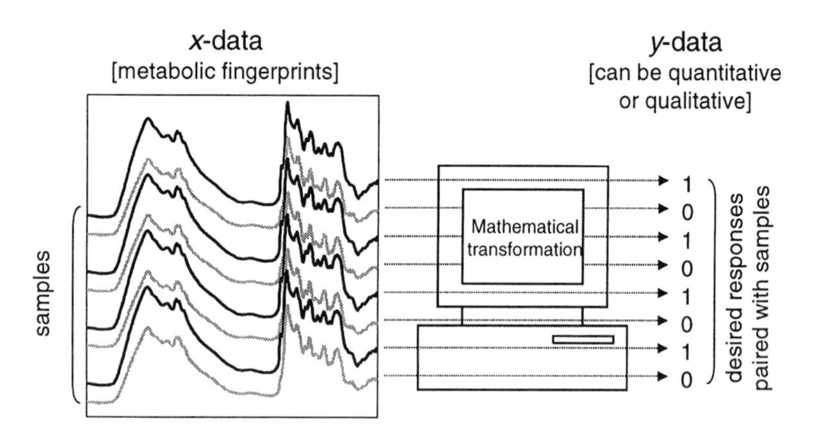

Figure 1. Supervised learning: When we know the desired responses (*y*-data, or targets) associated with each of the inputs (*x*-data, or metabolic fingerprints) then the system may be supervised. The goal is to find a mathematical transformation (model) that will correctly associate all or some of the inputs with the targets. In its conventional form this is achieved by minimizing the error between the known target and the model's response (output).

Many different algorithms perform supervised learning. Among the most common are (a) discriminant function analysis (DFA), which is a qualitative (categorical), cluster analysis-based method that involves projection of test data into cluster space (Manly, 1994; Radovic *et al.*, 2001), (b) partial least squares (PLS) which is a quantitative linear regression method (Martens and Næs, 1989) and (c) discriminant PLS, a qualitative (categorical) linear regression method (Martens and Næs, 1989; Alsberg *et al.*, 1998). However, arguably the most popular supervised learning methods are based on artificial neural networks (ANNs) which can learn non-linear as well as linear mappings. The most popular varieties are multilayer perceptrons (Werbos, 1994) and radial basis functions (Broomhead and Lowe, 1988; Saha and Keller, 1990; Bishop, 1995). In these supervised learning techniques there are minimally 4 data sets to be studied, as follows. The "training data" consist of (i) a matrix of *s* rows and *n* columns in which *s* is the number of objects/samples and *n* the number of variables (the *y*-data referred to above), and (ii) a second matrix, again consisting of *s* rows and typically 1 to *i* columns, in which the columns represent the variable(s) whose value(s) it is desired to know (the *y*-data or targets) and which for the training set have actually been previously determined by some existing "benchmark" method. The *x*-data (ii) are always paired with the patterns in the same row in the *y*-data (i). The "test data" also consist of two matrices, (iii) and (iv), corresponding to those in (i) and (ii) above, but the test set contains different samples. As the name suggests, this second pair is used to test the accuracy of the system; alternatively (and better) they may be used to cross-validate the model. That is to say, after construction of the model using the training set (i, ii) the test data (iii) are then used to challenge the calibration model so as to obtain the model's prediction of results, and these are then compared with the known or expected responses (iv). Once these are within acceptable ranges for the test data then the model is considered to be calibrated and ready to use.

2.3 The Inductive / Mining Variety

These algorithms are also based on *supervised* learning and seek to answer the question 'What have I measured in my metabolic fingerprint that makes samples in class A different from samples in class B?'

The problem with the supervised learning algorithms detailed above is that the mathematical transformation from multivariate data to the target question of interest is often largely inaccessible. DFA, PLS, and ANN methods are often perceived as 'black box' approaches to modeling spectra. It is known from the statistical literature that better (*i.e.* more robust) predictions can often be obtained when only the most relevant input variables are considered (Seasholtz and Kowalski, 1993; Kell and Sonnleitner, 1995; Bø and Jonassen, 2002). Thus the best machine learning techniques should not only give the correct answer(s), but also identify a subset of the variables with maximal explanatory power. This can provide an interpretable description of what, in biological terms, is the basis for that answer. Such explanatory modeling methods do exist and are based on rule induction (Breiman *et al.*, 1984; Harrington, 1991; Quinlan, 1993; Alsberg *et al.*, 1997), inductive logic programming (Lloyd, 1987; Muggleton, 1990; King *et al.*, 1992; Lavrac and Dzeroski, 1994), and, in particular, evolutionary computation (Holland, 1992; Koza, 1992; Bäck *et al.*, 1997).

3. EVOLUTIONARY COMPUTATION

Evolutionary computational-based algorithms are particularly popular inductive reasoning and optimization methods (Corne *et al.*, 1999; Michalewicz and Fogel, 2000). They are based on concepts of Darwinian selection (Bäck *et al.*, 1997) to generate and to optimize a desired computational function or mathematical expression that will yield explanatory 'rules'. These techniques include genetic algorithms (GAs; (Goldberg, 1989; Holland, 1992; Michalewicz, 1994; Mitchell, 1995)), evolution strategies (Schwefel, 1995; Beyer, 2001), evolutionary programming (Fogel, 1995, 2000) genetic programming (GP; (Koza, 1992, 1994; Banzhaf *et al.*, 1998; Koza *et al.*, 1999)) and genomic computing (GC; (Kell *et al.*, 2001; Kell, 2002A, 2002b)), and because the models are in English, and can penalize complex expressions, they may be made to be comparatively simple and easily interpreted.

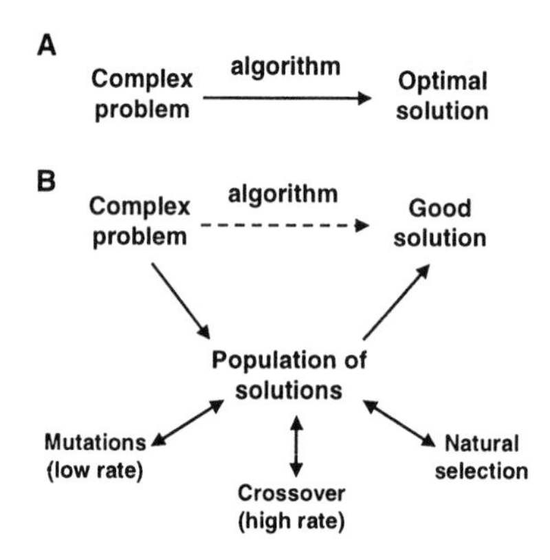

Figure 2. (A) The complex problem we wish to solve but cannot, and (B) the GA strategy.

If we consider the generic "Traveling Salesman Problem" where the object of the exercise is to find the shortest route between 20 cities, with the caveat that one may only visit each city once, we could (a) write down every possible order, (b) compute the distance for each, and (c) pick the shortest one. But is this really feasible? The number of possible orders is factorial and immense, $20! = 2.4 \times 10^{18}$, this number is so big that if your computer could check 1 million orderings every second it would still take 77,000 years to check them all! Thus even though we know how to solve the Traveling Salesman Problem we still cannot do it. This is true for identifying a subset of the variables from a metabolic fingerprint with the globally maximal explanatory power. For example, if we have measured only a modest 200 variables an exhaustive search of all possible permutations (where a variable is either used or not) is $2^{200} = 1.6 \times 10^{60}$. These problems are NP complete (see Garey and Johnson, 1979); that is to say to find the global optimum requires exhaustive search and this is computationally impossible. Thus route A in Fig. 2 is unfeasible as no algorithm can do this and an alternative strategy needs to be found. The premise is that a 'good' solution is acceptable and so we need an alternative method to search the huge spaces of possible solutions. Importantly, however, if the search space is large but the solution space is small, *i.e.* we can solve the problem with just a small number of variables, the effective search space becomes much narrower. Thus the number of permutations of 4 variables from 200 is just 6.47×10^7. GAs offer such an approach.

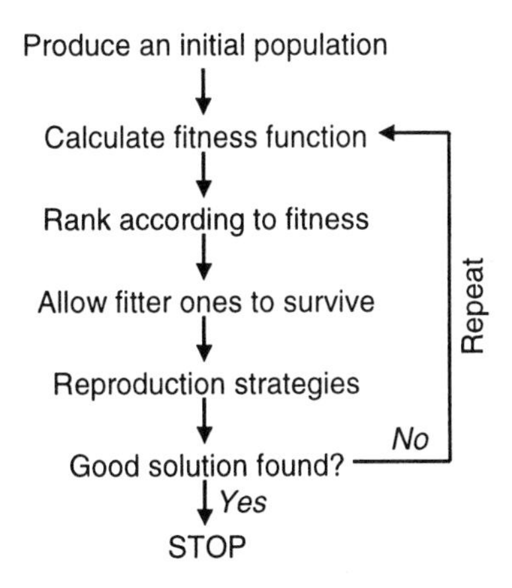

Figure 3. The overall procedure employed by GAs and GP. The criterion for a good solution will be based on setting a threshold error between the known target and the GAs' response.

In a GA a population of individuals, each representing the parameters of the problem to be optimized as a string of numbers or binary digits, undergoes a process analogous to evolution in order to derive an optimal or near-optimal solution (Fig. 2B). The parameters stored by each individual are used to assign it a *fitness*, a single numerical value indicating how well the solution using that set of parameters performs. New individuals are generated from members of the current population by processes analogous to asexual and sexual reproduction (Fig. 3).

Asexual reproduction, or *mutation*, is performed by randomly selecting a parent with a probability related to its fitness, then randomly changing one or more of the parameters it encodes. The new individual then replaces a less-fit member of the population, if one exists. Sexual reproduction, or *crossover*, is achieved by selecting two parents with a frequency related to their fitnesses, and generating two new individuals by copying parameters from one parent, and switching to the other parent after a randomly-selected point. The two new individuals then replace less fit members of the population as before. The above procedure is repeated, with the overall fitness of the population improving at each generation, until an acceptably fit individual is produced.

For variable selection prior to some supervised learning method, whether it is linear regression or ANNs, the state of each variable (in GA terminology a gene) is represented by a '1' (selected to be in the model) or a '0' (not selected) (Horchner and Kalivas, 1995; Broadhurst *et al.*, 1997). Together theses sets of variables are called a chromosome, this GA string would be of length m (where m = number of x-data input variables in the metabolic fingerprint). For example, in a variable selection problem starting with 7 variables, one possible chromosome would be 1101001. This can be translated such that variables 1, 2, 4, and 7 are to be used in the modeling process and variables 3, 5, and 6 are to be omitted. Other GA variants based on the selection of spectral windows for FT-IR and Raman spectroscopy are also popular (Williams and Paradkar, 1997; Taylor *et al.*, 1998; Roger and Bellon-Maurel, 2000; Leardi *et al.*, 2002; McGovern *et al.*, 2002).

However, whilst GAs are very successful search algorithms for tackling NP-hard problems, the disadvantage is that with the GA variable selection approach the relationship between one variable and another is not evident, only whether they contribute to a model or not. Therefore, a richer language is needed.

3.1 Genetic Programming

A GP is an application of the GA approach to derive mathematical equations, logical rules or program functions automatically (Koza, 1992, 1994; Gilbert *et al.*, 1997; Langdon, 1998; Koza *et al.*, 1999; Langdon and Poli, 2002). Rather than representing the solution to the problem as a string of parameters, as in a conventional GA, a GP usually (*c.f.* Banzhaf *et al.*, 1998) uses a tree structure. The leaves of the tree, or *terminals*, represent input variables or numerical constants. Their values are passed to *nodes*, at the junctions of branches in the tree, which perform some numerical or program operation before passing on the result further towards the root of the tree (Fig. 4). Genomic Computing (GC; Kell *et al.*, 2001; Kell, 2002a, 2002b) (www.abergc.com) is a variant on a GP.

The overall evolutionary procedure employed by GP is essentially identical to that of GAs. An initial (commonly random) population of individuals, each encoding a function or expression, is generated and their fitness to produce the desired output is assessed. In the second population three reproduction strategies are adopted (see Fig. 5 for pictorial details).

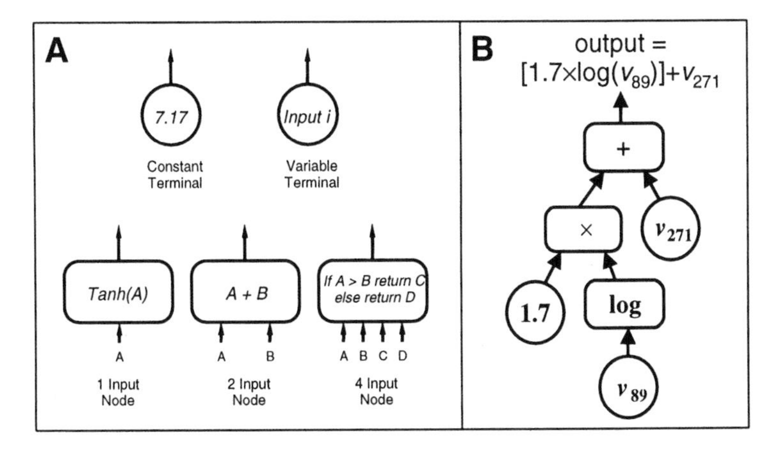

Figure 4. The richer language structure of a tree-encoded GP: (A) the building blocks and (B) a typical function tree.

(1) *Cloning:* some of the original individuals are allowed to survive unmodified.

(2) New individuals are generated by *mutation* where one or more random changes to a single parent individual are introduced. This can be when a node is randomly chosen, and modified either by giving it a different operator with the same number of arguments, or it may be replaced by a new random sub-tree. Terminals can be mutated by slightly perturbing their numerical values, or randomly choosing an input variable.

(3) Alternatively new children are generated by *crossover* where random rearrangement of functional components between two or more parent individuals takes place. Two parents are chosen with a probability related to their fitness. A node is randomly chosen on each parent tree, and the selected sub-trees are then swapped. At each reproduction stage because of the use of these trees to encode mathematical equations the new trees are still syntactically correct. The fitness of the new individuals in population 2 is assessed and the best individuals from the total population become the parents of the next generation. An individual's fitness is usually assessed as the root mean squared error of the difference between expected values and the GP's estimated values for the training set. In order to reduce 'bloat', a phenomenon in which the GP function trees gets so huge that it lacks explanatory power (Langdon and Poli, 1998), penalties to the number of nodes and depth of the tree in the individual's function tree can be applied.

This overall process is repeated until either the desired result is achieved or the rate of improvement in the population becomes zero. It has been shown (Koza, 1992) that if the parent individuals are chosen according to their fitness values, the genetic method can approach the theoretical optimum efficiency for a search algorithm, and EAs generally are guaranteed to find the global optimum provided the best individuals are retained between generations ('elitism') (Rudolph, 1997).

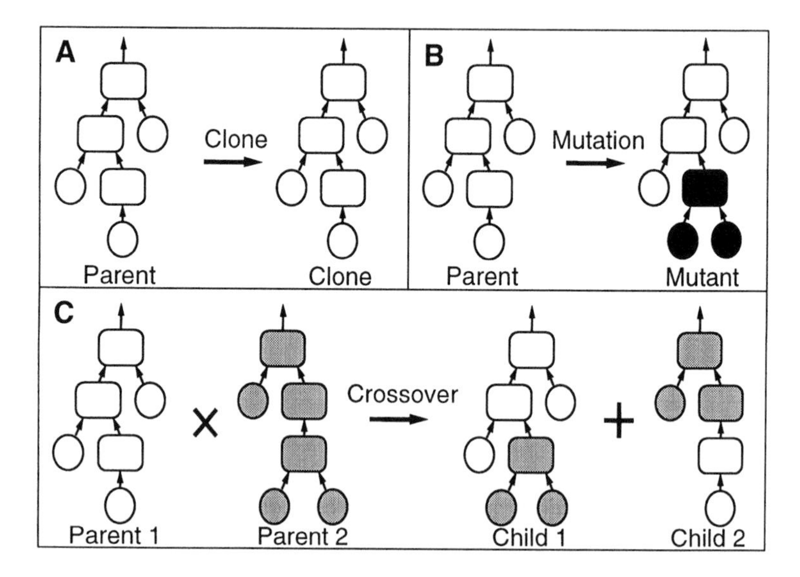

Figure 5. The GP reproduction processes, showing examples of (A) cloning, (B) mutation and (C) crossover events.

4. APPLICATION OF EVOLUTIONARY COMPUTATION-BASED METHODS TO METABOLOMICS

GAs and GPs are very efficient search algorithms and can be used to produce models that allow the deconvolution of metabolome data in chemical terms. Detailed below are five published examples illustrating this.

Example 1 (Goodacre *et al.*, 2000). Members of the genus *Bacillus* are widely distributed in soil, water, and air, and because their spores are so resistant their control is of considerable importance in the food processing industry and in the preparation of sterile products (Doyle *et al.*, 1997). In addition, the rapid identification of *Bacillus anthracis* spores is of importance because of its potential use as a biological warfare agent (Dando, 1994; Barnaby, 1997). Therefore, there is a need for a generic characterization system that can be used to carry out large-scale and rapid detection of bacterial spores. GP was used to analyze metabolic fingerprints generated from vegetative biomass and spores using Curie-point pyrolysis-MS (Py-MS) and FT-IR. Both fingerprinting approaches could be used to differentiate successfully between vegetative biomass and spores. GP produced mathematical rules that could be interpreted in simple biochemical terms. It was found that for Py-MS, a peak at *m/z* 105 was characteristic and attributable to a pyridine ketonium ion derived from the pyrolysis of pyridine-2,6-dicarboxylic acid (dipicolinic acid), a metabolite found in spores but not in vegetative cells. In addition, FT-IR analysis of the same system showed that a pyridine ring vibration at 1447-1439 cm^{-1} from the same metabolite, dipicolinic acid, was highly characteristic of spores. Thus, although the original datasets recorded hundreds of spectral variables from whole cells simultaneously, a simple biomarker was detected that can be used for the rapid and unequivocal detection of spores of these organisms.

Example 2 (Johnson *et al.*, 2000). Samples from tomato fruit grown hydroponically under both high- and low-salt conditions were analyzed by FT-IR, with the aim of identifying biochemical features linked to salinity in the growth environment. Examination of the GP-derived trees showed that there were a small number of spectral regions that were consistently used. In particular, the spectral region containing absorbances potentially due to a cyanide/nitrile functional group was identified as discriminatory. Cyanide is formed in plants during ethylene biosynthesis, and ethylene production is enhanced in plants subjected to stress conditions. Therefore, one may propose that plants grown under saline conditions may therefore have enhanced levels of cyanide as a result of enhanced ethylene biosynthesis. Thus inductive reasoning *via* GP has allowed the significance of a pathway turned on under tomatoes exposed to salinity to be highlighted as potentially important. This pathway can now be subjected to conventional biochemical analysis.

Example 3 (McGovern *et al.*, 2002). The previous two examples have been qualitative (*i.e.* the outputs were categorical variables). This example now demonstrates how GA and GP can be used in a quantitative fashion. The ability to control industrial bioprocess is paramount for product yield optimization, and it is imperative therefore that the concentration of the fermentation product (the determinand) is assessed accurately. Whilst IR and Raman spectroscopies have been used for the quantitative analysis of fermentations (McGovern *et al.*, 1999; Shaw *et al.*, 1999; Vaidyanathan *et al.*, 1999) the transformation of spectra to determinand concentration(s) has usually been undertaken by PLS and ANNs, and so one can not be sure whether the model is detecting the product itself, an increase in bi-products or decrease in substrates. By contrast, GA and GP have recently been used to analyse IR and Raman spectra from a diverse range of unprocessed, industrial fed-batch fermentation broths containing the fungus *Gibberella fujikuroi* which produces the gibberellic acid. The models produced allowed the determination of those input variables that contributed most to the models formed, and it was observed that those quantitative models were predominately based on the concentration of gibberellic acid itself.

Example 4 (Ellis *et al.*, 2002). Whilst a number of studies have applied FT-IR to the discrimination and adulteration of meats (Al-Jowder *et al.*, 1999; Downey *et al.*, 2000) its application to the rapid detection of microbial spoilage in meats has only very recently been demonstrated. A particularly robust and reproducible form of this method is attenuated total reflectance (ATR) where the food sample is placed in intimate contact with a crystal of high refractive index and an IR absorbance spectrum, a *metabolic snapshot*, collected in just a few seconds. It has been shown (Ellis *et al.*, 2002) that FT-IR with PLS allowed accurate estimates of bacterial loads (from 10^6 to 10^9 cm^{-2}) to be calculated directly from the chicken surface in 60s, and that GA and GP indicated that at levels of 10^7 bacteria.cm^{-2} the main biochemical indicator of spoilage as measured by FT-IR was the onset of proteolysis, a finding in agreement with the literature (Dainty, 1996; Nychas and Tassou, 1997).

Example 5 (Kell *et al.*, 2001). Within functional genomics the potential power of evolutionary methods has been shown for the analysis of metabolites from transgenic tobacco plants. Tobacco is a model organism for the study of salicylate biology in plant defense, but despite a considerable amount of research, little is known regarding its synthesis, catabolism, and mode of action. Six week old control plants and a transgenic expressing a bacterial gene encoding the enzyme salicylate hydroxylase (SH-L), which is known to block salicylic acid accumulation in transgenic tobacco (Darby *et al.*, 2000) were inoculated with tobacco mosaic virus and leaf samples were analyzed by HPLC. Genomic Computing analysis of these metabolome profiles identified 3 peaks as highly discriminatory for detecting the presence of the SH-L genotype in the transgenic. One of the peaks was indeed salicylate, but the other two were unknown and are now the subject of further investigation.

5. CONCLUSION

As scientists we are all aware of the cycle of knowledge (Fig. 6) (Kell, 2002b). One has some preconceived notions about the problem domain, experiments are designed to test these hypotheses, the observations from these experiments are recorded and by deductive reasoning the observations considered to be consistent or inconsistent with the hypotheses (Oldroyd, 1986). Actually, although this part is normally only implicit, by a process of induction these observations are synthesized or generalized to refine our accepted wisdom. The cycle then repeats itself until one is happy with the solution to a given problem. However, in the early stages of functional genomics programs we have a scenario where our knowledge is minute, that is to say we have no ideas about the role of an orphan open reading frame and there are few if any hypotheses to test (Brent, 1999; Brent, 2000; Kell and King, 2000). However, we can design experiments based, for example, on gene knockouts and controlled over-expression and observe the effect on the phenotype of the organism.

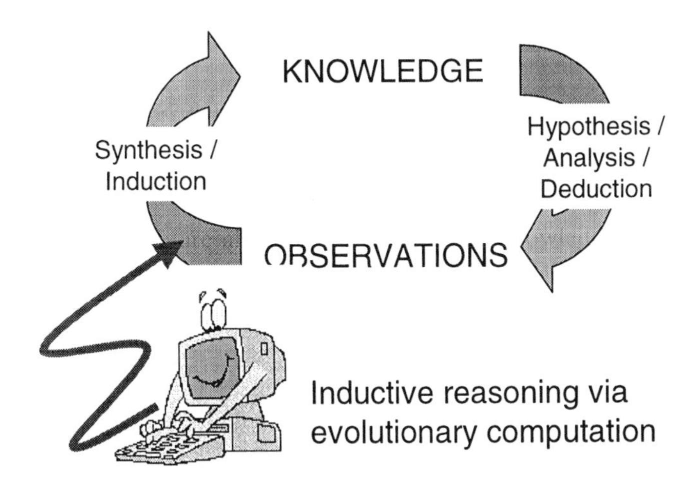

Figure 6. The cycle of knowledge showing where rule induction will play its part.

Metabolomics is one 'omics approach with which one can generate data floods from these genetic manipulations (as indeed are transcriptomics and proteomics, and the same general conclusions given here apply equally to these methods). Thus we are then positioned at the bottom of Fig. 6 where we have collected a great many observations and the trick is to drive the cycle round *via* inductive reasoning to generate new hypotheses. Evolutionary computing methods can be considered to be rule induction methods that are entirely data-driven and are thus especially appropriate for problems that are data-rich but hypothesis/information-poor. Rule induction by GP and GC can be used to generate rules and hence hypotheses from suitable examples. Of course these new theories will not necessarily be correct, but by testing them new knowledge will be generated which will lead to an increased understanding of the function of the orphan gene. In the new post-genomic biology, then, we shall need good databases (Mendes, 2002), very good data, and even better algorithms, with which to turn our data into knowledge.

ACKNOWLEDGEMENTS

The authors are indebted to the UK BBSRC (Engineering and Biological Systems Committee), the UK EPSRC and the Royal Society of Chemistry for financial support.

REFERENCES

Al-Jowder O, Defernez M, Kemsley EK, Wilson RH. Mid-infrared spectroscopy and chemometrics for the authentication of meat products. *J Agric Food Chem* 47: 3210-3218 (1999).

Allen JK, Davey HM, Broadhurst D *et al.* Metabolic footprinting: a high-throughput, high-information approach to cellular characterisation and functional genomics. *Nature Biotechnol* submitted (2002).

Alsberg BK, Goodacre R, Rowland JJ, Kell DB. Classification of pyrolysis mass spectra by fuzzy multivariate rule induction - comparison with regression, k-nearest neighbour, neural and decision-tree methods. *Anal Chim Acta* 348: 389-407 (1997).

Alsberg BK, Kell DB, Goodacre R. Variable selection in discriminant partial least squares analysis. *Anal Chem* 70: 4126-4133 (1998).

Altshuler D, Daly M, Kruglyak L. Guilt by association. *Nature Genet* 26: 135-137 (2000).

Bäck T, Fogel DB, Michalewicz Z. *Handbook of Evolutionary Computation*. Oxford University Press, Oxford (1997).

Banzhaf W, Nordin P, Keller RE, Francone FD. *Genetic Programming: An Introduction.* Morgan Kaufmann, San Francisco (1998).

Barnaby W. *The Plague Makers: The Secret World of Biolgoical Warfare*. Vision Paperbacks, London (1997).

Beavis RC, Colby SM, Goodacre R *et al.* Artificial intelligence and expert systems in mass spectrometry. In *Encyclopedia of Analytical Chemistry*. Meyers RA (Ed) pp. 11558-11597, John Wiley and Son, Chichester (2000).

Beyer H-G. *The Theory of Evolution Strategies*. Springer, Berlin (2001)

Bishop CM. *Neural Networks for Pattern Recognition*. Clarendon Press, Oxford (1995).

Bø TH, Jonassen I. New feature subset selection procedures for classification of expression profiles. http://genomebiologycom/2002/3/4/research/00171 3: research0017.1-0017.11 (2002).

Breiman L, Friedman JH, Olshen RA, Stone CJ. *Classification and Regression Trees.* Wadsworth Inc, Pacific Grove (1984).
Brent R. Functional genomics: learning to think about gene expression data. *Curr Biol* 9: R338-R341 (1999).
Brent R. Genomic biology. *Cell* 100: 169-183 (2000).
Broadhurst D, Goodacre R, Jones A *et al.* Genetic algorithms as a method for variable selection in PLS regression, with application to pyrolysis mass spectra. *Anal Chim Acta* 348: 71-86 (1997).
Broomhead DS, Lowe D. Multivariable function interpolation and adaptive networks. *Complex Sys* 2: 321-355 (1988).
Chatfield C, Collins AJ. *Introduction to Multivariate Analysis.* Chapman and Hall, London (1980).
Corne D, Dorigo M, Glover F (Ed). *New Ideas in Optimization.* McGraw Hill, London (1999).
Dainty RH. Chemical/biochemical detection of spoilage. *Int J Food Microbiol* 33: 19-33 (1996).
Dando M. *Biological Warfare in the 21st Century.* Brassey's Ltd., London (1994).
Darby RM, Maddison A, Mur LAJ *et al.* Cell specific expression of salicylate hydroxylase in an attempt to separate localised HR and systemic signalling establishing SAR in tobacco. *Plant Mol Pathol* 1: 115-124 (2000).
Downey G, McElhinney J, Fearn T. Species identification in selected raw homogenized meats by reflectance spectroscopy in the mid-infrared, near-infrared, and visible ranges. *Appl Spectr* 54: 894-899 (2000).
Doyle MP, Beuchat LR, Montville TJ (Ed) *Food Microbiology: Fundamentals and Frontiers.* American Society of Microbiology Press, Washington DC (1997).
Duda RO, Hart PE, Stork DE. *Pattern Classification.* 2nd Edn. John Wiley and Sons, London (2001).
Ellis DI, Broadhurst D, Kell DB *et al.* Rapid and quantitative detection of the microbial spoilage of meat using FT-IR spectroscopy and machine learning. *Appl Env Microbiol* 68: 2822-2828 (2002).
Everitt BS. *Cluster Analysis.* Edward Arnold, London (1993).
Fell DA. *Understanding the Control of Metabolism.* Portland Press, London (1996).
Fiehn O. Metabolomics – the link between genotypes and phenotypes. *Plant Mol Biol* 48: 155–171 (2002).
Fiehn O, Kloska S, Altmann T. Integrated studies on plant biology using multiparallel techniques. *Curr Opin Biotechnol* 12: 82-86 (2001).
Fiehn O, Kopka J, Dörmann P *et al.* Metabolite profiling for plant functional genomics. *Nature Biotechnol* 18: 1157-1161 (2000a).
Fiehn O, Kopka J, Trethewey RN, Willmitzer L. Identification of uncommon plant metabolites based on calculation of elemental compositions using gas chromatography and quadrupole mass spectrometry. *Anal Chem* 72: 3573-3580 (2000b).
Fogel DB. A comparison of evolutionary programming and genetic algorithms on selected constrained optimization problems. *Simulation* 64: 397-404 (1995).
Fogel DB. *Evolutionary Computation: Toward a New Philosophy of Machine Intelligence.* IEEE Press, Piscataway (2000).
Garey M, Johnson D. *Computers and Intractability: A Guide to the Theory of NP-Completeness.* Freeman, San Francisco (1979).
Gilbert RJ, Goodacre R, Woodward AM, Kell DB. Genetic programming: a novel method for the quantitative analysis of pyrolysis mass spectral data. *Anal Chem* 69: 4381-4389 (1997).

Goldberg DE. *Genetic Algorithms in Search, Optimization and Machine Learning*. Addison-Wesley, Reading (1989).

Goodacre R, Neal MJ, Kell DB. Quantitative analysis of multivariate data using artificial neural networks: a tutorial review and applications to the deconvolution of pyrolysis mass spectrtra. *Z Bakteriol* 284: 516-539 (1996).

Goodacre R, Shann B, Gilbert R *et al.* The detection of the dipicolinic acid biomarker in *Bacillus* spores using Curie-point pyrolysis mass spectrometry and Fourier transform infrared spectroscopy. *Anal Chem* 72: 119-127 (2000).

Goodacre R, Timmins ÉM, Burton R *et al.* Rapid identification of urinary tract infection bacteria using hyperspectral, whole organism fingerprinting and artificial neural networks. *Microbiol* 144: 1157-1170 (1998).

Harrington PB. Fuzzy rule-building expert systems: minimal neural networks. *J Chemometrics* 5: 467-486 (1991).

Hastie T, Tibshirani R, Friedman J. *The Elements of Statistical Learning: Data Mining, Inference and Prediction.* Springer-Verlag, Berlin (2001).

Heinrich R, Schuster S. *The Regulation of Cellular Systems*. Chapman and Hall, New York (1996).

Holland JH. *Adaption in Natural and Artificial Systems*. MIT Press, Cambridge (1992).

Horchner U, Kalivas JH. Further investigation on a comparative study of simulated annealing and genetic algorithm for wavelength selection. *Anal Chim Acta* 311: 1-13 (1995).

Johnson HE, Gilbert RJ, Winson MK *et al.* Explanatory analysis of the metabolome using genetic programming of simple, interpretable rules. *Genet Program Evolv Mach* 1: 243-258 (2000).

Jolliffe IT. *Principal Component Analysis*. Springer-Verlag, New York (1986).

Kell DB. Defence against the flood: a solution to the data mining and predictive modelling challenges of today. *Bioinformatics World* (part of Scientific Computing News) Issue 1: 16-18 (2002a) http://www.abergc.com/biwpp16-18_as_publ.pdf.

Kell DB. Genotype-phenotype mapping: genes as computer programs. *Trends Genet* in press (2002b).

Kell DB, Darby RM, Draper J. Genomic computing. Explanatory analysis of plant expression profiling data using machine learning. *Plant Phys* 126: 943-951 (2001).

Kell DB, King RD. On the optimization of classes for the assignment of unidentified reading frames in functional genomics programmes: the need for machine learning. *Trends Biotechnol* 18: 93-98 (2000).

Kell DB, Mendes P. Snapshots of systems: metabolic control analysis and biotechnology in the post-genomic era. In *Technological and Medical Implications of Metabolic Control Analysis*. Cornish-Bowden A, Cárdenas ML (Ed) pp. 3-25, Kluwer Academic Publishers, Dordrecht (2000) (see http://qbab.aber.ac.uk/dbk/mca99.htm).

Kell DB, Sonnleitner B. GMP - Good Modelling Practice: an essential component of Good Manafacturing Practice. *Trends Biotechnol* 13: 481-492 (1995).

Kell DB, Westerhoff HV. Towards a rational approach to the optimization of flux in microbial biotransformations. *Trends Biotechnol* 4: 137-142 (1986).

King RD, Muggleton S, Lewis RA, Sternberg MJE. Drug design by machine learning - the use of inductive logic programming to model the structure-activity-relationships of trimethoprim analogs binding to dihydrofolate-reductase. *Proc Natl Acad Sci USA* 89: 11322-11326 (1992).

Koza JR. 1992. *Genetic Programming: On the Programming of Computers by Means of Natural Selection.* MIT Press, Cambridge (1992).

Koza JR. *Genetic Programming II: Automatic Discovery of Reusable Programs*. MIT Press, Cambridge (1994).
Koza JR, Bennett FH, Keane MA, Andre D. *Genetic Programming III: Darwinian Invention and Problem Solving*. Morgan Kaufmann, San Francisco (1999).
Langdon WB. *Genetic Programming and Data Structures: Genetic Programming + Data Structures = Automatic Programming*! Kluwer Academic Publishers, Boston (1998).
Langdon WB, Poli R. Fitness causes bloat: mutation. In *Proc First European Workshop on Genetic Programming*. Vol. 1391. Banzhaf W, Poli R, Schoenauer M, Fogarty TC (Ed) pp. 37-48, Springer-Verlag, Berlin (1998).
ftp://ftp.cwi.nl/pub/W.B.Langdon/papers/WBL.euro98_bloatm.ps.gz.
Langdon WB, Poli R. *Foundations of Genetic Programming*. Springer-Verlag, Berlin (2002).
Lavrac N, Dzeroski S. *Inductive Logic Programming: Techniques and Applications*. Ellis Horwood, Chichester (1994).
Leardi R, Seasholtz MB, Pell RJ. Variable selection for multivariate calibration using a genetic algorithm: prediction of additive concentrations in polymer films from Fourier transform-infrared spectral data. *Anal Chim Acta* 461: 189-200 (2002).
Lindon JC, Nicholson JK, Holmes E, Everett JR. Metabonomics: metabolic processes studied by NMR spectroscopy of biofluids. *Concepts Magn Reson* 12: 289-320 (2000).
Lloyd JW. *Foundations of Logic Programming*. Springer-Verlag, Berlin (1987).
Manly BFJ. *Multivariate Statistical Methods: A Primer*. Chapman and Hall, London (1994).
Martens H, Næs T. *Multivariate Calibration*. John Wiley and Sons, Chichester (1989).
McGovern AC, Broadhurst D, Taylor J *et al.* Monitoring of complex industrial bioprocesses for metabolite concentrations using modern spectroscopies and machine learning: application to gibberellic acid production. *Biotechnol Bioeng* 78: 527-538 (2002).
McGovern AC, Ernill R, Kara BV *et al.* Rapid analysis of the expression of heterologous proteins in *Escherichia coli* using pyrolysis mass spectrometry and Fourier transform infrared spectroscopy with chemometrics: application to α2-interferon production. *J Biotechnol* 72: 157-167 (1999).
Mendes P. Emerging bioinformatics for the metabolome. *Briefings Bioinformat* 3: 134-45 (2002).
Mendes P, Kell DB, Westerhoff HV. Why and when channeling can decrease pool size at constant net flux in a simple dynamic channel. *Biochim Biophys Acta* 1289: 175-186 (1996).
Michalewicz Z. *Genetic Algorithms + Data Structures = Evolution Programs*. Springer-Verlag, Berlin (1994).
Michalewicz Z, Fogel DB. *How to Solve It: Modern Heuristics*. Springer-Verlag, Heidelberg (2000).
Mitchell M. *An Introduction to Genetic Algorithms*. MIT Press, Boston (1995).
Mitchell TM. *Machine Learning*. McGraw Hill, New York (1997).
Muggleton SH. Inductive logic programming. *New Generation Comput* 8: 295-318 (1990).
Nychas GJE, Tassou CC. Spoilage processes and proteolysis in chicken as detected by HPLC. *J Sci Food Agric* 74: 199-208 (1997).
Oldroyd D. *The Arch of Knowledge: An Introduction to the History of the Philosophy and Methodology of Science*. Methuen, New York (1986).
Oliver SG. Proteomics: guilt-by-association goes global. *Nature* 403: 601-603 (2000).
Oliver SG, Winson MK, Kell DB, Baganz F. Systematic functional analysis of the yeast genome. *Trends Biotechnol* 16: 373-378 (1998).
Quinlan JR. *C4.5: Programs for Machine Learning*. Morgan Kaufmann, San Mateo (1993).

Raamsdonk LM, Teusink B, Broadhurst D *et al.* A functional genomics strategy that uses metabolome data to reveal the phenotype of silent mutations. *Nature Biotechnol* 19: 45-50 (2001).

Radovic BS, Goodacre R, Anklam E. Contribution of pyrolysis mass spectrtrometry (Py-MS) to authenticity testing of honey. *J Anal Appl Pyrolysis* 60: 79-87 (2001).

Roger JM, Bellon-Maurel V. Using genetic algorithms to select wavelengths in near-infrared spectra: application to sugar content prediction in cherries. *Appl Spectr* 54: 1313-1320 (2000).

Rudolph G. *Convergence Properties of Evolutionary Algorithms*. Verlag Dr Kovac, Hamburg (1997).

Saha A, Keller JD. Algorithms for better representation and faster learning in radial basis functions. In *Advances in Neural Information Processing Sytems*. Vol. 2. Touretzky D (Ed) pp. 482-489, Morgan Kaufmann, San Mateo (1990).

Schwefel H-P. *Evolution and Optimum Seeking*. John Wiley and Sons, New York (1995).

Seasholtz MB, Kowalski B. The parsimony principle applied to multivariate calibration. *Anal Chim Acta* 277: 165-177 (1993).

Shaw AD, Kaderbhai N, Jones A *et al.* Non-invasive, on-line monitoring of the biotransformation by yeast of glucose to ethanol using dispersive Raman spectroscopy and chemometrics. *Appl Spectr* 53: 1419-1428 (1999).

Tukey JW. *Exploratory Data Analysis*. Addison-Wesley, Reading (1977).

Vaidyanathan S, Kell DB, Goodacre R. Flow-injection electrospray ionization mass spectrometry of crude cell extracts for high-throughput bacterial identification. *J Am Soc Mass Spectrom* 13: 118-128 (2002).

Vaidyanathan S, Macaloney G, McNeill B. Fundamental investigations on the near-infrared spectra of microbial biomass as applicable to bioprocess monitoring. *Analyst* 124: 157-162 (1999).

Vaidyanathan S, Rowland JJ, Kell DB, Goodacre R. Rapid discrimination of aerobic endospore-forming bacteria via electrospray-ionisation mass spectrometry of whole cell suspensions. *Anal Chem* 73: 4134-4144 (2001).

Werbos PJ. *The Roots of Back-Propagation: From Ordered Derivatives to Neural Networks and Political Forecasting*. John Wiley and Sons, Chichester (1994).

Westerhoff HV, Kell DB. What BioTechnologists knew all along...? *J Theor Biol* 182: 411-420 (1996).

Wilkinson L. *The Grammar of Graphics*. Springer-Verlag, New York (1999).

Williams RR, Paradkar RP. Correcting fluctuating baselines and spectral overlap with genetic regression. *Appl Spectr* 51: 92-100 (1997).

Winson MK, Goodacre R, Woodward AM *et al.* Diffuse reflectance absorbance spectroscopy taking in chemometrics (DRASTIC). A hyperspectral FT-IR-based approach to rapid screening for metabolite overproduction. *Anal Chim Acta* 348: 273-282 (1997).

Chapter 14

DYNAMIC PROFILING AND CANONICAL MODELING

Powerful Partners in Metabolic Pathway Identification

Eberhard O. Voit[1] and Jonas Almeida[1,2]
[1]Department of Biometry and Epidemiology [2]Department of Biochemistry and Molecular Biology, Medical University of South Carolina, Charleston, SC 29435, USA

1. INTRODUCTION

Biological research attempts to answer the question: How do organisms function? Once we can answer this question, we can explain our natural surroundings and begin to change them in a targeted fashion that offers us benefit, may it be in medicine, agriculture, biotechnology, or a responsible exploitation of the environment. The challenge is that we were not provided with a blueprint of the inner workings of organisms. We have very many observational data, but they are almost always mere snapshots of some parts of some organisms under some more or less controlled conditions. Often these snapshots are clustered in some interesting corner of the biological universe, but more often they are separated by gaping holes in our knowledge. Our task is then to interpolate between rather scarce data in order to construct a picture that matches the observations and, more interestingly, explains what lies between and beyond.

Aside from mere description, biological discovery consists of two components, namely the collection of data and the application of procedures to evaluating them. Data collection currently demands the lion share of biological research, and it is not the purpose of this chapter to discuss it in any detail. The evaluation of data in the above sense of interpolating between snapshots requires a mental construct, a model. Such a model may be very simple and conceptual. For instance, looking at data on the size of a bacterial population, graphed as dots over a series of time points, we "automatically" construct a conceptual model that connects the dots in a smooth fashion. This model allows us to predict population sizes that occur between those observed, but were not measured *per se*. The model also may give us insights that are not deducible from individual data points. For instance, we may conclude that the growth process eventually saturates at a finite level, and this in turn may provide hypotheses for future experimentation. Every scientific experiment is based on some explicit or implicit hypothesis that has its roots in a formal or intuitive model.

For simple phenomena like the growth trend in a bacterial population, a conceptual model is often sufficient, at least for an initial exploration However, as soon as a phenomenon involves a variety of factors and influences, our intuition is prone to failure. As an example, imagine a physiological process that is affected by several activators and inhibitors. Without a structured, quantitative approach it is difficult to predict whether this process will increase or decrease in activity under conditions that had not been measured before. More generally, if many constituents, all with their own features, interact within a system, the responses of the system to new inputs are difficult to explain or predict. We discuss in this chapter how a mathematical model with proper features can be a powerful aid for efficient and reliable data evaluation.

Models and data have always been complementing each other on the path of biological discovery, but their roles have changed repeatedly. At times, plenty of data were driving the development of models, but at other times, models were constructed faster than data could be obtained. As an example of the former, consider again the growth of organisms or populations. Growth phenomena often exhibit a quasi-exponential initial phase, which is followed by decreasing growth rates and, finally, saturation, thereby giving the appearance of an S-shaped curve. Data collected from the three characteristic phases do not contain much specific information, especially if they are corrupted by measurement error. As a consequence, many alternative models emerged in the 19th and 20th centuries in the form of differently formulated growth *laws* (*e.g.* Savageau, 1980). Although they lacked a mechanistic basis and were entirely data dependent, these *models-of-data* (DiStefano, 1985; Jacquez, 1996; Voit, 2002) were very useful, and they are still employed for interpolations, predictions, for instance in forestry and actuarial science, and the definition of normalcy, for example in growth patterns of children. The generic limitation of data models is that they are insufficient for predictions under untested conditions.

As an example for models running out of data, consider the brave attempts of Garfinkel (*e.g.* 1968, 1980, 1985), who constructed *models-of-processes* (DiStefano, 1985; Jacquez, 1996; Voit, 2002) of large metabolic networks with the goal not only to interpolate values between observed data, but also to provide explanations and new knowledge. Garfinkel based these models on mechanistic models of individual enzyme-catalyzed processes and used for their mathematical formulation the well-established kinetic rate laws of Henri, Michaelis and Menten, Hill, and their many extensions. Garfinkel's models were theoretically unlimited in size and complexity, but it turned out that they were not as successful as had been expected. The two main reasons were that model results were difficult to interpret (see Heinrich *et al.*, 1977; Torres and Voit, 2003) and that one needed far more data to populate the models than could be obtained from experiments.

The step from singular kinetic processes to metabolic networks does not constitute simple scaling, because the networks do not consist of loose collections of processes that can be studied exhaustively in isolation. Often different processes use and thus compete for the same substrates, some inhibit each other, and many of them exhibit some sort of synergistic or antagonistic connectivity. Therefore, in addition to characterizing individual processes through data, interactions between processes must be identified qualitatively and quantitatively, and this leads to a huge demand for experimental measurements.

To summarize the interplay between data and models, models-of-data are driven by the data themselves, and it is up to the ingenuity of the modeler to find a suitable mathematical function that captures the data. Models-of-processes, by contrast, are conceived in the minds of modelers and based on some mechanistic understanding; to test them, work with them, or apply them to real-world problems, they have to be quantified with a lot of data. In the past, data were not available in sufficient quantity and quality, and this paucity became the most significant limitation to applying models-of-processes to real-world problems. However, modern molecular biology, genomics, proteomics, and especially intracellular monitoring and metabolic profiling are in the process of changing this situation. Novel experimental techniques are not just producing enormous volumes of data, but some of these data are highly quantitative, accurate, and of a nature that is ideally suited for metabolic modeling. The premier example of interest here is metabolic profiling, which has reached a level of sophistication where hundreds or even thousands of metabolite concentrations can be measured simultaneously (*e.g.* Goodenowe, 2001 and Chapter 8). This enormous "metabolic density" is unprecedented in the history of biomathematical modeling. In a different dimension of the space of biochemical dynamics, it is becoming feasible to record concentrations of select metabolites in individual cells in very short time sequence (*e.g.* Almeida *et al.*, 1995; Neves *et al.*, 1999). Some of the methods of genomic or metabolic characterization are still crude, and the accuracy of their results is to be considered with caution, but this is surely only a matter of time.

The merging of data that are densely spaced both metabolically and temporally is bound to usher in a new era of biomathematical modeling. We demonstrate in this chapter how this type of dense metabolic profiling provides ideal data for powerful "canonical" modeling and pathway identification.

2. DYNAMIC METABOLIC PROFILES

The past years have seen the development of several experimental techniques that show extraordinary potential for the identification of the structure of pathways, if combined with the right type of mathematical modeling. We will refer to these techniques categorically as *metabolic profiling*. What is common to these methods is that they permit simultaneous measurements of metabolites directly *in vivo* or under conditions that are similar to those in the intact organism. While it is not our intention to review these experimental methods, it is useful at least to mention mass spectrometry and nuclear magnetic resonance, which appear to be particularly promising.

Mass spectrometry has been refined to a point where even very similar masses can be separated reliably. For example, it allows the distinction between one carbon atom plus two hydrogen atoms and one nitrogen atom. Without accurate determination, both would share a peak at mass 14, but modern mass spectrometry is able to distinguish the true weights of CH_2 ($12.011 + 2 \times 1.008 = 14.027$) and N (14.008). Modern spectrometry is not only very accurate in mass determination, it is also executable with very small quantities of material and still allows measurements of a large array of metabolites. For example, Goodenowe (2001) was able to measure over 6,000 metabolites in ripening strawberries.

A mass spectrum of this quality is the paradigm of a snapshot of the metabolic state of the cell under investigation. While a great advancement, it does not in itself provide much information about the dynamics of the underlying pathways. This is a crucial limitation, because the dynamics captures how a cell responds to an external or internal stimulus. Furthermore, it is known from theoretical analyses (*e.g.* Sorribas and Cascante, 1994; Voit, 2000: pp. 219 and 452) that different dynamical models may have exactly the same steady state. Thus, in terms of metabolic pathway identification from data, steady-state information alone is insufficient for the full characterization of a pathway.

The real power of mass spectrometry in the context of profiling and pathway identification will come with simultaneous measurements of many metabolites at a series of relatively dense time points. As a generic example for this type of *dynamic metabolic profiling*, one may think of a cell that is operating at its normal steady state and, at time t_0, is exposed to some stimulus, like the addition of some chemical to the medium, a sudden change in pH, or an abrupt change in temperature. Suppose the normal steady state before t_0 had been characterized with a metabolic profile, and further profiles are taken at time points t_1, t_2, ..., t_n, whose temporal density depends on cell type and stimulus, but may be imagined for illustration at the order of a few seconds. Thus, stratifying the data by metabolites, these measurements give discrete time courses of absolute or relative concentrations of all metabolites of interest, over the time period between t_0 and t_n.

Mass spectrometry is not the only method for generating such time courses. Conceptually similar experiments and results are becoming available with a variety of other methods, such as nuclear magnetic resonance that can be used *in vivo*, probing intracellular concentrations for indefinite periods of time as physiological processes unfold. This technique rewards the investment of expensive isotopes with the valuable distinction between intracellular and extracellular metabolic profiles (Sibjesma *et al.*, 1996) and with a temporal resolution as short as a few seconds (*e.g.* Neves *et al.*, 2000). Beyond the organism level, this ability to probe active cultures has been used to unveil elusive metabolic pathways that are collectively maintained by complex microbial consortia (Santos *et al.*, 1999).

Dynamic metabolic profiling provides us with time courses, that is, with measurements of metabolic concentrations at (many) sequential time points, and the biomathematical challenge is to deduce from these data information about the structure and topology of the underlying pathway. In contrast to the earlier example of fitting a growth function, where the goal was to determine any suitable function that would fit the data with sufficient accuracy, the goal in metabolic pathway identification is to determine a detailed, valid mathematical model whose dynamic output is consistent with the observed metabolic profiles. In the case of growth functions, the modeling result allowed us to interpolate between measured data points and make limited extrapolations, as long as the experimental conditions did not change. A valid model of a metabolic pathway, in comparison, allows predictions of responses to new, natural or artificial conditions, provides explanations for the observed design and operation of the pathway, and suggests means of altering and optimizing the network according to desired objectives (Torres and Voit, 2003).

Dynamic metabolic profiles contain enormous information about the flux distribution and regulation of metabolic pathways *in vivo*. This information is not immediately explicit though, but requires adequate analytical and computational methods of retrieval and interpretation. The challenge of deducing a mathematical model from metabolic profiles is twofold. On the biological side, the information about the pathway is often incomplete or not existing. Secondary pathways or regulatory signals may not be known in sufficient detail, and the data themselves may be sketchy or noisy. Thus, any identification attempt is faced with enormous numbers of structural possibilities and uncertainties. On the mathematical side, there are no *a priori* guidelines for the types of functions or equations to be used. For a case in point, the interested reader is referred to a diverse collection of mathematical functions that were all formulated with the same goal of adequately describing nutrient uptake by tree roots (Voit and Sands, 1996). For metabolic pathways, one might be tempted to resort to the tried-and-true generalized Michaelis-Menten formulations, but they have the significant drawback for pathway identification in that they change their mathematical structure every time a modulator is included. For instance, the mathematical form of an inhibited process depends strongly on the type of inhibition, be it competitive, non-competitive, allosteric, or of an entirely different type. This diversity in structure poses problems for any estimation algorithm, because if the type of inhibition is unknown, a separate search must be executed for any candidate function, and an objective comparison among the different data fits is difficult.

A particularly promising set of theorems and methods for the purpose of metabolic pathway analysis and identification is collectively known as *Biochemical Systems Theory* (*BST*; Savageau, 1969a, 1969b, 1976; Voit, 1991, 2000; Torres and Voit, 2003). According to BST, traditional rate laws are approximated with power-law functions. The resulting models have the crucial advantage that they always have the same mathematical structure. They also have been shown to be sufficiently accurate in a rich variety of applications. For pathway identification, we suggest here to combine BST models with computational methods of parameter estimation and machine learning that are based on regression, *genetic algorithms (GAs)* and, possibly, *artificial neural networks* (*ANNs*) (Almeida, 2002). Preliminary analyses indicate that this combination of powerful modeling methods and structures is promising but not always sufficient for efficient metabolic pathway identification. We introduce the general procedures next and discuss possible refinements and extensions in a later section.

3. BST—A POWERFUL MODELING FRAMEWORK FOR METABOLIC NETWORKS

BST captures the dynamics of metabolic networks by focusing on temporal changes in each of the metabolites. If metabolite X_i receives input from several sources, and if these inputs are affected by other metabolites or factors outside the system, all these sources and effectors are collected in a production term for X_i. Analogously, all processes leading to degradation or use of X_i are collected in one term, so that the changes of X_i over time are represented as the difference between the production and the degradation terms. Rigorous results from numerical analysis and rich biological validation has shown that it is legitimate to approximate these two terms by products of power-law functions that contain as variables all metabolites and factors that affect a given term. Because the construction of equations in BST always follows these simple rules, this type of modeling is referred to as *canonical modeling* (Savageau and Voit, 1987; Voit, 1991).

As an example, consider the dynamics of metabolite X_1 in the simple generic pathway in Fig. 1.

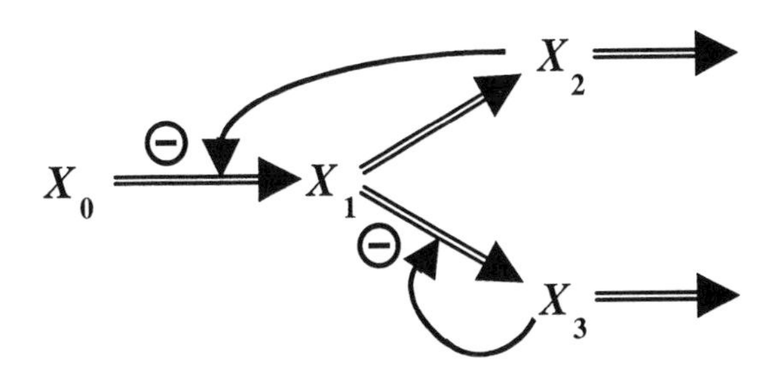

Figure 1. Generic branched pathway with one source, X_0, and two feedback loops.

The production of X_1 depends on the source variable X_0 and is also affected by inhibition exerted by X_2. Thus, X_0 and X_2 enter the production term V_1^+, which can be formulated symbolically without any further information. The straightforward result is $V_1^+ = \alpha_1 X_0^{g_{10}} X_2^{g_{12}}$. Now consider the degradation of X_1, which is affected by X_1 itself, because it serves as substrate for the reaction, and by the inhibition from X_3. The corresponding representation of the term of the canonical model is thus $V_1^- = \beta_1 X_1^{h_{11}} X_3^{h_{13}}$. Note that this term does not depend on X_2. X_2 is a "passive" recipient of material flowing out of X_1 but does not directly affect this process. The change in X_1, which is indicated by the dot notation $\dot{X}_1 = dX_1/dt$, is thus directly obtained as

$$\dot{X}_1 = \alpha_1 X_0^{g_{10}} X_2^{g_{12}} - \beta_1 X_1^{h_{11}} X_3^{h_{13}} .$$

The symbolic equations for X_2 and X_3 are set up in exactly the same fashion. Furthermore, these simple rules apply to pathways of arbitrary size and complexity. Numerous examples of canonical models are found in Voit (2000).

The intriguing consequence of the power-law representation is that every kinetic order and every rate constant has a uniquely defined meaning and role. Each kinetic order g_{ij} specifically captures quantitatively the direct effect that the variable or factor X_j has on the production of X_i. Similarly, the kinetic order h_{ij} specifically represents the corresponding direct effect of X_j on the degradation of X_i. The rate constants α_i and β_i simply quantify the speed or turnover rate of the production and the degradation of X_i, respectively. Because of the specificity of the roles of all parameters, identification of a particular process or effect is quasi synonymous with the quantification of the corresponding parameter. For instance, if the result of the identification procedure assigns a value of 0 to the kinetic order g_{62}, one may deduce that X_6 does not directly affect the growth or production of X_2. Alternately, if g_{62} had a negative value, the effect of X_6 on X_2 would be interpreted as negative, *i.e.* inhibiting production, while a positive value would signal a substrate or positive effector. Summarizing these properties of canonical models, the identification of the structure of a pathway is largely reduced to the simpler, though still challenging problem of identifying optimal parameter values from metabolic profiles. Of course, the direct mapping between parameter values and effects has been known since the inception of BST and constitutes the basis for the method of controlled mathematical comparisons (Irvine and Savageau, 1985; Savageau, 1985; Alves and Savageau, 2000), which has been used extensively for analyzing design and operating principles of biological systems.

In the example of Fig. 1, we constructed the equations from the known structure of the pathway. For identification of the structure from data, the process has to be turned around. That is, time course data are available for X_0, X_1, X_2, and X_3, and the task is to identify from these data a graph as in Fig. 1. Generally this is accomplished by setting up S-system equations that allow for all possible interactions among the variables and finding out, with the help of an optimization algorithm, which parameters have values to zero and which not. Thus, under ideal conditions, the equation for changes in X_1 would be formulated in general symbolic terms as

$$\dot{X}_1 = \alpha_1 X_0^{g_{10}} X_1^{g_{11}} X_2^{g_{12}} X_3^{g_{13}} - \beta_1 X_0^{h_{10}} X_1^{h_{11}} X_2^{h_{12}} X_3^{h_{13}}$$

and the optimization algorithm would identify g_{11}, g_{13}, h_{10} and h_{12} as zero and also bring forth the accurate numerical values of the remaining parameters.

4. IDENTIFICATION OF METABOLIC PATHWAYS FROM DYNAMIC PROFILES

As outlined in the previous section, the identification of the structure of a metabolic pathway is by and large a matter of parameter estimation. At first glance, the problem thus seems to be solved: Subject the data and the symbolic model (with all parameter values unspecified) to a nonlinear regression routine, obtain optimal parameter values, and interpret them as demonstrated before. In fact, several analyses can be found where this procedure was successful. The oldest may be an analysis of ethanol production in a yeast culture (Voit and Savageau, 1982a), where the question was investigated whether the generated alcohol could be used as a secondary substrate of whether it inhibited growth. Irvine and Savageau (1985) studied the structure of an immune cascade, though they did not use time series data. In a theoretical study, Sorribas and Cascante (1994) investigated to what degree the structure of a metabolic pathway could be deduced solely from steady-state data. Zhang *et al.* (1996) used a genetic algorithm to estimate parameters in a simple S-system describing oil production in palm trees, and several Japanese groups (Okamoto *et al.*, 1997; Akutsu *et al.*, 2000; Kikuchi *et al.*, 2001; Maki *et al.*, 2001; Sakamoto *et al.*, 2001) recently extended methods of estimating parameters in S-systems with genetic algorithms for much larger gene-regulatory systems. Pursuing the related problem of optimizing the structure of a pathway, Hatzimanikatis *et al.* (1996) developed a method based on mixed integer linear programming. Other recent approaches of pathway identification, though not with BST models, were proposed by Samoilov (2001) and Oliveira (2001).

While simple in theory, all of these approaches face significant challenges when the underlying pathway contains more than just a few variables and/or if the available data are scarce. These challenges fall into three categories. First, nonlinear regression is prone to failure if the parameter space has local minima or if the solution resides at the center of a "banana-shaped valley" with an almost flat and curved valley floor and steep sides. In the former case, the solution may terminate early, but at a local and not the global minimum. Given non-optimal parameter values, the corresponding interpretation in terms of pathway structure may be faulty. For example, a small negative value of g_{ij} would indicate an inhibitory signal, whereas $g_{ij}=0$ would suggest the absence of such a signal. The latter problem often leads the optimization algorithm to cycle, or to make no progress toward the optimal solution, and to spend excessive iterations without greatly improving the current solution. Genetic algorithms avoid some of these problems, but they do not necessarily converge to the optimal solution for other reasons.

The second challenge is the fact that the optimization algorithm, whether based on regression or a genetic algorithm, must solve the system of differential equations during every iteration. Informal estimates suggest that this numerical integration may easily account for 90% or more of the total estimation time. In some cases, the solution may not even be obtainable because of stiffness or other numerical problems.

The third challenge is that distinct solutions (*i.e.*, sets of parameter values) sometimes have essentially the same residual error with respect to a given data set. This may have different reasons. It may be a matter of convergence of the optimization algorithm, but it may also happen that different BST models actually lead to the same or very similar time courses. This may be due to true underlying redundancies (Sands and Voit, 1996), the existence of transformation groups that lead to equivalent time courses (Voit, 1991:Ch.12; Voit, 1992; Hernández-Bermejo and Fairén, 1997), or to the flexibility of canonical models, which allows for differently parameterized models with rather similar time courses that are difficult to distinguish from data, especially if they contain significant experimental error.

Preliminary studies suggest that all three challenges can be overcome or at least ameliorated. The studies also suggest that it is unlikely that the process will be fully automated in the near future and produce the best solution without human "curation." The first challenge of convergence is one of numerical analysis and computer science. Our current insights indicate that a mixture of nonlinear regression and a genetic algorithm or artificial neural network may be the method of choice. For instance, one may use a genetic algorithm to obtain a set of many "coarse" solutions, of which none may be optimal, but which may all serve as different starting points of a nonlinear regression (see Appendix).

One way of dealing with the second challenge, namely the need of repeatedly solving differential equations, is the approximation of the derivatives dX_i/dt with slopes of the variable X_i at all measured time points. The slopes may be obtained from the data directly by hand (Voit and Savageau, 1982a, 1982b; Voit, 2000) or with some smoothing algorithm, or one may employ the "three-point formula," a popular technique in biochemical engineering that uses the values immediately before and after a given position to determine the expected tangent angle of the underlying function at this position (Burden and Faires, 1993). The crucial advantage of substituting the derivatives with slopes is that the differential equations are thereby replaced by sets of algebraic equations, which are optimized with incomparably greater speed. We are in the process of fine-tuning a web-based tool, http://bioinformatics.musc.edu/webmetabol/, that combines these methods to identify metabolic pathways from dynamic metabolic profiles (see Appendix). This tool allows the user to submit data, for instance in

Excel format, select boundaries for parameter values, and obtain parameter estimates in an interactive fashion.

The third challenge of non-uniqueness probably requires the highest degree of human interaction. Two approaches or their combination may be useful for this challenge. First, many practical cases may not only rely on the metabolic profiles alone, but additional knowledge about the pathways may be available. For instance, one may know from other sources that X_3 is a precursor of X_4 or that X_8 inhibits the degradation of X_5. This information can and should be translated either into hard or soft constraints. A hard constraint may be the definition that two parameter values (for instance those describing the degradation of X_3 and those describing the production of X_4) must be the same. Hard constraints of this type can drastically reduce the search effort for the optimization algorithm. Soft constraints may place boundaries on parameter values. For example, if it is known that X_8 inhibits the degradation of X_5, one should posit that the corresponding parameter value h_{58} is negative. Experience suggests that the value of such a kinetic order is seldom below –1, which further limits the numerical search for the optimal solution.

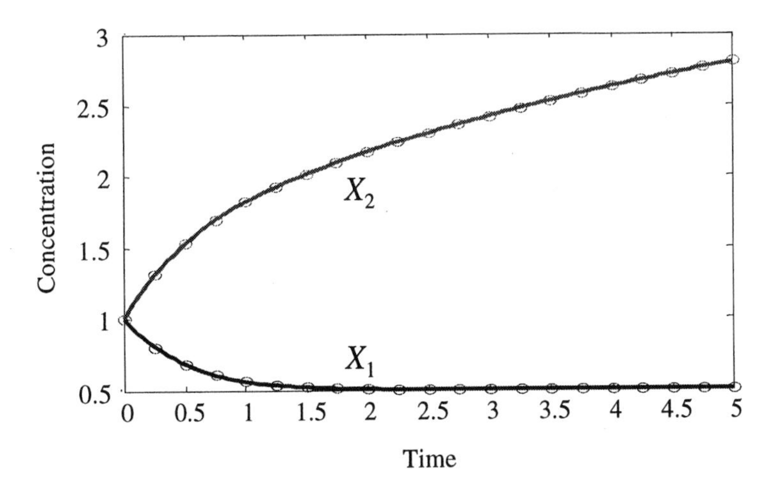

Figure 2. Time courses of two variables of a simple generic system upon perturbation.

The second human interaction occurs at the back end of the process. It is useful to obtain several solutions and to compare their features in an attempt to differentiate noise from reality. As an illustration, we estimated parameter values from simulated time courses of a very simple two-variable system, which however nicely illustrates the need for human curation. The time courses show how the two variables return to their steady-state values after a perturbation (Fig. 2). Six sample solutions are given in Table 1.

It is emphasized that all solutions in Table 1 have essentially the same small residual error, even though there is quite a variety in parameter values. In fact, any of these solutions fits the data as shown in Fig. 2 (solid lines). An interpretation of results may proceed as follows. One notes first that g_{11} always has small values and that some of them are positive and some negative. This may suggest that in fact g_{11} may be zero, which would mean that X_1 does not affect its own production, which is often the case. Secondly, g_{12} and h_{12} are very similar in all solutions, even though their values vary quite a bit from one solution to the next. The kinetic orders g_{12} and h_{12} capture the effects of X_2 on the production and degradation of X_1, respectively, and since they are negative, X_2 would inhibit both processes at the same rate. Since they have the same value, the effects in some sense cancel out and merely affect the overall speed of the change in X_1. Furthermore, since one of these pairs is close to zero, one might cautiously think about eliminating both kinetic orders and to conclude that X_2 has no effect on the dynamics of X_1. To test this hypothesis further, one could fix g_{12} and h_{12} at zero and rerun the optimization routine to see whether solutions can be obtained that satisfy $g_{12}=h_{12}=0$ and still have a residual error similar to that obtained before, in which case g_{12} and h_{12} could indeed be insignificant. It must be emphasized that there is no guarantee that X_2 has no effect on the dynamics of X_1, but that it appears that the time courses can be generated without this effect. Further analysis of this question is only possible with more, independent data on X_1 and X_2, for instance, following a different type of perturbations.

Table 1. Kinetic Orders of Six "Optimal" Solutions Obtained from the Time Courses in Fig 2 by Means of a Genetic Algorithm

Solution	g_{11}	g_{12}	h_{11}	h_{12}	g_{21}	g_{22}	h_{21}	h_{22}
1	0.13	-0.07	0.73	-0.08	0.55	-1.40	0.17	0.36
2	0.07	-0.15	0.69	-0.15	0.47	-0.54	-0.40	-0.19
3	-0.19	-0.18	0.63	-0.16	0.58	-0.36	0.41	0.01
4	0.01	-0.22	0.76	-0.24	0.63	-0.48	-0.31	-0.03
5	0.09	-0.20	0.69	-0.22	0.54	-0.79	0.17	-0.41
6	-0.03	-0.43	0.41	-0.44	0.37	-0.79	-0.15	-0.47

A third observation from Table 1 is that the values of g_{21} are always similar. This indicates a consistent and rather strong effect of X_1 on the production of X_2. Furthermore, the value of h_{11} shows a similar pattern, and taken together these observations might suggest the hypothesis that X_1 is the precursor of X_2. Again, there is no guarantee, but equating the degradation term of X_1 with the production term of X_2 and accounting for the observations described above leads to hypotheses that can be tested against the observed data. Indeed, the data were generated from a model of the pathway in Fig. 3. It is noted that, at least for this example, a single optimization would not have been sufficient. It is also worth considering whether data of the same system, following different perturbations, would provide a stronger database.

Figure 3. Pathway structure deduced from time courses in Fig. 2 and results in Table 1.

5. CONCLUSION

Biological discovery requires good data and methods for their analysis. It is indicated here that novel methods of metabolic profiling, if extended to dynamic profiling, have unprecedented potential for metabolic pathway identification, if they are accompanied by efficient mathematical and computational tools. We suggest that these tools come from BST which provides a proven unique theoretical framework for this purpose. Reasons why BST is particularly well suited include its successful application to numerous data throughout biology and, maybe more important, the homogeneous structure of its canonical models, which maps parameters essentially uniquely onto structures of the metabolic pathway under study. This mapping has the important consequence that structure identification is reduced to the much simpler, yet still challenging task of parameter estimation.

Pathway structures have been obtained from time courses for some while, but recent interest in this task has grown immensely, because modern methods of metabolic profiling are on the verge of generating data in a quantity and quality never available before. It is suggested here that "metabolic density," resulting in very many measurements of metabolite concentrations, be merged with "temporal density," which is already achievable, though at a smaller scale with respect to the number of metabolites measured. This merger will result in dynamic metabolic profiles, for which mathematical methods of canonical analysis are available or imminent.

ACKNOWLEDGMENTS

This work was supported in part by a Quantitative Systems Biotechnology grant (BES-0120288; E.O. Voit, PI) from the National Science Foundation, a Cancer Center grant from the Department of Energy (C.E. Reed, PI), a Department of Energy grant to Oak Ridge National Laboratory (ERKP280; C. Brandt, PI) and an Interdisciplinary USC/MUSC grant (E.P. Gatzke, PI). Any opinions, findings, and conclusions or recommendations expressed in this material are those of the authors and do not necessarily reflect the views of the sponsoring institutions.

REFERENCES

Abu-Mostafa YS. The Vapnik-Chervonenkis dimension: information versus complexity in learning. *Neural Computat* 1: 312-317 (1989).

Akutsu T, Miyano S, Kuhara S. Inferring qualitative relations in genetic networks and metabolic pathways. *Bioinformatics* 16: 727-734 (2000).

Almeida JS. Predictive non-linear modeling of complex data by artificial neural networks. *Curr Opin Biotechnol* in press (2002).

Almeida JS, Reis MAM, Carrondo MJT. Competition between nitrate and nitrite reduction in denitrification by *Pseudomonas fluorescens*. *Biotechnol Bioeng* 46: 476-484 (1995).

Alves R, Savageau M. Extending the method of mathematically controlled comparison to include numerical comparisons. *Bioinformatics* 16: 786-798 (2000).

Burden RL, Faires JD. *Numerical Analysis*. 5th Edn. pp. 156-167, PWS Publishing Co, Boston (1993).

DiStefano III JJ. The modeling methodology forum: an expanded department. *Am J Physiol* 248: C187-C188 (1985).

Garfinkel D. The role of computer simulation in biochemistry. *Comp Biomed Res* 2: 31-44 (1968).

Garfinkel D. Computer modeling, complex biological systems, and their simplifications. *Am J Phys* 239: R1-R6 (1980).

Garfinkel D. Computer-based modeling of biological systems which are inherently complex: problems, strategies, and methods. *Biomed Biochim Acta* 44: 823-829 (1985).

Goodenowe DB. Metabolic network analysis: integrating comprehensive genomic and metabolomic data to understand development and disease (abstract). Cambridge Healthtech Institute Conference on Metabolic Profiling: Pathways in Discovery, Chapel Hill (2001).

Hatzimanikatis V, Floudas CA, Bailey JE. Optimization of regulatory architectures in metabolic reaction networks. *Biotechnol Bioeng* 52: 485-500 (1996).

Heinrich R, Rapoport SM, Rapoport TA. Metabolic regulation and mathematical models. *Prog Biophys Mol Bio* 32: 1-82 (1977).

Hernández-Bermejo B, Fairén V. Lotka-Volterra representation of general nonlinear systems. *Math Biosci* 140: 1-32 (1997).

Irvine DH, Savageau MA. Network regulation of the immune response. *J Immunol* 134: 2100-2130 (1985).

Irvine DH, Savageau MA. Efficient solution of nonlinear ordinary differential equations expressed in S-system canonical form. *SIAM J Numer Anal* 27: 704-735 (1990).
Jacquez, JA. *Compartmental Analysis in Biology and Medicine*. 3rd Edn. Thomson-Shore, Inc, Dexter, MI (1996).
Kantz H, Schreiber T. *Nonlinear Time Series Analysis*. Cambridge University Press, Cambridge (1997).
Kikuchi S, Tominaga D, Masanori A, Tomita M. Pathway finding from given time-courses using genetic algorithm. *Genome Informat* 12: 304-305 (2001).
Maki Y, Tominaga D, Okamoto M et al. Development of a system for the inference of large scale genetic networks. *ProcPacific Symposium on Biocomputing*. pp. 446-458, World Scientific, Singapore (2001).
Michel, M. *An Introduction to Genetic Algorithms*. MIT Press, Cambridge (1998).
Neves AR, Ramos A, Nunes MC *et al. In vivo* nuclear magnetic resonance studies of glycolytic kinetics in *Lactococcus lactis*. *Biotechnol Bioeng* 64: 200-212 (1999).
Neves AR, Ramos A, Shearman C *et al. In vivo* nuclear magnetic resonance studies of glycolytic kinetics in *Lactococcus lactis*. *Eur J Biochem* 267: 3859-3868 (2000).
Okamoto M, Morita Y, Tominaga D *et al.* Design of virtual-labo-system for metabolic engineering: development of biochemical engineering system analyzing tool-kit (BEST KIT). *Comp Chem Engng* 21: S745-S750 (1997).
Oliveira JS, Bailey CG, Jones-Oliveira JB, Dixon DA. An algebraic-combinatorial model for the identification and mapping of biochemical pathways. *Bull Mathem Biol* 63: 1163-1196 (2001).
Sakamoto E, Iba H. Inferring a system of differential equations for a gene regulatory network by using genetic programming. *Proc 2001 Congress on Evolutionary Computing, CEC2001*. pp. 720-726, IEEE Press, Piscataway, NJ, (2001).
Samoilov M, Arkin A, Ross J. On the deduction of chemical reaction pathways from measurements of time series of concentrations. *Chaos* 11: 108-114 (2001).
Sands PJ, Voit EO. Flux-based estimation of parameters in S-systems. *Ecol Model* 93: 75-88 (1996).
Santos MM, Lemos PC, Reis MA, Santos H. Glucose metabolism and kinetics of phosphorus removal by the fermentative bacterium *Microlunatus phosphovorus*. *Appl Environ Microbiol* 65: 3920-3928 (1999).
Savageau MA. Biochemical Systems Analysis, I. Some mathematical properties of the rate law for the component enzymatic reactions. *J Theor Biol* 25: 365-369 (1969a).
Savageau MA. Biochemical Systems Analysis, II. The steady-state solutions for an n-pool system using a power-law approximation. *J Theor Biol* 25: 370-379 (1969b).
Savageau MA. *Biochemical Systems Analysis. A Study of Function and Design in Molecular Biology*. Addison-Wesley, Reading (1976).
Savageau MA. Growth equations: a general equation and a survey of special cases. *Math Biosci* 48: 267-278 (1980).
Savageau MA. A theory of alternative designs for biochemical control systems. *Biomed Biochim Acta* 44: 875-880 (1985).
Savageau MA, Voit EO. Recasting nonlinear differential equations as S-systems: a canonical nonlinear form. *Math Biosci* 87: 83-115 (1987).
Sibjesma WFH, Almeida JS, Reis MAM, Santos H. Evidence for uncoupling effect of nitrite during of denitrification by *Pseudomonas fluorescens: in vivo* ^{31}P-NMR study. *Biotechnol Bioeng* 52: 176-182 (1996).
Sorribas A, Cascante M. Structure identifiability in metabolic pathways: parameter estimation in models based on the power-law formalism. *Biochem J* 298: 303-311 (1994).

Torres NV, Voit, EO. *Pathway Analysis and Optimization in Metabolic Engineering.* Cambridge University Press, Cambridge, in press (2003).
Voit EO (Ed). *Canonical Nonlinear Modeling. S-System Approach to Understanding Complexity.* Van Nostrand Reinhold, New York (1991).
Voit EO. Symmetries of S-systems. *Math Biosci* 109: 19-37 (1992).
Voit EO. *Computational Analysis of Biochemical Systems. A Practical Guide for Biochemists and Molecular Biologists.* Cambridge University Press, Cambridge (2000).
Voit EO. Models-of-data and models-of-processes in the post-genomic era. *Mathem. Biosci* in press (2002).
Voit EO, Sands PJ. Modeling forest growth. I. Canonical approach. *Ecol Model* 86: 51-71 (1996).
Voit, EO, Savageau MA. Power-law approach to modeling biological systems; II. Application to ethanol production. *J Ferment Technol* 60: 229-232 (1982a).
Voit EO, Savageau MA. Power-law approach to modeling biological systems; III. Methods of analysis. *J Ferment Technol* 60: 233-241 (1982b).
V'Yugin, VV. Algorithmic complexity and stochastic properties of finite binary sequences. *Comp J* 42: 294-317 (1999).
Zhang Z, Voit EO, Schwacke LH. Parameter estimation and sensitivity analysis of S-systems using a genetic algorithm. In *Methodologies for the Conception, Design, and Application of Intelligent Systems.* YamakawaT, Matsumoto G (Ed) World Scientific, Singapore, (1996).

APPENDIX: NUMERICAL DETAILS OF S-SYSTEM-BASED PATHWAY IDENTIFICATION

The identification of S-system parameters from dynamic metabolic profile data represents a complex of three challenges: 1) global minimization; 2) numerical integration; and 3) evaluation of multiple solutions. This appendix summarizes some of the numerical tools we use to tackle these challenges. For clarity of representation, it is beneficial to discuss the challenges in reverse order.

Identification of an objective function to discriminate multiple solutions

The last numerical step of the entire process is the identification and minimization of an appropriate objective function, *ObjFunc*, that uses S-system parameters as independent variables, $Ssys(\alpha,\beta,g,h)$ and, by optimizing them, minimizes the residual error between data and S-system model. The typical objective function for this purpose is the sum of squared errors, *SSE*, that is defined as follows:

$$ObjFunc\,(\alpha,\beta,g,h) = SSE(X,\alpha,\beta,g,h) = \sum (X^{*} - X)^2 = \sum_t \left(\int_{t_0}^{t} Ssys(\alpha,\beta,g,h) \cdot dt - X_t \right)^2 \quad \text{(A1)}$$

where X represents the vector of metabolites and the integral indicates numerical solution of all S-system differential equations.

The observation that multiple combinations of the parameter values α, β, g, h and the initial condition X_0 may correspond to the same or a very similar minimum value of SSE suggests that the error function might have to be compounded with a measure of complexity of the solution. The rationale for this step is that one would probably prefer a solution with fewer interactions between metabolites over one with more interactions, if both solutions fit equally well. It must be cautioned though, that this criterion of minimal complexity has its justification more in philosophy than mathematics. Furthermore, there are no generally accepted expressions quantifying algorithmic (Kolmogorov) complexity even of simple binary sequences (V'Yugin, 1999), let alone of time series. Another possibility would be to use the concept of entropy in time series, as discussed by Kantz (1997), or, as suggested by Abu-Mostafa (1989), complexity associated with artificial learning, which is quantified by the so-called Vapnik-Chervonenkis dimension. Unfortunately, neither is practically achievable for complexity-based rankings of S-systems. As an alternative, one may consider an S-system model as less complex if the totality of the magnitudes of its kinetic orders is small. Such a criterion can be modeled by adding the weighted absolute values of all the kinetic orders, g and h, to SSE (Eq. A2):

$$ObjFunc(\alpha,\beta,g,h) = SSE(X,\alpha,\beta,g,h) + W\left(\sum|g| + \sum|h|\right) \quad \text{(A2)}$$

Similarly, one could maximize, as a secondary objective, the number of kinetic orders with a value of zero, which would minimize the number of interactions among the metabolites of the system. Again, it must be cautioned that nature may not favor simpler solutions of this type over more complex solutions.

Numerical Integration of S-systems

The straightforward implementation would be the determination of predicted metabolite concentrations, X^*, by numerical integration. However, this path is computationally very costly because many intermediate parameter solutions generated by the minimization of $ObjFunc(\alpha,\beta,g,h)$ are stiff or even impossible to integrate in practice. A more sensible solution, as discussed in the text, is numerically to estimate the slopes $S_i(t_j)$ of all variables X_i at all time points t_j from the experimental data and to minimize the difference between these estimated slopes and the right-hand sides of the S-system equations, which in the optimal solution are equivalent to the true slopes $\dot{X}_i(t_j) = dX_i/dt$ at t_j. The residual error to be minimized then becomes

$$SSE(X,\alpha,\beta,g,h) = \sum_{i,j}\left(S_i(t_j) - \dot{X}_i(t_j)\right)^2 = \sum\left(S(t) - Ssys(\alpha,\beta,g,h)\right)^2 \quad \text{(A3)}$$

An additional transformation has been shown to increase the effectiveness of this approach further. It is based on expressing all S-system variables in terms of logarithms $y_i = \ln(X_i)$. Since the derivative of y_i is simply the derivative of X_i divided by X_i itself, the transformed S-system equations read

$$\dot{y}_i = \alpha_i \cdot \exp(\sum_{j=1}^{n} g'_{ij} y_j) - \beta_i \cdot \exp(\sum_{j=1}^{n} h'_{ij} y_j), \tag{A4}$$

where

$$\begin{cases} g'_{ij} = g_{ij} & \text{for } i \neq j \\ g'_{ij} = g_{ij} - 1 & \text{for } i = j \end{cases} \quad \begin{cases} h'_{ij} = h_{ij} & \text{for } i \neq j \\ h'_{ij} = h_{ij} - 1 & \text{for } i = j \end{cases}$$

as discussed in detail in the literature (Savageau, 1976, Irvine and Savageau, 1990). This formulation requires fewer costly computational steps than the original S-system representation in Cartesian coordinates.

Global minimization

Conventional minimization algorithms are based on nonlinear regression and rely on the local gradient of the objective function with regard to the parameters, which steers the solution iteratively toward the best-fitting solution. This approach requires starting values for the parameters, in this case $[\alpha_i, \beta_i, g_{ij}, h_{ij}, X_i(t_0)]$, and is prone to failure if a local minimum happens to occur on the path toward the global solution. Two global search solutions have been proposed as alternatives, namely *simulated annealing* (SA) and *genetic algorithms* (GA). For a number of reasons, which are beyond the scope of this chapter, genetic algorithms appear to be the preferable choice for canonical models. In a process inspired by the mechanism of natural selection, these algorithms generate populations of individual solutions that are allowed to mate, recombine and compete for selection; the fittest individuals, according to the objective function, are allowed to enter the next iterative round of mating and recombination (Michel, 1998).

Implementation

We assembled a preliminary web-based application that accounts for the steps outlined above; it is accessible at http://bioinformatics.musc.edu/webmetabol/ with Microsoft's Internet Explorer. The logical steps of this tool are:

Execution of a genetic algorithm with *SSE* as objective function and the discrete S-system with algorithmically estimated slopes, according to Eqs. (A3) and (A4).

Nonlinear gradient regression, initiated with the results of the genetic algorithm, using again Eqs. (A3) and (A4).

Nonlinear gradient regression with the continuous S-system in logarithmic coordinates (Eq. A4)

Nonlinear gradient regression with the continuous S-system in Cartesian coordinates, *i.e.*, in the space of metabolite concentrations.

The genetic algorithm library used for this analysis, *GAOT*, was developed at Oak Ridge National Laboratory, TN, and is maintained by the Department of Industrial Engineering at

North Carolina State University [http://www.ie.ncsu.edu/mirage/GAToolBox/gaot/]. Constrained regression was implemented using the reference function *FMINCON* of the numerical library of MATLAB's Optimization Toolbox (Mathworks Inc). The interpretation of results is currently done by human intelligence, but will be aided in the future with an additional algorithm.

The sequential regression procedure is illustrated in Figure A1 for a two-variable time series and an objective function that simply minimizes the residual error, *i.e.*, without compounding the cost function with a measure of complexity as described in Eq. (A2). The left panel shows the concentrations of variables A and B, plotted against time, and the right panel the corresponding slopes, as estimated originally by the three-point method. Symbols represent artificial data, which contain "experimental error." The lines show intermediate solutions of the sequential process, as described above and in the figure legend.

The relative computational costs of each regression step, for the example in Figure A1, are represented in Figure A2. The cost distribution indicates the importance of using a genetic algorithm for the identification of effective starting values for subsequent gradient regression steps. Even though the genetic algorithm alone leads to reasonable parameter estimates, gradient methods that refine the results incur most of the computational costs. This observation reflects the complicated local properties of the error surfaces encountered in identifications of S-system structures from experimental time series.

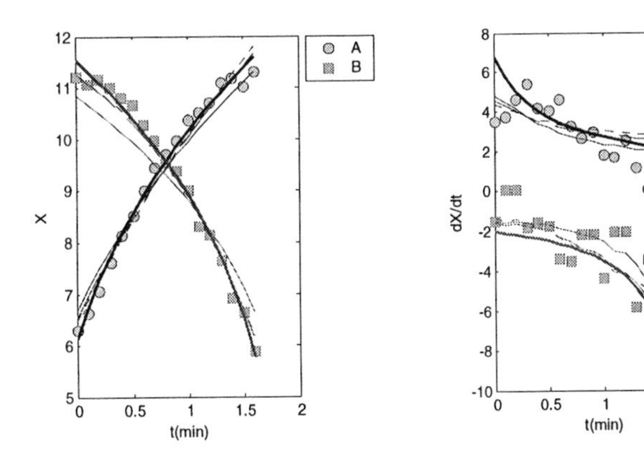

Figure A1. Sequential regression results for a didactic two-variable time series with experimental noise. Left panel: Concentrations *X* (named A and B here); right panel: first derivatives *dX/dt*. Symbols correspond to raw data and to derivatives estimated by the three-point formula, respectively. Short dashed lines are estimates obtained solely from the genetic algorithm, applied to estimated slopes *dy/dt* (Eq. A4); long dashed lines resulted from subsequent regression using a gradient method; thin solid lines were obtained with the same method but targeting *dX/dt*; finally, thick solid lines show improvements by additional regression of *X* values, using costly but efficacious numerical integration of the differential equations, according to Eq. (A1).

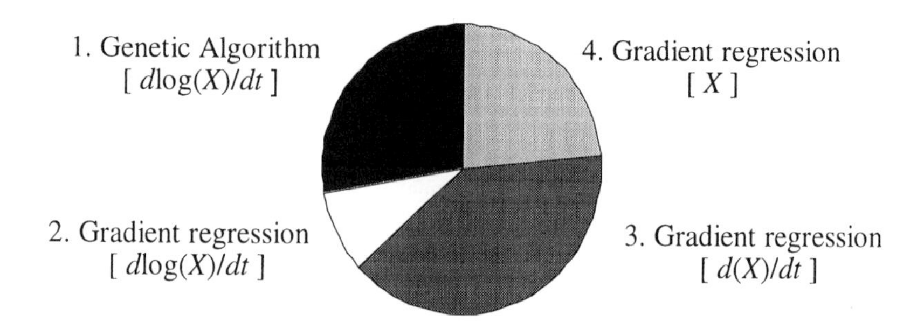

Figure A2. Relative distribution of computational costs of the four-step regression procedure described above. Total running time, on a Dell PowerEdge 2550 1.4 GHz 2 GbRAM was 83 sec.

Chapter 15

DATABASES, DATA MODELING AND SCHEMAS

Databse Development in Metabolomics

Nigel Hardy and Helen Fuell
Department of Computer Sciences, University of Wales, Aberystwyth, SY23 3DB, UK

1. INTRODUCTION

1.1 What is a Database?

A database is a collection of data. The term carries with it a number of additional implications that are useful to consider.

There is a clear implication that the collection will be well structured and well organized for continual retrieval of data and the addition, removal and updating of data. Databases are therefore rarely constructed for long term archival of rarely used, static data sets.

There is an implication that data in a database will be logically related, in two senses. They will all concern one organization, concept or project. More fundamentally, elements of the data are internally related to one another in ways analogous to textual cross-referencing.

The concept of a database system carries with it an implication of shared access. Private databases exist but many benefits accrue from simultaneous controlled access to the data by a number of users. There are many technical problems arising from this but here we note the practical benefits, particularly following the advent of the Internet.

A database is expected to provide a range of data checking functions. It will improve consistency - not least by reducing duplication of data items. It will help ensure the integrity of the data as a whole by enforcing constraints on the values of individual data items and on the combinations of values that are plausible in the real world at one time. By a combination of checking and of controlled access to data, a database can enforce (or at least support) a wide range of standards for operating procedures and other processes.

Placing an entire collection of possibly valuable data in one place (a database) can seem to be a case of "all your eggs in one basket". In practice, specialized backup and recovery mechanisms mean that the risk of accidental loss of data (due to human error and to machine failure) can be reduced to very low levels. A database implies a relatively safe place to store data, particularly when protection against unauthorized access is available.

The collection of data in a database is expected to be "self describing". This is achieved through mechanisms variously described as *metadata,* a *system catalog* or a *data dictionary.* This means that information about the data, as well as the data itself, can be retrieved. It leads to the concept of *data independence*: that is, programs (or people) using the data need not be strictly tied to a shared and fixed understanding of the data. Changes to the structure of the data, which do not directly affect a particular use, need not cause a problem. Consider the trivial data set (suggestive of gas chromatography-mass spectrometry (GC-MS) data) in Fig. 1.

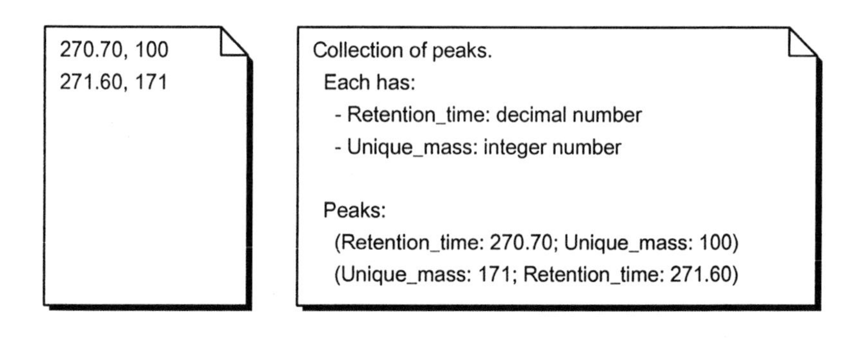

Figure 1. Self description.

Access to the plain data on the left requires instructions of the form: "I want the second number of each row, numbers are separated by commas". For the self-describing set on the right this could be: "I want the Unique_mass". If Peak_area were added to the description of each peak, the instructions for accessing the unique mass from self-describing data will continue to work. The instructions for the plain data may or may not work, depending on where the Peak_area "column" is inserted and whether the reader is resilient to changes in the number of columns. The verbosity of the self-describing data in Fig. 1 suggests a huge overhead but real database systems achieve self-description in more effective ways than this textual illustration suggests.

1.2 Obtaining the Benefits of a Database

This range of facilities (and others) which are now seen as synonymous with databases give ample justification for the use the technology in many areas of scientific research, not least in metabolic profiling, and many workers in the field recognize this.

The benefits of database technology do not follow automatically and they do not come cheaply. To develop a database for a particular application requires a significant design and implementation effort. Users (in our case including experimental biologists, technicians and statisticians) must expect to become involved in part of the process of designing a database. This comes under the heading of *requirements analysis* and is often characterized as *"developing a database schema"*. The data to be stored, the constraints that should be applied and the manipulations and queries that are required must be carefully identified and characterized.

In this chapter we outline the principles behind this process.

2. DATABASE DEVELOPMENT

Study of this field, particularly for the beginner, is bedeviled by confusion of terminology. This arises, at least in part, from the re-discovery of the same good ideas in several communities, each with its own vocabulary, and from the use of common phrases with increasingly specific, but divergent, meanings. The word *model* and more specifically *data model* is a particular instance of this problem. This chapter attempts to use consistent terminology, and to give definitions for important terms. The reader should note that other texts will differ.

Data independence (*q.v.*) is supported by the *three-level architecture* (American National Standards Institute, 1975; Tsichritzis and Klug, 1978) that has long been recognized as a necessary framework for the development of systems that protect the users (the external level) from implementation details (the internal level) by use of a conceptual layer. This latter can be thought of as an independent or community view of the database that is shared and understood both by users and by those developing and maintaining the database. Establishment of this view is where database development typically begins.

A database schema is a description of the structure of the database. This description can be at any particular level in the architecture, and it is typical for schemas at different levels to co-exist. The description at the internal level may be called the *physical* schema. The users' view is called the *external schema* (there may be several of these, known as *subschemas,* suiting the needs of different user groups). In the middle lies the *conceptual* schema. These are all different descriptions of the same concepts. In the world of architecture, a small-scale 3-D model, an artist's impression and a full set of plans in various projections are all conceptual descriptions of a new planned building. Each serves a purpose in communicating ideas, checking for completeness, structural integrity etc. and each can be related to the others and shown to be describing the same concept. Some are closer to the needs of users; some are closer to the needs of those who will construct the building. To continue the analogy, different people can follow the plans to produce a number of identical buildings. A database schema can similarly be used more than once to produce the container for different data that has the same fundamental structure. A database therefore has a schema, which it may share with other databases, and data, which are specific to it. All copies of a contacts and appointments program, for example, will have the same schema, but each will be a database containing data specific to a person.

In a plant metabolomics example, we can imagine a range of users. For illustration, consider the greenhouse technician and the statistician, in addition to the senior scientist who plans the work. The view that each of these has on a plant (in terms of the data about it which they use, create and alter) will differ, but it must be ensured that they are all working with the same plant. The greenhouse technician will not be concerned with data about the machines used to analyze samples, though these will be part of the understanding of the world shared by the other two users. We therefore wish to understand and support the views of the world appropriate to each user, in terms of which parts of the world they work with and which data about it they handle. These views must be rationalized into one conceptual schema, which supports them all. All parts of this schema must be understandable to relevant users and also to the engineers who are building the database system. The engineers will be responsible for supporting it by realization as a physical schema using techniques and technologies that need not trouble the users. Users see the benefits of this physical schema as a system that conforms to the conceptual schema (or at least to the parts of it relevant to them). The greenhouse technician and the statistician need not be concerned with how data about a plant are stored and managed on computer discs, only that they can each work in a complementary way on an appropriate set of data.

Users are much less likely to be involved in parts of the development process other than conceptual modeling and we review them only briefly here. The conceptual model must be carefully checked and validated. This will typically be interleaved with discussions with the users. When all parties believe they are satisfied with the model, the developers will make design decisions for the building phase (see section 5) and proceed with building and testing. Prototypes or partial implementations may well be shared with users to check earlier decisions. Deployment and user training follow and, depending on circumstance, there may be a phase when pre-existing or standard data are entered. Once the database system is in service, tuning may be necessary and appropriate since as the usage patterns emerge, features of the system can be adjusted to improve performance. No matter how faithful to the real world the original model was, the world will change and leave the model out of step. Maintenance and enhancement are therefore an inevitable part of database systems development.

Though it is largely beyond the scope of this chapter, we now briefly consider the software for accessing a database. A database system is typically provided as a service. In computing terms this means a resource that can be called on, possibly across networks, by means of a well-defined protocol to perform some service for a client agent. (A database may or may not be provided as a service in commercial and managerial terms – that is a separate issue). Client software can take many forms. Special purpose, bespoke interfaces can be developed, supporting the view of a particular type of user, for example the greenhouse technician, in an appropriate way. Alternatively, very general-purpose access can be provided, to the statistician for example, giving a very free and unstructured interface, often at the cost of some technical complexity. Access can be via special purpose programs, via web browsers or even through dedicated hardware (with embedded software) offering direct submission of data from collection devices. Database development must therefore include development (or at least customization) of a range of client programs. In data terms, the conceptual schema defines the service that can be offered. Client programs can only offer, to the user, facilities based on the schema (and computations based on data from it). A suitable schema supports the required range of clients. We note that this is where data independence (see section 1.1) becomes most valuable.

The crucial importance of establishing a good common view of the required database early in the development process cannot be overstated. Many will know of the problems of working in a building that was designed for another purpose or was designed with a poor understanding of the intended purpose. Similarly, a database designed with a poor understanding of the data it should hold or of the way in which those data are accumulated, used and changed will not be effective. "Demolition" may be the best course of action in such a case. We now consider how to establish a common view.

3. DATA MODELING

Data represent aspects of the real world. That part of the real world has a particular structure and operates in a particular set of ways. When we commit data to a computer system, what shape and style of storage structures should we use? One answer is that we should build structures that closely mimic the structures of the real world. In this way, we would know that the data would "fit" – a place for everything and everything in its place. No new untried system has to be invented and external verification of the adequacy of the database system is easier. Users (assuming they are familiar with that part of the real world) will find the database system more natural to use and easy to learn about. As the real world changes and develops, the aspects of the database system that must be altered for maintenance and enhancement will more easily identifiable. If (as database designers typically do) we take this approach, we are building a simplified replica of part of the real world inside the computer. We are building and using a model.

The community view of a database at the conceptual level will similarly be a model. Establishment of this view is therefore commonly called *data modeling* and more specifically *conceptual modeling*. A conceptual data model can be defined as *"A detailed model that captures the overall structure of organizational data while being independent of any database management system or other implementation considerations*" (Hoffer *et al.*, 2002).

An architect's model of a building might be built of plastic and wood. What will we use to build our data models? To build a specific model of data for some application area we will typically use one of the available paradigms, which, confusingly, are also informally called "data models". A data model in this generic sense can be defined as *"An integrated collection of concepts for describing and manipulating data, relationships between data, and constraints on the data in an organization"* (Connolly and Begg, 2002).

4. "E-R" DIAGRAMS FOR CONCEPTUAL DATA MODELING

The most common and longest standing generic model used for conceptual modeling is the Entity-Relationship (E-R) model. Originally introduced by Chen (1976) this model has relatively few concepts and provides a reasonably intuitive environment for development of models. It has subsequently been extended with more modern concepts, which can offer additional expressivity sometimes at the cost of intelligibility for non-technical users.

E-R modeling is essentially graphical. An E-R model can be expressed in other ways but diagrams are an attractive and effective medium for most people involved in database design. The concepts in the generic model must each have a graphical representation and there will be rules about their composition. Over the past 30 years a number of graphical notations have been proposed and used but they are all broadly equivalent. Here we use UML, The Unified Modeling Language (Booch *et al.*, 1999) which is becoming a common choice of notation for E-R modeling. UML is a more recent development and encompasses modeling of a wide range of aspects of software systems. It supports a number of types of diagram, including the one used here, which is entirely suitable for E-R modeling.

The E-R model has three basic concepts: *entity, relationship* and *attribute*. These are the "raw materials" from which we build our conceptual model and for which there must be a graphical notation. We will describe these concepts and then illustrate them with simplified examples.

Figure 2. Examples of entity set notation.

An **entity** corresponds to a thing or object in the real world. It may or may not be a physical thing. It is expected to have an independent existence and be distinguishable from all other similar things in the world. A useful informal description of an entity is "something which must be recorded and tracked". We record the values of certain properties of an entity. These are called **attributes** and the values for some subset of an entity's attributes serve to distinguish it from all other entities. Typically, we find in the world sets of similar entities. All members of each set have the same attributes (with different values for at least some of them). These are variously called *entity sets, entity types* or *entity classes*. We use here the term **entity set** and will use the term **entity instance** to emphasize the contrast with an entity set where necessary. An E-R diagram is built with entity sets. (Useful design diagrams can be drawn using entity instances. These may be called *instance diagrams*, or *occurrence diagrams.* We will not consider them here.)

Terminological confusion can arise from the common shorthand of referring to the items on an E-R diagram as "entities". For a clear understanding it is important to recognize that they represent sets. To illustrate entity sets consider Fig. 2. The boxes represent entity sets and their names are given in the top inner box. (By convention, the names are singular nouns). The items in the middle inner box are the attributes for which all entity instances will have a value. Our example is for a world where experimental plants of various species are grown under controlled conditions. An entity instance for Species might be represented by the attribute values (data) *"Arabidopsis thaliana (L)* Heynh*"*. For Plant, an entity instance might be represented by *"5764837, Greenhouse 2, 14-May-2002"*.

A **relationship** is an association between two or more entities. More precisely this is a *relationship instance* or *relationship occurrence*. On an E-R diagram we again represent *relationship sets* but commonly call them just relationships. Relationships involving more that two entities are rare and often confusing to deal with. We omit them from this discussion and concentrate on relationships involving just two entities. These are called **binary relationships**. There is a relationship between an instance of a plant and an instance of a species. There is therefore a set of relationships between the set of species and the set of plants. This is shown in Fig. 3.

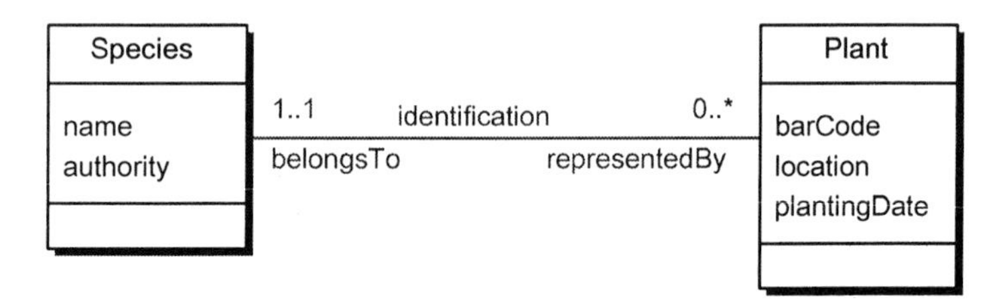

Figure 3. Example of relationship set notation.

The line represents the relationship set. The name of this set (i.e. of the relationship) is `identification`. All relationship sets must have a name. The other two words (phrases) are optional. They represent the relationship from the two ends. Thus, a species is represented by plants and a plant belongs to a species. Where no single relationship name is appropriate these additional "roles" may improve intelligibility of the model. The "1..1" and "0..*" represent constraints on the relationship set and are known as **cardinalities**. They are probably best understood, however, from the perspective of entity instances. They constrain how may relationships an entity instance may be involved in. We may "read" the relationship in Fig. 3 as "a plant belongs to one and only one species, while a species is represented by zero or more plants". At each end of the relationship, we can state the minimum and maximum number of relationships. Though we can conceptually use any values to do this, it is common to state the constraints using only 0, 1 or * (i.e. many). The notation is always *min..max*. Another way to consider the minimum values is as "participation"; a value of zero represents optional participation; a value of 1 represents mandatory participation. Thus, a particular species need not take part is any of these relationships while a plant must always take part in one (i.e. be identified). At this point we note that a major difference between alternative graphical E-R notations is the representation of cardinalities. Unless the notation is correctly understood the diagram can easily be misinterpreted.

So far, we have seen attributes of entity sets. Relationship sets can also have attributes (Fig. 4).

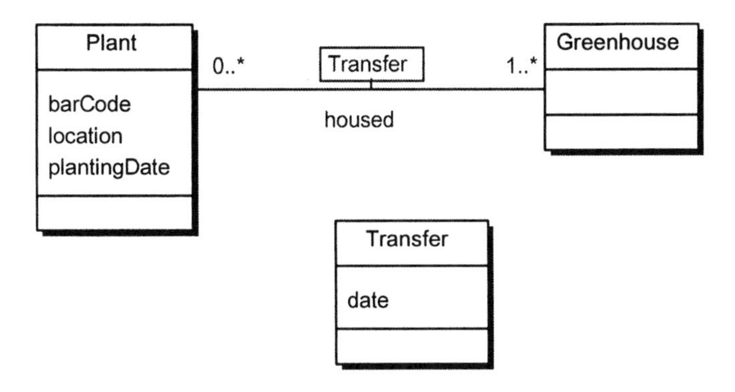

Figure 4. Attributes of a relationship set.

The relationship shows that a plant may be housed in 1 or more greenhouses (over its life). Conversely, a greenhouse may house many plants (at various times). If we are to store the date on which a plant is moved to a particular greenhouse, where do we do that. Is it an attribute of `Plant`? No, since we cannot then identify which greenhouse it refers to. Is it an attribute of `Greenhouse`? No, since we cannot identify which plant is concerned. The date is an attribute of the relationship between a particular instance of the plant entity and a particular instance of the greenhouse entity. Attributes of relationship sets are notated in UML, by creating a special purpose collection of attributes (`Transfer` in this case) and then connecting that to the relationship.

Modern versions of the E-R model (the so-called "extended" model) include a number of other concepts. We consider here just one more: *specialization/generalization*. This is the concept that in other modeling techniques may be called *subtyping*. Entity sets that are specializations of other entity sets are a common way of viewing the world and therefore provide a powerful modeling tool. We note above that all entities in a set have the same attributes. Specialization models the situation where the entities in a set have attributes in common, but subsets have additional attributes not relevant to the rest. Fig. 5 shows a simplified model for greenhouses. All greenhouses have a set of common attributes. Three specialized types of greenhouse are shown: temperature-controlled, light-controlled and humidity-controlled. Each has additional attributes that are not relevant to other types in addition to the attributes from the general type. Attributes of a temperature-controlled greenhouse are, therefore, {`name`, `technician`, `minTemp`, `maxTemp`}. Put another way, a temperature-controlled greenhouse is still a greenhouse – all `TCGreenhouse` entities are also `Greenhouse` entities.

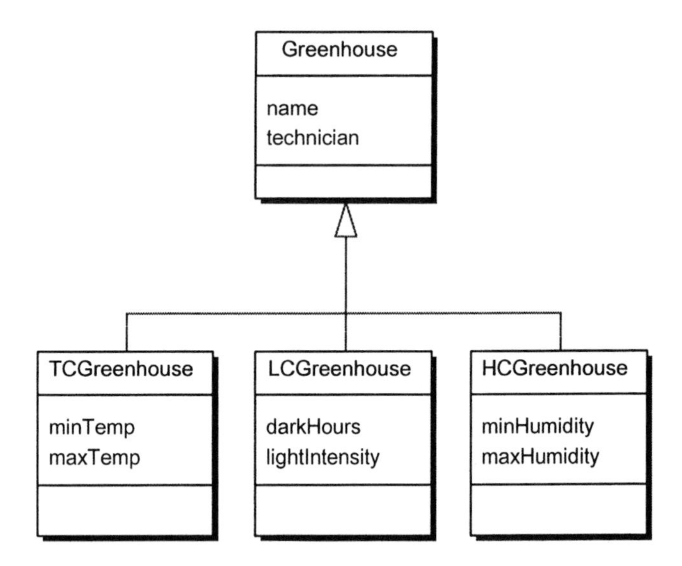

Figure 5. Specialization in extended Entity-Relationship modeling.

Must all greenhouses have some form of control? Can a particular greenhouse be both temperature and light controlled? How do we model the answers to these questions? Two additional aspects of specialization take the place of cardinality when we use this modeling technique. The first is *participation*: is it mandatory for all generalized entities to be specialized, or is it optional? In our example, there probably exist greenhouses with no controls and therefore participation is optional. The second aspect of specialization asks whether specialized entity sets are *disjoint.* The sets of specialized greenhouses will not be disjoint (some greenhouses are in more than one). It is common to describe this latter concept with an *"and"* or *"or"* between the specializations. The specialization of greenhouses should therefore be constrained by *{optional, and}*. As a contrast, consider specialized types of (GC-MS) equipment. All will have some attributes in common. Specializations might include time-of-flight and quadrupole systems. No system can be without such a specialization and no system could have both. The constraint would therefore be *{mandatory, or}*.

A final aspect of specialization/generalization modeling is that specialized entities may have relationships which other specializations or the generalized entity cannot have. A plant entity set might be specialized into genetically modified (GM) and non-GM sets because GM plants will have a relationship with a modification technique entity that is not appropriate for non-GM entities.

At this point we imagine (hope?) that the reader is starting to argue with, for example, Fig. 3. *"How do I represent cultivars? How do I represent ecotypes? (What would the relationship between species, cultivars, ecotypes and plants be?) I need to be more precise about plant location, etc. etc."* Such arguments do not represent a problem with E-R modeling. They represent its strength. The clarity and precision of the model will usually elicit in the prospective user a range of such questions, which would often not arise if a more discursive and less formal method of capturing the situation were used. We delay further consideration of this point to Section 7 below.

5. BUILDING SYSTEMS

The choice of a suitable data model for use in conceptual modeling is determined by expressivity and intelligibility for all the people involved. It must be sufficiently expressive to allow all the pertinent aspects of the enterprise to be precisely captured. It must be intelligible for efficiency and so that no ambiguity or uncertainty creeps in due to misunderstanding. These concerns suggest "high level" modeling in the sense that it may not be possible to directly implement on a computer a system that supports it. (At least, it may not be possible to efficiently and safely implement it). The database developer must therefore take the conceptual model and translate it to an implementation model.

Building and maintaining a database on a computer is a time consuming and intricate task. Very early in the development of computer technology it was recognized that there were common tasks and facilities used in most database systems. As with programming languages and operating systems this led to the development of major pieces of software that could be re-used for many applications. These are called *Database Management Systems (DBMS).* A DBMS is *"A software system that enables users to define, create, maintain and control access to the database"* (Connolly and Begg, 1999). A database could therefore be built with any technology, but we choose not to re-invent the wheel and to benefit from an available DBMS. Oracle (Abbey *et al.*, 2002), Microsoft Access (Irwin *et al.*, 2002) and AceDB (www.acedb.org) are examples of DBMS. If we are willing to operate within the paradigms and constraints of a DBMS we can benefit greatly from its facilities. Those paradigms and constraints are also the fruits of much scholarship and experience and so we do well to follow them, at least in the first instance.

Each DBMS implements support for a particular implementation data model. Thus we commonly hear of a *relational* database management system (RDBMS) or of an object-based or *object-oriented* database management system (OODBMS). These refer to the data models in which a conceptual model may be built.

The relational model is the longest established (Codd, 1970) data model currently used for DBMS. Almost all commercial database systems today are built using a RDBMS. Though the model is simple, it has a strong mathematical foundation and offers good support for data integrity. A conceptual model built using the relational model can be directly implemented, using a language for expressing it and a RDBMS. Almost all RDBMS use SQL as the language of expression, and more specifically the *Data Definition Language* (DDL) part of SQL. An E-R model cannot be directly implemented. There is no E-R language or "ERDBMS". We can, however, translate an E-R model to a relational model and implement that. The translation is reasonably direct (indeed the E-R model for conceptual modeling and the relational model for implementation grew up side by side) and engineers can be reasonably confident that the translation will not result in inadvertent changes to the agreed model. It should be noted that conceptual modeling could accurately (and appropriately) capture real aspects of the world that are fundamentally hard to represent in a computer-based system. Such problems are very likely to be identified at the translation stage, which represents a positive aspect to it.

It should be noted that a conceptual model could also be translated to other implementation models. More generally, parts of the conceptual model may be realized in a DBMS while other parts are automated using other technologies or are left as parts of the overall system that are performed by people. XML (Bray *et al.*, 2000) technology is of growing importance. It is described as a model for *semi-structured data* and it is a possible candidate as an implementation model. Though developed from SGML, which was largely applied to the structuring and storage of documents, XML is now widely used for the transmission of data of many other types. Database management systems whose implementation model is XML are possible and are beginning to appear, but it is most unlikely that they will challenge relational DBMS in the near future. A conceptual schema might therefore be converted to an XML schema to enable data interchange between systems, including relational databases, which support the same schema. This approach is being taken with MAGE (European Bioinformatics Institute, 2001) for microarray work (Brazma *et al.*, 2001) based on MIAME (Minimum Information about a Microarray Experiment).

6. SCHEMAS FOR METABOLOMICS

A number of challenges face the designers of databases to serve the metabolomics community. In this section we concentrate on large multi-user databases to hold data on plant metabolomic experiments and their results and highlight some of the issues that surround the design of such databases.

The data necessary to describe plant metabolomic experiments are diverse. For example, data are required:

- to describe the plants that provided the material for the experiments, their provenance, growth, harvest, sampling and preparation for metabolic analysis;
- to describe the analyses of the plant material and the setup of each of the range of machines (chromatographic, spectroscopic etc.) on which the analyses are carried out;
- to record the results of the analyses.

Such information is necessary for repeatability of experiments and for identification of sources of experimental variation during statistical and data mining analysis.

When developing a schema for such a wide range of diverse data the question of flexibility arises. There will be a desire to remain generic to as many different experimental methods and analytical technologies as possible and to be easily expandable to handle emerging methods and technologies as they are developed. Users will wish to store as much pertinent information as possible. A flexible design can come at the cost of utility. A highly complex structure can be hard to use consistently and to search reliably. It will be necessary to strike a balance between flexible structures, which appear supportive of storage, and rigid structures, which support retrieval.

The raw results files produced by the analytical technologies are often very large and noisy. The software packages provided with the analytical technologies usually provide some processing functions to clean and pre-process the data, perhaps providing metabolite identification. These processes are also sometimes carried out (at least in part) by hand. During database design, decisions need to be made about the type of results data to be held in the database. Arguably pre-processed data is of more use in statistical analysis and data mining. However, raw data is the less subjective result of the analysis. If the decision is made to store raw data, attention must be given during database design to appropriate structures to make such data usable in an efficient manner. If the decision is made to store processed data, then archiving strategies for the raw data must be considered as must tagging the processed data with its provenance and processing details; the latter so that third party users of the data can ascertain which datasets are comparable and the level of confidence that they can assign to information within a dataset given its processing history. Each manufacturer of analytical machines is inclined to have their own proprietary software packages, which produce a wide range of different datasets that vary significantly in content. Access to truly "raw" data can be a problem. Consideration must be given to storing results from analytical technologies that produce metabolic fingerprint data along with those from others that produce data on identified metabolites.

Many fields have identified the need for controlled vocabularies to enable effective deposition an retrieval of data from the database by a number of different users. Metabolomics will be no different. Database systems in the field must address this problem and can support and enforce the use of standard terminology.

7. CONCLUSIONS: THE ROLE OF THE USERS

Developers of database systems for metabolomics are unlikely to be top-flight practicing biologists. Indeed it is probably more appropriate and effective if they are not. Similarly, the biologists are unlikely to be experienced systems analysts or software engineers. It is through the effective collaboration of specialists from across the spectrum that effective and usable systems will arise. The process of developing conceptual models can and should be one that clarifies, codifies and even simplifies the processes and events that go to make up a particular activity. As such, it will be challenging. A user who constantly simplifies the story or provides sanitized versions of real world procedures will not help in the production of an effective schema. (Similarly, a software engineer who accepts the story as first recited and does not question ambiguities and contradictions or follow up on hints of hidden complexity will not satisfy the users in the end). The users' role is therefore to engage in detailed exposition, questioning and confirmation of the ways in which their world operates – E-R techniques can support this. They will need to engage in this process without fear of uncovering problems. (Many systems analyses have uncovered flaws in long established working practices).

Above all the users (as well as the engineers) will need to exercise patience. Building models of the world is not a quick process. It is often repetitive (returning to a solution which was discounted earlier in the process) and tedious (requiring careful consideration of enormous detail). The benefits can be great.

REFERENCES

Abbey M, Corey M, Abramson I. *Oracle9i: A Beginner's Guide.* Oracle Press, Berkeley (2002).

American National Standards Institute. ANSI/X3/SPARC Study Group on Data Base Management Systems. Interim Report, *FDT - Assoc Comp Machin SIGMOD Bulletin* 7:1-140 (1975).

Booch G, Rumbaugh J, Jacobson I. *Unified Modelling Language User Guide.* Addison Wesley, Reading (1999).

Bray T, Paoli J, Sperberg-McQueen CM, Maler E (Ed) *Extensible Markup Language (XML) 1.0.* 2nd Edn. World Wide Web Consortium (2000). http://www.w3.org/TR/2000/REC-xml-20001006.

Brazma A, Hingamp P, Quackenbush J *et al.* Minimum information about a microarray experiment (MIAME)-towards standards of microarray data. *Nature Genet* 29:365-371 (2001).

Chen PP. The Entity-Relationship Model: towards a unified view of data. *Assoc Comp Machin Trans Database Sys* 1: 9-36 (1976).

Codd EF. A relational model of data for large shared data banks. *Comm Assoc Comp Machin* 13:377-387 (1970) (reprinted in *M D Computing* 15:162-166 (1998).

Connolly T, Begg C. *Database Systems: A Practical Approach to Design, Implementation and Management.* 3rd Edn. Addison-Wesley, Reading (2002).

European Bioinformatics Institute, Rosetta Inpharmatics. *Gene Expression RFP Response.* Object Management Group Document (2001). http://cgi.omg.org/cgi-bin/doc?lifesci/01-10-01.

Hoffer JA, George JF, Valacich JS. *Modern Systems Analysis and Design.* 3rd Edn. Prentice Hall, New Jersey (2002).

Irwin MR, Prague CN, Reardon J. *Microsoft Access 2002 Bible.* JohnWiley and Sons, Chichester (2001).

Tsichritzis DC, Klug A. The ANSI/X3/SPARC DBMS Framework: Report of the Study Group on Data Base Management Systems. *Information Sys* 3:173-191(1978).

Chapter 16

DATABASES AND VISUALIZATION FOR METABOLOMICS

X. Jing Li, Olga Brazhnik, Aejaaz Kamal, Dianjing Guo, Christine Lee, Stefan Hoops, and Pedro Mendes
Virginia Bioinformatics Institute, Virginia Tech (0477), 1880 Pratt Dr., Blacksburg, VA 24061, USA

1. INTRODUCTION

Genomics has revolutionized research in biological sciences. The reductionist approach of single-molecule analysis is being slowly, but steadily, replaced by global views of the entire cellular machinery. This started with genetics, which turned to determining complete DNA sequences of organisms and is resulting in global comparisons of these sequences to infer their evolutionary past. With the availability of the first complete genome sequences, however, it became obvious that we know very little about how cells work. The problem is that a large number of genes in any genome have functions that are yet unknown, as judged by the lack of obvious phenotype of the corresponding mutants. These "orphan" genes are also not similar at the DNA or protein sequence levels to other genes of known function. The first complete eukaryotic genome, of the yeast *Saccharomyces cerevisiae* (Goffeau *et al.*, 1996), perhaps the species for which we know most biochemistry, revealed a staggering 40% of genes to which no function could be assigned. Based on this, Oliver called for a systematic approach to the discovery of gene function (Oliver, 1996), which since became known as *functional genomics*. The word function itself is a matter for debate (Kell and King, 2000): it is often simply a label of what cellular process the gene is involved in (Riley, 1993) but could also be a more complete, maybe even mechanistic, description of the molecular processes that its products are involved in at the molecular level (Casari *et al.*, 1996). We favor the latter because the classification in terms of "cellular processes" is itself rather artificial (What are cellular processes? Are they fixed or dynamic?). Indeed, function would be better described by some form of biochemical network that the gene is part of (Brazhnik *et al.*, 2002) or even by a mathematical model.

An approach to characterizing apparently silent phenotypes is to determine detailed molecular profiles of the organisms under specific physiological conditions (Oliver *et al.*, 1998). Such molecular phenotypes would include levels of mRNA, proteins and metabolites[1]. It is in this context of functional genomics that the term *metabolome* was coined, as a synonym to *metabolite profile*. More recently Fiehn (2002) defines the metabolome as the set of metabolites synthesized by an organism, but leaving the term *metabolomics* to describe comprehensive analyses that aim at identifying and quantifying all the metabolites present in a specific physiological state. Finally, *metabolic fingerprints* are measurements that reflect the levels of a partial (or full) set of metabolites in a physiological state, but not through their explicit concentrations (*e.g.* by a spectrum (Oliver *et al.*, 1998)).

For a long time, some biochemists have been using computers to simulate the behavior of biochemical networks (Chance *et al.*, 1960; Kibby, 1969; Garfinkel *et al.*, 1970; Hulme, 1971; Park and Wright, 1973; Garfinkel, 1981; Hofmeyr, 1986; Fell and Sauro, 1990; Mendes, 1997). However, these efforts have always been limited by the availability of information about enzyme properties, metabolite concentrations, or even what reactions occur in cells. Metabolomics promises to facilitate model construction by providing a great deal of data about metabolite concentrations at various physiological states. This can then be used to infer those details needed for constructing quantitative (kinetic) models of biochemistry. But this interaction between modeling and metabolomics (or functional genomics, in general) is reciprocal. Biochemical models are needed to make sense of the data and to put hypotheses to test. Although this may seem like a vicious circle, it is more like a spiral that is converging to more correct models of the cellular physiology. A hypothesis is generated from observed (metabolomic) data and put to test through computer simulation. Most often simulation results will not be able to predict new observations, refuting the hypothesis. An improved hypothesis/model is reformulated that result in new simulated data. That is then compared to yet more new data and the cycle continues, hopefully converging to a model that is capable of predicting experimental

[1] By metabolites we refer to non-DNA encoded organic molecules, irrespective of their molecular mass (*i.e.* everything other than polypeptides and nucleic acids) (see also Beecher, Chapter 17).

results and thus provide explanations about the mechanisms of the phenomenon of interest.

Metabolomics provides a systems view of metabolism through production of massive data sets (Mendes, 1997). The deluge of data coming out of mass spectrometers is not conveniently disseminated through paper articles, instead metabolomic experimental results are best archived and accessed through online databases (Mendes, 2002). Humans are notoriously incapable of visualizing more than 4 dimensions, yet metabolomic data can easily contain thousands. Data visualization is then a limiting factor for the interpretation of experimental results. Metabolomics has created new challenges for bioinformatics. While the organization of data generates little knowledge *per se*, without it there would be no knowledge at all and metabolomics would be reduced to an expensive and laborious way to overwhelm biochemists with numbers.

2. DATABASES FOR METABOLOMICS

Several types of databases are needed to organize and interpret metabolomic data. Obviously, they are primarily needed to store and retrieve the measured metabolite levels. This is commonly referred to as "the data", however it is only a portion of the data. Crucial to interpreting these data, is information about the experimental details, such as the nature of the biological material, the environmental conditions in which the experiment was carried out, or the equipment and methodologies used. These details are usually referred to as "metadata" (meaning data about data). Finally, it is also important to relate the data to existing knowledge and to use unambiguous vocabulary. This is achieved through the adoption of widely accepted ontologies (Schulze-Kremer, 1997), and through relating the data to established information.

We have recently classified the types of databases useful to metabolomics research (Mendes, 2002). These are:

1. databases storing detailed metabolite profiles, including raw data and detailed metadata;
2. databases storing metabolite profiles obtained for a single biological species;
3. databases collecting diverse metabolite profile data from many biological species and at many different physiological states;
4. databases listing all known metabolites (metabolome) for each biological species;
5. databases representing established biochemical facts.

Actual implementations do not necessarily belong to only one of these classes. However it is useful to consider explicitly this classification, as it covers all of the current needs. Currently there are available databases that fulfill the roles of 4 and 5 but they are sub-optimal as they were devised for different applications.

2.1 Laboratory Metabolite Profile Databases

Laboratories involved in metabolomic research are producing large numbers of metabolite profiles. These laboratories have a need to store all of the data related with the experiments in an efficient way. Traditionally, laboratories kept results in books, which served as long-term registers of the details of experiments, including much more information than published journal articles. Because of the sheer volume of data that metabolomics experiments produce, and the fact that the data are collected already in electronic format, it is much more convenient to keep the results in databases. Laboratory metabolite profile databases will eventually become the means of publishing data.

A laboratory metabolite profile database should contain a great deal of details about the experiments, as it would be a primary source of data. These details are extremely important for analysis of results, without them it would not be possible to repeat the experiments nor would one later be able to re-process the data, when better algorithms become available. It is also important that they store the raw data. By "raw" one refers to the first level of data that is produced by the instrumentation. In metabolomics, this is usually composed of chromatograms, electrophoretograms, mass or electromagnetic spectra, but it could be the result of any other analytical method. Usually these raw data are too detailed to be interpreted easily, and are processed with several algorithms until they result in a list of molecular species (known or unknown), and a measure of their concentrations. Again, it is important that all the intermediary steps be kept, as well as details on how they were obtained. Most researchers will be interested only in the last level of processed data, and this is what is most likely to be included in generic metabolomic databases (see below).

In our laboratory we are constructing one such database to fulfill the needs of projects we are involved in. Because we are combining metabolomic with proteomic and transcriptomic data, its scope is wider than discussed here. Nevertheless, it includes all of the characteristics discussed in the previous paragraphs. Fig. 1 represents a high-level data model for the metabolic part of our database. It will be made available to other researchers once it is in a working state.

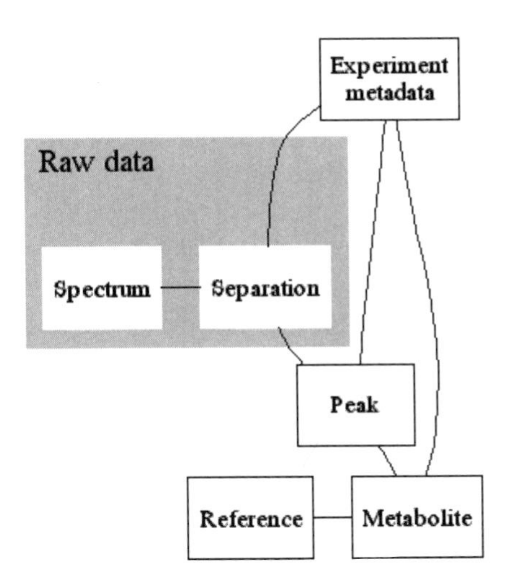

Figure 1. High-level data model of a laboratory metabolic profile database. The raw data conforms to a separation-spectrum technique (*e.g.* gas chromatography-mass spectrometry). The reference is important to provide context to the metabolites.

Metabolite profile databases of different laboratories should make their data available in a standard format. Bioinformatics is already plagued with a plethora of different file formats, which are an obstacle for data exchange and interoperability. Given that bioinformatic support for metabolomics is only now emerging, there is a good opportunity to establish standardization early. The Extensible Markup Language, XML, is currently the most appropriate vehicle for creation of standards (Barillot and Achard, 2000; Achard *et al.*, 2001). The metabolomics community has an excellent opportunity to achieve this goal early if it can agree on open standards based on XML to exchange data.

2.2 Generic Metabolite Profile Databases

Metabolic profiles and metabolomes will be useful to a range of studies and disciplines, much like sequences and gene expression profiles. In order to facilitate such studies it will be important that a few databases collect varied metabolic profiles and metabolomes. Similarly to what happens already in other areas of bioinformatics, one sees utility in a generic repository and also in species-specific ones.

A number of species-specific databases are already in existence, such as SGD for *Saccharomyces cerevisiae* (Cherry *et al.*, 1998) and TAIR for *Arabidopsis thaliana* (Huala *et al.*, 2001), which are intended as one-stop-shops for all related information about that species. Obviously, these databases should soon also incorporate metabolomic data. As happens for other data types, they would not be the primary data repositories, but rather collect information obtained from the primary sources, whether they are journal articles or other databases. This could be achieved in essentially two models: data warehouse or portal. The first copies data from the sources to its own local databases periodically, while the second is mostly an access point which links to the original data sources. In either case, they become the preferential data access point for those concentrating mostly on that species. These species-specific databases are an excellent means to organize a scientific community and must be proactive in finding the active research groups and their data. These species-specific databases would benefit from the establishment of data standards, as they would be constantly retrieving data from the primary sources.

While species-specific databases are important for many researchers, others need to compare data from various organisms and therefore would be best served by a central repository containing all metabolite profile and metabolomes studied, irrespective of biological species. The existence of such a resource will certainly be as important as those for sequences or protein structure are already. This GenBank of metabolomics would collect all metabolite profiles if researchers had to upload the relevant data before publication, much like what happens with sequences. Again, data standards would be major facilitators for data transfer. Given the nature and size of metabolomic data, the central repository would probably only store metabolite levels and the essential associated metadata (details on the biological species, environmental conditions, and other essential experimental details, but not instrument settings, operator names, and other purely technical details).

2.3 B-Net: A Reference Biochemical Database

Another class of biochemical databases useful to metabolomic research are reference databases. This group is the only one that already exists and with a fairly large number of members. Reference databases are required to provide biological and chemical context to the measurements made by metabolomic technologies. These databases catalogue the known biochemical compounds, reactions, enzyme activities, proteins, and genes for each organism. Examples of existing databases that could serve this purpose are KEGG (Kanehisa *et al.*, 2002), EcoCyc (Karp *et al.*, 1996), EMP (Selkov *et al.*, 1996), UM-BBD (Ellis *et al.*, 2001), BRENDA (Schomburg *et al.*, 2002), and PathDB (Mendes *et al.*, 2000). Two recent reviews analyzed them and concluded that they all overlap to a large extent, but then each has some characteristic that makes it unique (Wittig and De Beuckelaer, 2001; Wixon, 2001). For the purpose of metabolomics, such reference biochemical databases would need to display the following properties:

1. All chemical compounds represented should be specific molecules. This means that they should be associated with a singular molecular mass and be completely specified in terms of stereochemistry.
2. (Bio)chemical reactions should only list specific compounds.
3. All enzymes represented should be single entities, in the sense that they were translated from a unique mRNA species. Thus, each isoenzyme should be represented as a single entity; likewise ortholog enzymes should be represented as different entities.
4. All facts stored in the database should be appropriately documented and substantiated. No entities should be included in the database for which there is no evidence.
5. The database should represent the relations between the various entities of different natures.

All of the above properties are intended for clarity and precision. Indeed the function of this database would be essentially to serve as ontology, a way to explicitly specify the meaning of and relation between the fundamental concepts (Schulze-Kremer, 1997). Arguably, several ontologies already exist (Baker *et al.*, 1999; Karp, 2000; Consortium, 2001) and it would be ideal if they could be adopted for the current purpose. Unfortunately, they do not have sufficiently good coverage of the chemistry, which is essential for metabolomics applications. Juty *et al.* (2001) described a set of XML formats and data files for *Escherichia coli* that appear to fulfill most of the desired properties, but unfortunately their data set is not widely available.

It is important to stress some issues related with the five properties listed above. Part of the activity of metabolomics is to identify organic molecules present in living systems, thus it is very important that each one listed in the reference database be a unique entity. This extends to the second point, that reactions be composed only of specific compounds. This is a particular deficient aspect of existing biochemical databases, which, following the specifications of the Nomenclature Committee of the International Union of Biochemistry and Molecular Biology (NC-IUBMB), contain several reactions that have classes of compounds as members (rather than single compounds). Precision in representing chemical reactions is needed because the reference database will be used for computation of networks, which are required for identification of points of action. The same applies to proteins and nucleic acids, being very important that paralogue genes and their products be uniquely identified. Most of the existing biochemical databases assume existence of most central metabolism reactions or enzymes for all organisms, although it is remarkable that, in many cases, no evidence exists for that. An important role of metabolomics, as well as proteomics and gene expression, is to establish which molecular species and reactions do really exist. In summary, the reference database (ontology) is required to relate measurements to biological entities. An example would be to represent the fact that D-glucose was observed in roots of *Zea mays*, that it is a substrate for the enzyme activity hexokinase, and that the protein hexokinase1 was also observed in the same tissue, but not the protein hexokinase2. Another example would be the lack of evidence for the occurrence of the reaction between D-fructose 6-phosphate and ATP in *A. thaliana*, with very important consequences in the annotation of the genome of this organism (no gene encoding the EC 2.7.1.11 enzyme activity is expected in the genome).

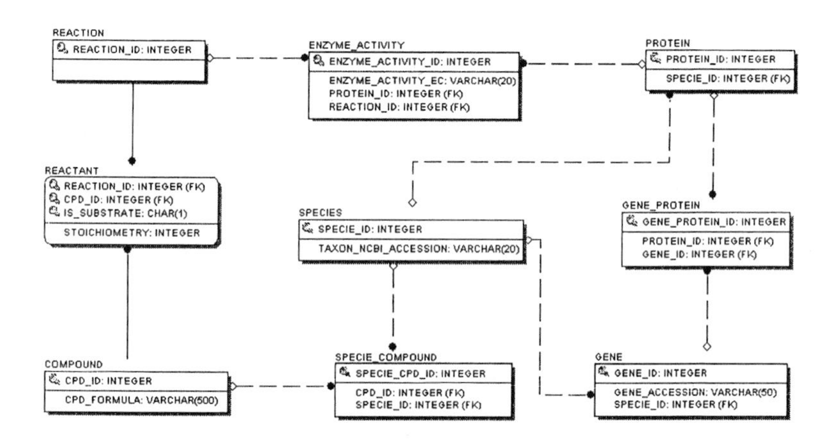

Figure 2. High-level data model of the B-Net biochemical reference database.

While constructing our metabolomics laboratory database (see above) it became obvious that a reference database with the above properties was sorely needed. We have since then embarked on creating one such reference complying with all the requirements above, and we named it B-Net (for *B*iochemical *NET*work). Our intention is not to create *the* reference biochemical database, but rather to give context to our data sets. Given that our other databases are relational, we have then implemented B-Net in a relational framework. The data model (Fig. 2) has been kept to the simplest possible but which reflects the biology as we understand it. Importantly, biological facts in B-Net will be substantiated by a reference (bibliographic or other) and by evidence codes. We have adopted the evidence codes defined by the Gene Ontology™ (Consortium, 2001) and adopted the procedure that each fact is supplemented with as many evidences as we can find. This should allow one to make judgments about the plausibility of these "facts". B-Net has the following major data types: compounds, reactions, enzyme activities, proteins, genes and biological species. The database is able to represent such things as genes that are alternatively spliced to produce several proteins, proteins that are composed of polypeptides encoded in several genes, proteins that contain several enzyme activities, enzyme activities that are contained in several proteins, enzyme activities that include many reactions (non-specific ones, such as EC 1.1.1.1). Finally the database is also able to capture what metabolites, proteins and genes, are included in a biological species, and thus can be used as a metabolome database (indeed we are currently collecting information about the metabolome of *Medicago truncatula*, which will soon become available at http://medicago.vbi.vt.edu).

To construct B-Net we started with public data sets and have trimmed them down to what conforms to the five points above. For compounds we used the LIGAND database of the Kyoto group (Goto *et al.*, 1998), but eliminated entries that represent classes of compounds, polymers, proteins other than redox pairs, and adducts. We used the reaction and enzyme activity data from the NC-IUBMB, but only retained reactions that had compounds in the previous list (*i.e.* refused reactions with classes of compounds, *etc.*). We are currently adding data on proteins and genes for *M. truncatula* that have been reported in the literature, in proteomic (Mathesius *et al.*, 2001; Bestel-Corre *et al.*, 2002), metabolomic (Huhman and Sumner, 2002), and gene expression (Bell *et al.*, 2001) studies. We will make this reference database available to others in XML format, together with Oracle and PostgreSQL scripts. Hopefully this will allow others to not have to repeat this work and perhaps this will become a nucleus for an ontology for functional genomics studies. We will also make efforts to create links between the objects in our reference database and those of the Gene Ontology™.

3. VISUALIZATION OF METABOLOMIC DATA

Functional genomic data has the characteristic of being of very high dimensions. This applies equally to gene expression, proteomics or metabolomics data. The high dimensionality is an obstacle for humans to gain intuition from their observation. Visualization occurs mostly in two dimensions, although often we represent 3- or even up to 5-dimensional data in 2-D projections (using perspective, animation and color). Visualization is a major area of development for functional genomic informatics. Currently two strategies are extensively used to represent these data, including metabolomics. One is to use hierarchical (or other forms of) clustering to group the many dimensions. A particularly popular visualization is produced by software developed by Michael Eisen (Eisen *et al.*, 1998), which summarizes gene expression data for thousands of genes in a compact way (one or two pages). It would be very simple to apply this software to metabolomic data as well. A second type of visualization is based on data reduction, usually done through principal components analysis (PCA), and which has been applied to metabolomic data (Oliver *et al.*, 1998; Fiehn *et al.*, 2000; Roessner *et al.*, 2001). PCA is a method of choosing a plane to project the data onto, which retains most of the variance. Other alternatives for this are the use of projection pursuit (Friedman, 1987), or the visualization technique "Grand Tour" (Asimov, 1985), both included in the XGobi software (Friedman, 1987). Fiehn and co-workers (Kose *et al.*, 2001) describe a method based on correlations between metabolite levels to infer portions of the metabolic network. Meyer and Cook (2000) have recently reviewed data visualization for biotechnology.

The methods from multivariate statistics enumerated above visualize the data without any *a priori* knowledge. Such methods are mostly useful when one is searching for patterns in the data, and in one case the results are hypotheses about the underlying biochemical networks (Kose *et al.*, 2001). However, there are also advantages from visualizations that embed the data in pre-existing knowledge. This can be done primarily using biochemical network diagrams, and examples already exist for gene expression data (Nakao *et al.*, 1999; Wolf *et al.*, 2000).

3.1 Metabolic Pathways

We have become interested in using metabolic network diagrams for visualizing metabolomic data, but also to integrate it with gene expression data. Currently we have finished a Java-based software to view metabolomic data produced with the Met-Ex technology from Phenomenome Discoveries Inc. (see Goodenowe, Chapter 8). Our software combines metabolomic data with gene expression or proteomic data in metabolic pathway maps. We have used the maps from the KEGG system (Kanehisa *et al.*, 2002) as the backbone, given that these are now popular views of metabolism, though it could easily use other maps. Ratios of metabolite levels are translated into a color and the circles representing metabolites in the map are colored accordingly. A similar approach is used to represent gene expression ratios, which are painted on the boxes representing reactions. Here we associate the genes with reactions through the enzyme activity of their protein products. This visualization is useful for comparing two states of biological material, as it uses ratios of concentrations between two states. The software provides further details on the data (such as details about the mass spectrometry used, see Goodenowe, Chapter 8) when the user right-clicks the metabolite or reaction locations. Fig. 3 illustrates this visualization.

3.2 Metabolite Neighborhoods

While using metabolic pathways to give visualize metabolomic data in context is very powerful, some reservations could be raised. The concept of metabolic pathways is not well defined, and there is considerable disagreement between biochemists on the concept.

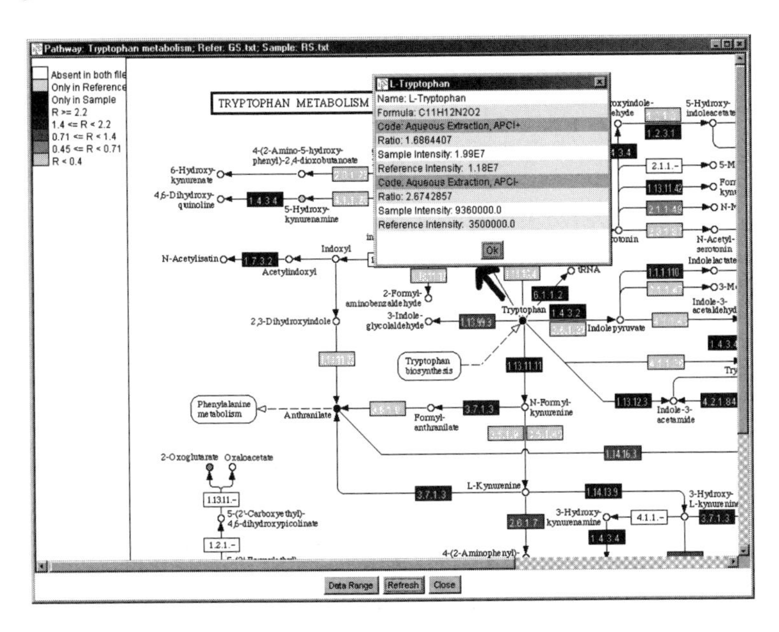

Figure 3. A metabolic map view of metabolomic and transcriptomic data. Filled circles represent metabolite concentration ratios, filled boxes represent gene expression ratios. The inset box was activated by a mouse click on tryptophan.

We have entered in casual and even formal discussions on this subject (http://www.sun.ac.za/biochem/btk/program.html) and have heard many definitions (from "a linear set of reactions without branches" to "the titles of chapters of biochemistry text books"). To complicate things, the fact that many null mutants reveal no phenotype (in organisms ranging from yeast to mice) may be a consequence of redundancy in the metabolic network – several paths usually connect two metabolites. Metabolic pathways in general, including the ones from KEGG, usually ignore side reactions that their intermediate metabolites may participate. Co-substrates are also often not represented in such maps (and certainly not in KEGG). The absence of these elements obscures any intuition or conclusions that one may expect to arise from the visualizations described in the previous section. The whole metabolic network of a cell type is also too complicated (dense) to be of any value to visualization.

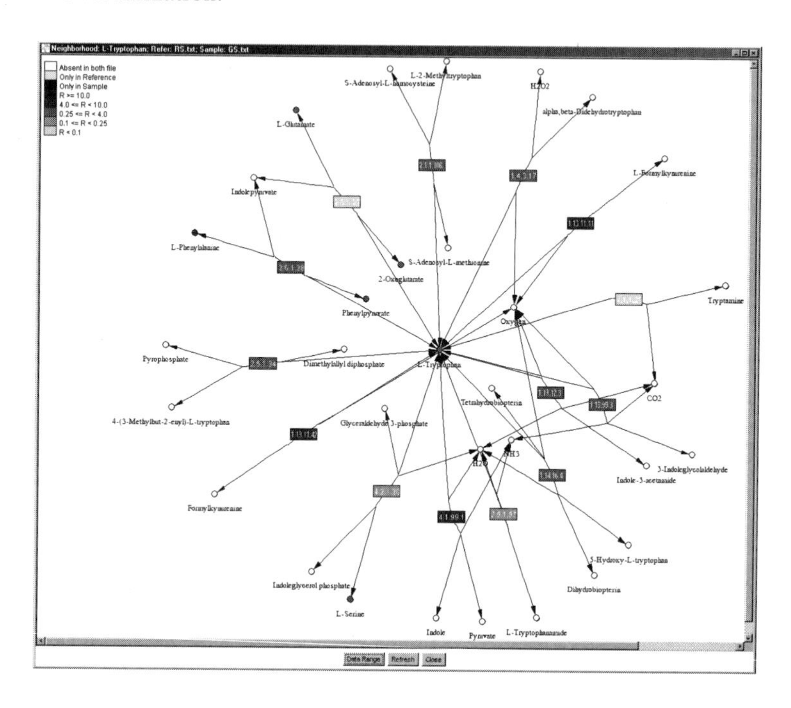

Figure 4. The metabolite neighborhood of L-tryptophan.

In response to this need for accurate but intelligible representations of metabolic networks, we have developed a novel concept that we named *metabolite neighborhoods*. Metabolite neighborhoods are local views of the network and are defined as *the set consisting of a central metabolite, all the reactions that include it as substrate or product, plus all metabolites that take part in those reactions*. Metabolite neighborhoods are generally simple enough that their diagrams are intelligible. Their number is, of course, the same as the number of metabolites. In our B-Net database this is currently *circa* 2000, which makes impractical to compose these diagrams by hand. The alternative is to use graph layout software to automatically construct such diagrams. We have created an automated process to construct metabolite neighborhood diagrams for the metabolites in B-Net. The first step is to query the database to extract each neighborhood as sets of reaction strings. Then a program converts the metabolites into vertices and the reactions into edges and adds a number of hidden edges and vertices in order to constrain the graph to what biochemists expect of metabolic diagrams. Finally the graph is laid out using the GraphViz engine by from the AT&T Labs (Gansner and North, 1999). The final result is a bitmap that uses the same elements used in the KEGG maps (circles for metabolites and rectangles for reactions), plus a file with the coordinates of each element. Alternatively we can produce other types of output, such as SBML (Systems Biology Markup Language, an XML format for specifying biochemical dynamic models, Hucka *et al.*, 2002).

Our software Met-ExViewer is also able to use the metabolite neighborhood diagrams to visualize metabolomic and transcriptomic data. Such views of the data have the advantage that all the reactions that the central metabolite is involved in are represented in the same diagram, allowing one to inspect which of the reactions may have been slowed down.

Currently we are building a program similar to Met-Ex Viewer but that will be able to display data from technologies, such as gas- or liquid chromatography-mass spectrometry. That software will be made freely available at our web site (http://www.vbi.vt.edu/~mendes).

4. CONCLUSION

Metabolomics is becoming an important piece of functional genomics. Metabolite profiles are the molecular profiles that are closer to phenotype and thus are expected to be extremely important to identifying gene function (Oliver *et al.*, 1998; Trethewey *et al.*, 1999; Fiehn *et al.*, 2000; Cornish-Bowden and Cardenas, 2001; Raamsdonk *et al.*, 2001). In addition, metabolite profiles can be important resources for construction of mathematical computer models of cellular function (Mendes, 2002).

Like other genomic technologies, metabolomics is capable of producing large amounts of data that form a challenge to organize and make sense of. Indeed, if one is interested in metabolomics as a part of functional genomics, it is important to combine these data with those of proteomics and gene expression (microarray, SAGE, ESTs, *etc.*). This can be done through multivariate statistical analysis and by embedding the data into its biochemical context. Giving biochemical context to functional genomics data is done through identification of molecules (metabolites, proteins, mRNA) and by relating them to each other through known interactions. We briefly described B-Net, a reference biochemical database that is intended to do just so. B-Net acts as ontology for molecular cellular interactions and is implemented through a relational database. We are also developing a database system to store metabolomic raw and processed data linked to its experimental context. This will act as a laboratory functional genomics database.

Visualization of metabolomic data can be done by generic methods (Meyer and Cook, 2000; Kose *et al.*, 2001), or through the use of biochemical network diagrams. The latter is a very convenient way to put the data in context with previous biochemical knowledge (the network) and allows fusion of these data with those of transcriptomics or proteomics. This is akin to the use of geographical information systems to fuse data from different sources (*e.g.* atmospheric conditions with population densities). To overcome some limitations of classical metabolic (pathway) diagrams, we have developed the concept of metabolite neighborhoods, which are a local, but complete, view of the metabolic network. Metabolite neighborhoods provide an alternative view, used for the same purpose of data fusion and visualization. Metabolite neighborhoods are also useful for analyses of structural properties of the network (Jeong *et al.*, 2000; Wagner and Fell, 2001).

ACKNOWLEDGEMENTS

We thank Jennifer Weller and Lloyd Sumner for helpful discussions. We are grateful to the National Science Foundation (Grant DBI- 0109732) and Phenomenome Discoveries Inc. for financial support.

REFERENCES

Achard F, Vaysseix G, Barillot E. XML, bioinformatics and data integration. *Bioinformatics* 17: 115-125 (2001).

Asimov D. The grand tour: a tool for viewing multidimensional data. *SIAM J Sci Stat Comput* 6: 128-143 (1985).

Baker PG, Goble CA, Bechhofer S *et al.* An ontology for bioinformatics applications. *Bioinformatics* 15: 510-520. (1999).

Barillot E, Achard F. XML: A lingua franca for science? *Trends Biotechnol* 18: 331-333 (2000).

Bell CJ, Dixon RA, Farmer AD *et al.* The *Medicago* Genome Initiative: a model legume database. *Nucleic Acids Res* 29: 114-117 (2001).

Bestel-Corre G, Dumas-Gaudot E *et al.* Proteome analysis and identification of symbiosis-related proteins from *Medicago truncatula* Gaertn by two-dimensional electrophoresis and mass spectrometry. *Electrophoresis* 23: 122-137. (2002).

Brazhnik P, de la Fuente A, Mendes P. Gene networks: how to put the function in genomics. *Trends Biotechnol* in press (2002).

Casari G, De Daruvar A, Sander C, Schneider R. Bioinformatics and the discovery of gene function. *Trends Genet* 12: 244-245 (1996).
Chance B, Garfinkel D, Higgins J, Hess B. Metabolic control mechanisms. V. A solution for the equations representing interaction between glycolysis and respiration in ascites tumor cells. *J Biol Chem* 235: 2426-2439 (1960).
Cherry JM, Adler C, Ball C *et al.* SGD: *Saccharomyces* Genome Database. *Nucleic Acids Res* 26: 73-79 (1998).
Consortium, The Gene Ontology. Creating the gene ontology resource: design and implementation. *Genome Res* 11: 1425-1433 (2001).
Cornish-Bowden A, Cardenas ML. Functional genomics. Silent genes given voice. *Nature* 409: 571-572 (2001).
Eisen MB, Spellman PT, Brown PO, Botstein D. Cluster analysis and display of genome-wide expression patterns. *Proc Natl Acad Sci USA.* 95: 14863-14868 (1998).
Ellis LB, Hershberger CD, Bryan EM, Wackett LP. The University of Minnesota Biocatalysis /Biodegradation Database: emphasizing enzymes. *Nucleic Acids Res* 29: 340-343 (2001).
Fell DA, Sauro HM. Metabolic control analysis by computer - progress and prospects. *Biomed Biochim Acta* 49: 811-816 (1990).
Fiehn O, Kopka J, Dörmann P *et al.* Metabolite profiling for plant functional genomics. *Nature Biotechnol* 18: 1157-1161 (2000).
Fiehn O. Metabolomics - the link between genotypes and phenotypes. *Plant Mol Biol* 48: 155-171 (2002).
Friedman JH. Exploratory projection pursuit. *J Am Stat Assoc* 82: 249-266 (1987).
Gansner ER, North SC. An open graph visualization system and its applications to software engineering. *Software Practice Experience* 30: 1203-1233 (1999).
Garfinkel D, Garfinkel L, Pring M *et al.* Computer applications to biochemical kinetics. *Ann Rev Biochem* 39: 473-498 (1970).
Garfinkel D. Computer modeling of metabolic pathways. *Trends Biochem Sci* 6: 69-71 (1981).
Goffeau A, Barrell BG, Bussey H *et al.* Life with 6000 genes. *Science* 274: 546-567 (1996).
Goto S, Nishioka T, Kanehisa M. LIGAND: chemical database for enzyme reactions. *Bioinformatics* 14: 591-599 (1998).
Hofmeyr JHS. Steady-state modeling of metabolic pathways. A guide for the prospective simulator. *Computer Appl Biosci* 2: 5-11 (1986).
Huala E, Dickerman AW, Garcia-Hernandez M *et al.* The *Arabidopsis* Information Resource (TAIR): a comprehensive database and web-based information retrieval, analysis, and visualization system for a model plant. *Nucleic Acids Res* 29: 102-105 (2001).
Hucka M, Finney A, Sauro HM *et al.* The Systems Biology Markup Language (SBML): a medium for representation and exchange of biochemical network models. *Bioinformatics*: submitted (2002).
Huhman DV, Sumner LW. Metabolic profiling of saponins in *Medicago sativa* and *Medicago truncatula* using HPLC coupled to an electrospray ion-trap mass spectrometer. *Phytochemistry* 59: 347-360 (2002).
Hulme EC. Simulation of biochemical systems. *J Theoret Biol* 31: 131-137 (1971).
Jeong H, Tombor B, Albert R *et al.* The large-scale organization of metabolic networks. *Nature* 407: 651-654 (2000).
Juty NS, Spence HD, Hotz HR *et al.* Simultaneous modelling of metabolic, genetic and product-interaction networks. *Briefings Bioinformat* 2: 223-232 (2001).

Kanehisa M, Goto S, Kawashima S, Nakaya A. The KEGG databases at Genome Net. *Nucleic Acids Res* 30: 42-46 (2002).

Karp PD, Riley M, Paley SM, Pelligrinitoole A. Ecocyc - An encyclopedia of *Escherichia coli* genes and metabolism. *Nucleic Acids Res* 24: 32-39 (1996).

Karp PD. An ontology for biological function based on molecular interactions. *Bioinformatics* 16: 269-285 (2000).

Kell DB, King RD. On the optimization of classes for the assignment of unidentified reading frames in functional genomics programmes: the need for machine learning. *Trends Biotechnol* 18: 93-98 (2000).

Kibby MR. Stochastic method for the simulation of biochemical systems on a digital computer. *Nature* 222: 298-299 (1969).

Kose F, Weckwerth W, Linke T, Fiehn O. Visualizing plant metabolomic correlation networks using clique-metabolite matrices. *Bioinformatics* 17: 1198-1208 (2001).

Mathesius U, Keijzers G, Natera SH *et al.* Establishment of a root proteome reference map for the model legume *Medicago truncatula* using the expressed sequence tag database for peptide mass fingerprinting. *Proteomics* 1: 1424-1440 (2001).

Mendes P. Biochemistry by numbers: simulation of biochemical pathways with Gepasi 3. *Trends Biochem Sci* 22: 361-363 (1997).

Mendes P, Bulmore DL, Farmer AD *et al.* PathDB: a second generation metabolic database. In *Animating the Cellular Map*. Hofmeyr JHS, Rohwer JM, Snoep JL (Ed) pp. 207-212, Stellenbosch University Press, Stellenbosch (2000).

Mendes P. Emerging bioinformatics for the metabolome. *Briefings Bioinformat* in press (2002).

Meyer RD, Cook D. Visualization of data. *Curr Opin Biotechnol* 11: 89-96 (2000).

Nakao M, Bono H, Kawashima S *et al.* Genome-scale gene expression analysis and pathway reconstruction in KEGG. *Genome Informat* 10: 94-103 (1999).

Oliver SG. From DNA sequence to biological function. *Nature* 379: 597-600 (1996).

Oliver SG, Winson MK, Kell DB, Baganz F. Systematic functional analysis of the yeast genome. *Trends Biotechnol* 16: 373-378 (1998).

Park DJ, Wright BE. METASIM, a general purpose metabolic stimulator for studying cellular transformations. *Computer Prog Biomed* 3: 10-26 (1973).

Raamsdonk LM, Teusink B, Broadhurst D *et al.* A functional genomics strategy that uses metabolome data to reveal the phenotype of silent mutations. *Nature Biotechnol* 19: 45-50 (2001).

Riley M. Functions of the gene products of *Escherichia coli*. *Microbiol Rev* 57: 862-952 (1993).

Roessner U, Luedemann A, Brust D *et al.* Metabolic profiling allows comprehensive phenotyping of genetically or environmentally modified plant systems. *Plant Cell* 13: 11-29 (2001).

Schomburg I, Chang A, Schomburg D. BRENDA, enzyme data and metabolic information. *Nucleic Acids Res* 30: 47-49 (2002).

Schulze-Kremer S. Adding semantics to genome databases: towards an ontology for molecular biology. *Intel Sys Mol Biol* 5: 272-275 (1997).

Selkov E, Basmanova S, Gaasterland T *et al.* The metabolic pathway collection from EMP: the enzymes and metabolic pathways database. *Nucleic Acids Res* 24: 26-28 (1996).

Trethewey RN, Krotzky AJ, Willmitzer L. Metabolic profiling: a Rosetta stone for genomics? *Curr Opin Plant Biol* 2: 83-85 (1999).

Wagner A, Fell DA. The small world inside large metabolic networks. *Proc Royal Soc London B* 268: 1803-1810 (2001).

Wittig U, De Beuckelaer A. Analysis and comparison of metabolic pathway databases. *Briefings Bioinformat* 2: 126-142 (2001).
Wixon J. Website review: pathway databases. *Compar Funct Genom* 2: 391-397 (2001).
Wolf D, Gray CP, de Saizieu A. Visualising gene expression in its metabolic context. *Briefings Bioinformat* 1: 297-304 (2000).

Chapter 17

THE HUMAN METABOLOME

Chris W.W. Beecher
Paradigm Genetics, 108 TW Alexander Dr., Research Triangle Park, NC 27709, USA

1. INTRODUCTION

The human metabolome is best understood by analogy to the human genome, *i.e.* where the human genome is the set of all genes in a human, the human metabolome is the set of all metabolites in a human. Whereas the science of genomics is based upon a genome, the science of metabolomics is grounded in a metabolome. To continue the genome/metabolome analogy, it should be apparent that any published human genome is a statistical approximation as it is derived from a limited number of individuals, and that ultimately any individual has a unique genome. Similarly, the human metabolome is a statistical approximation of the total human metabolic potential. Furthermore, just as the human genome is differentiable from other genomes, for instance, the *Xenopus* or *Caenothus* genomes, there is a human metabolome that defines the human biochemical potential that is different from that of *Xenopus* or *Caenothus*. While the vitamins and essential amino acids are the most obvious unique characteristics of the human metabolome they are not, by any means, the only defining features. By its definition, the metabolome is the comprehensive set of all potential metabolites that can be expressed in any human under all conditions, the high-water mark of human biochemical potential. It is likely that most individuals vary in their biochemical potential, expressing only incomplete subsets of the metabolome, depending on their genetic makeup and health state. Indeed, many metabolic diseases and even the efficacy of most drugs are variable, due, at least in part, to individual variances in metabolism and the resulting biochemistry. Fundamentally, the concept of a metabolome is based in set theory. The metabolome is the total set, and all individuals, tissues, and conditions are subsets. What is unique about the metabolome is its ability to directly reflect physiological status when used as part of a metabolomic analysis. This manuscript will explore the implications of the human metabolome and explore ways for their generation and use.

2. METABOLOMICS

Metabolomics, also known as biochemical profiling, is an emerging "omics" science that has tremendous potential and many unique characteristics (Glassbrook *et al.*, 2000; Roessner *et al.*, 2001).

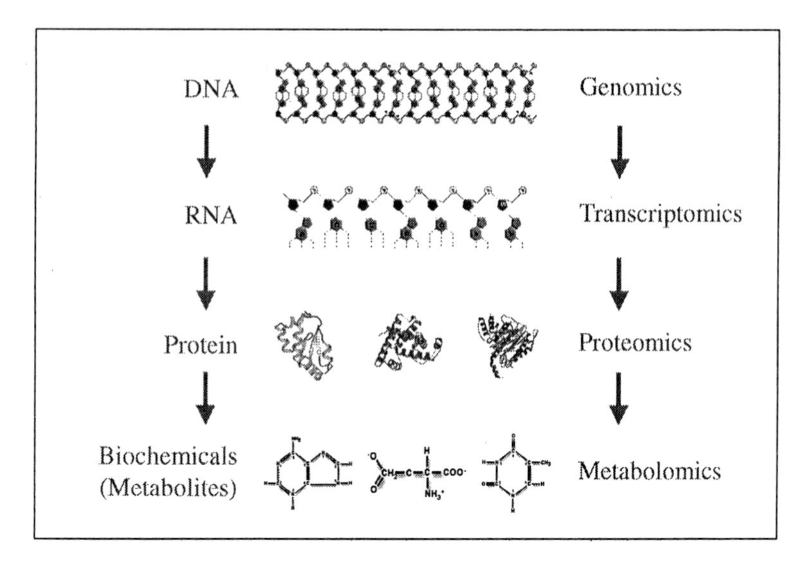

Figure 1. The "Omics" sciences are characterized by complex datasets of related phenomena each of which taken as a whole constitute a picture of an organism.

It is most obviously useful for the exploration of problems in which physiology is altered, *e.g.* through stress, disease, chemical or other insult. Like all "omics" sciences (Fig. 1) it takes a global view of an organism, *i.e.* attempting to understand the current physiological status of a sample or organism in light of its full physiologic potential. Thus, genomics looks at the total genome to understand the significance of an individual gene. Similarly, transcriptomics and proteomics attempt to understand change as a function of individual transcript expression or protein concentration in light of the total expression of RNA transcripts or proteins, respectively. Despite the strength of these technologies, deriving from their breadth of information, in all three cases, the current limitations to interpretation lie in the fact that the vast majority of the relevant entities (*i.e.* genes, transcripts and/or proteins) are only partially characterized. In other words, while their respective sequences may be known, their functional significance is often largely hypothetical, if understood at all. These sciences, responsible for huge advances in the past few years, are comparative in nature and, although understood through "analytical" technologies, they are not truly quantitative. They usually achieve their successes by comparing one or more states to determine where change has occurred. In this light, the "omics" sciences are generally hypothesis-generating, and form the basis for specific "non-omics" investigations that are supported by classical experiments.

Metabolomics is a parallel science in that it characterizes the physiological state of a sample by determining the "concentration" of all of the small molecules that constitute metabolism (Fig. 2).

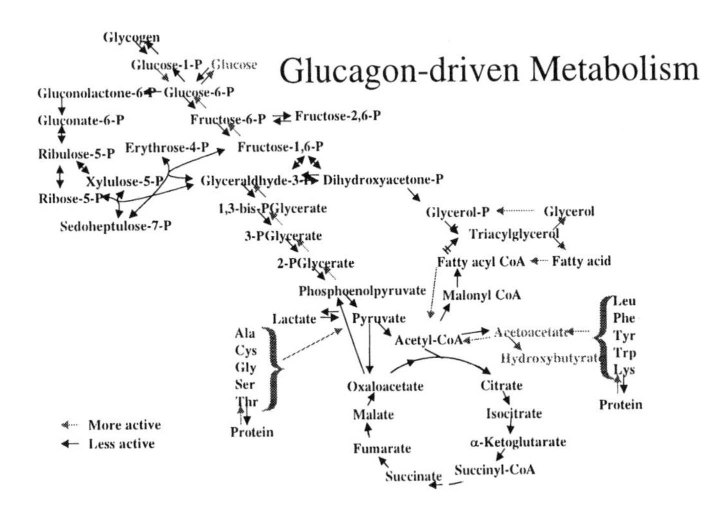

Figure 2. A simplification of a metabolic pathway that responds to a condition, in this case glucagon, that causes large-scale redirection of metabolites. Note that the relationship of almost all the biochemical metabolites are well understood. The enzymes (shown as arrows) are direct links to the proteomic, transcriptomic and genomic levels.

In contrast to the other "omics," however, the nature and the relationship of almost all of the metabolomic entities (*i.e.* biochemicals) have been thoroughly established through over a century's worth of biochemical investigations. As noted earlier, the establishment of a metabolome is fundamental to a full understanding of metabolomics.

2.1 Definitions

Human biochemistry is very complex, enough so that at the outset a restrictive definition is likely to be helpful. For purposes of this chapter (and to keep metabolomics from being all-encompassing), the metabolome should consist only of those native small molecules (definable, non-polymeric compounds) that are participants in general metabolic reactions and that are required for the maintenance, growth and normal function of a cell. A number of implications flow from this definition. Consider the following:

1. Enzymes, other proteins and most peptides are generally not small molecules and thus excluded. The fact that many members of this class participate in biochemical reactions with small molecules; for instance, isoprenylation, glycosylation, *etc*., may be problematic but the products of these reactions, modified proteins, are not small molecules and thus are better considered within proteomics. In their construction and degradation they consume or yield small molecules, and thus, have locatable and specific output or input points to the metabolome.
2. Genetic material (all forms of DNA and RNA) is also excluded according to size and function. Furthermore, these are better considered within their own disciplines of genomics and transcriptomics. In its construction and degradation it consumes or yields small molecules and, thus, it also has locatable and specific output or input points to the metabolome.
3. Structural molecules (glycosaminoglycans, and other polymeric units) similarly may be built up from and degraded to small molecules but do not otherwise participate in metabolic reactions.
4. Polymeric compounds, such as glycogen, are important participants in metabolic reactions but are not chemically defineable. Thus, like all the previous categories, they represent a source of metabolites, *i.e.* an input/output to metabolism, but are not metabolites themselves.
5. Metabolites of xenobiotics are not native, required for the maintenance, growth or normal function of a cell, and thus are not really part of the metabolome. Their significance, not to be underestimated, is the realm of toxicology.
6. Essential or nutritionally required compounds are not synthesized *de novo,* (*i.e.* not native), but are required for the maintenance, growth or normal function of a cell; therefore, they are part of the metabolome as inputs from nutrition.

While arguments could be made for the inclusion of these, or the exclusion of other categories, to do so does not serve metabolomics, as it would shift the focus away from metabolism, and, with it, from physiology. At some point a general catalog of all of the components of a cell, inclusive of genomic, transcriptomic, proteomic, metabolomic and other molecular entities will need to be collated and a metabolome may be considered a first step in this direction. Such a general catalog would begin to approach the natural ecology of all molecular space. It will undoubtedly be a powerful perspective that will build upon current ongoing efforts, of which the establishment of a metabolome will be but one first step.

Finally, despite efforts to be limiting in the previous paragraphs, biochemicals and metabolites are generally not viewed as congruent terms. As a matter of historical precedence, the term "metabolites" is often mistakenly taken to indicate only that subset of all metabolites that are part of degradation pathways. It is important to reconcile this by recognizing the broadest possible definition of the term as the basis of metabolomics and metabolomes.

2.2 Metabolome Construction

One way to deduce the human metabolome is by metabolic reconstruction from the human genome (Selkov, 1995; Selkov *et al.*, 2000) Such an effort would mine a variety of publicly accessible and proprietary databases to compile a listing of compounds that are within the potential of a human being to biosynthesize. Since the metabolome is developed from the human genome, it will not include xenobiotics or their metabolites.

Metabolic reconstructions have been widely used for the reconstruction of microbial genomes, but not yet on a more complex organism (Covert *et al.*, 2001). The reconstruction will first scan the annotations from the Human Genome for all enzymes (Fig. 3). This starting point, a preliminary list of enzymes, will be qualified to ascertain that they are all appropriate. Those enzymes that emerge will be informatically translated to a preliminary list of reactions expected to occur by reference to a metabolic database. This list of reactions defines a starting list of compounds that act as either substrates or products for these "assumed-to-be-present-in-humans" reactions. These compounds, as they are newly identified, are inserted into the Human Metabolome Database. At this point a return cycle begins which looks at the preliminary set of metabolites and asks in what other reactions they are known to participate. This examination suggests a list of proposed reactions that yields a list of proposed enzymes. The known sequences for these enzymes are then sought in the Human Genome yielding a continuation cycle. These continuation cycles are reiterated until as much of primary metabolism as possible can be uncovered by genomic probing. When there is sufficient understanding of the framework of metabolism, then a new strategy may be employed, namely the proposal of whole pathways that are suggested by the so-far-discovered collection of reactions. This yields additional proposed enzymes that become a source of additional continuation cycles.

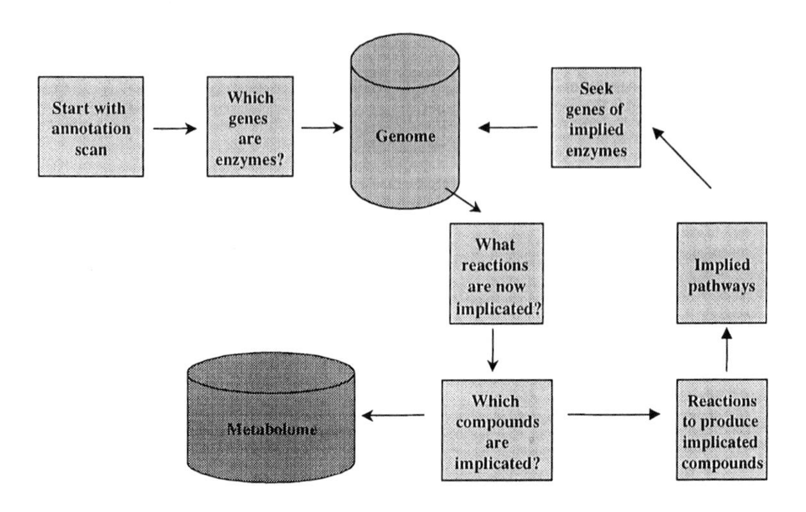

Figure 3. The logic flow of developing a metabolome from a genome. First determine all known enzymes in a genome. This implies a preliminary set of precursors and products. These each imply additional enzymes that are then sought in the genome. As enzymes are identified their associated chemistry may be assumed to be part of the metabolome of that organism. This process of metabolic reconstruction was championed by Selkov (see refs.).

In this way it is expected that the majority of human biosynthetic capacity may be understood from first principles; certainly, all of the core metabolic intermediates should be identified. Beyond this point additional methods may be needed to fill in more peripheral pathways. These will likely be added by literature-based curation, a more strenuous and often less reliable method. A benefit of this reaction-based approach will be that all of the metabolites in the metabolome will be associated with one or more enzymes, and will fit into known biosynthetic relationships. Other approaches have been proposed which are not reaction-based, but rather chemistry based. These suffer from the drawback of being just lists of disjointed compounds.

The completion of such a project will be the development in one place of a scientifically validated Human Metabolome, *i.e.* a list of compounds to be expected in a human, irrespective of location. Clearly every compound that is referenced in the Human Metabolome will need to be verified in the scientific primary literature. Furthermore, during the verification steps, not only will the presence of the compound in a human be verified, but any identified tissue specificity will be recorded.

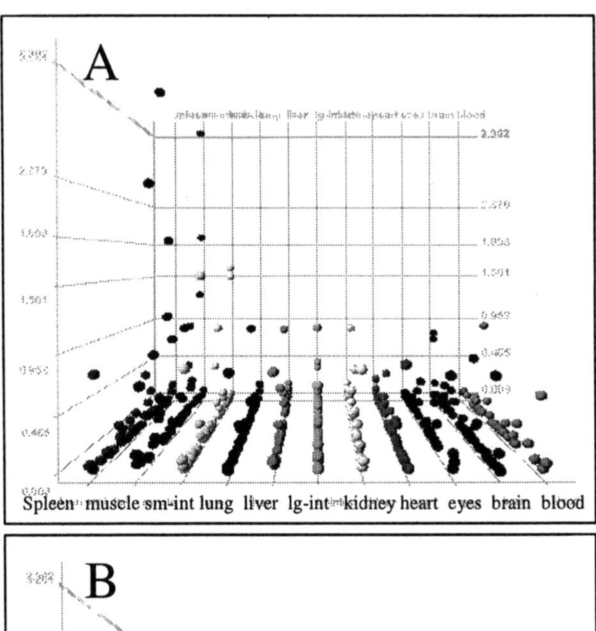

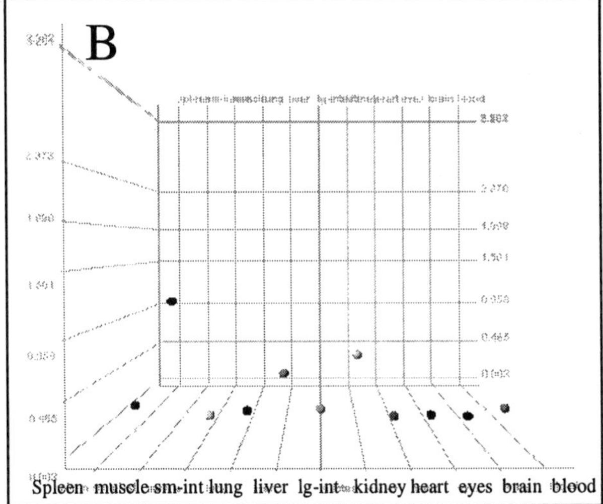

Figure 4. In Figure 4A is depicted the metabolomic analysis of a wide variety of compounds across 11 different tissues from a mouse. The height of each dot represents the relative concentration of each compound. The distribution of a single compound across all 11 tissues is depicted in Figure 4B.

With this as a guide, it will be possible to define what is not to be expected. For instance, a novel compound may be found to be an indication of an abnormal metabolic reaction. It is expected that there are some areas where such abnormal chemistry may be routinely found to be indicative of an altered metabolic state, such as in stressed, diseased or chemically-challenged tissues. Where this leads to a pathological condition the understanding of the root (*i.e.* biochemical) cause will be illuminating to the disease, and provide a hypothetical reference for a return of the biochemistry to a more normal state.

From experiments that have already been done in other species (Fig. 4) it is clear that, just like the relationship between the transcriptome and a gene array pattern, the number of all compounds likely to be expressed in any human tissue will be a subset of the full metabolome. Thus, the biochemical profile of a liver is different from the biochemical profile of a muscle (Fig. 5). Even though the biochemical profile of a tissue is generally stable and reproducible, and represents the baseline metabolic status of that tissue over time, there is some absolute drift in physiological status. By extension of this thought, when a tissue is diseased, the biochemical status of the diseased tissue must reflect the altered physiological status. The usefulness of the complete human metabolome in understanding these issues will be immense.

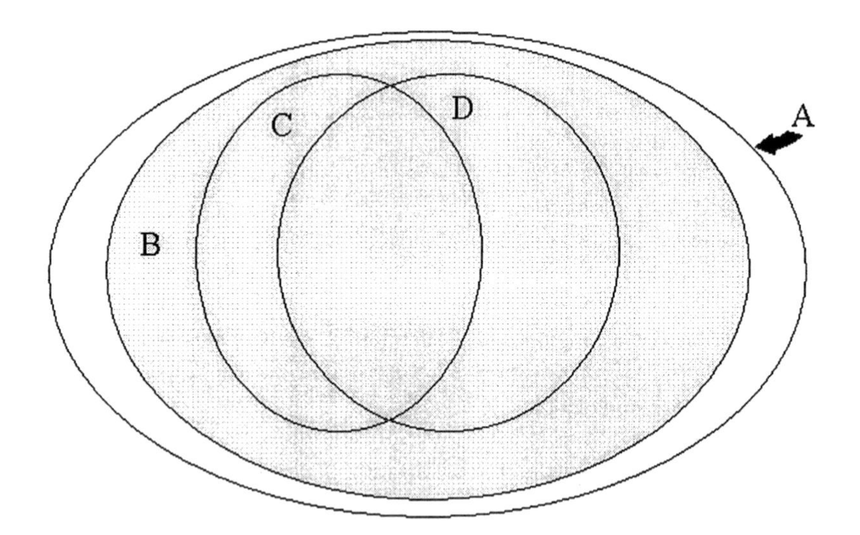

Figure 5. In terms of set theory, A represents the total human biochemical potential or the Metabolome. The subset B may represent a single individual, while subsets C and D each represent tissues within that individual. It is critical to realize that all individuals are likely to be represented by different subsets and that all tissues are in a constant state of flux and thus C and D likely vary, to some minimal extent, over time.

Preliminary estimates of the total number of compounds in a human have been varied. The standard wall-chart of metabolism, which lists reactions that are not present in humans, lists only about 800 compounds in core primary metabolism. Most biochemical textbooks extend this list to no more that 1200 to 1500 compounds, again drawing from all life forms. By extensive querying of publicly available databases we have been able to extend this list, limited now to humans, to no more than 2000 compounds. This number of compounds will form a very workable and firm foundation for the evolving science of metabolomics.

3. CONCLUSION

The Human Metabolome is a list of all compounds capable of being synthesized by humans. There is no expectation that any human tissues, or even individuals will express their total biochemical potential under any given circumstance, yet there is an expectation that there are a finite number of compounds that can be produced by humans, and that they can be discovered and documented. The human metabolome will be immensely useful in the development and support of the science of metabolomics, and through this medium should yield new perspectives on human physiology. The discoveries that will derive from understanding the full extent and diversity of the human metabolome are likely to be profound.

REFERENCES

Covert MW, Schilling CH, Famili I *et al.* Metabolic modeling of microbial strains *in silico*. *Trends Biochem Sci* 26: 179-186 (2001).

Glassbrook N, Beecher C, Ryals J. Metabolic profiling on the right path. *Nature Biotechnol* 18: 1142-1143 (2000).

Roessner U, Luedemann A, Brust D *et al.* Metabolic profiling allows comprehensive phenotyping of genetically or environmentally modified plant systems. *Plant Cell* 13: 11-29 (2001).

Selkov E. Reconstruction of metabolic networks using incomplete information. *Proc Int Conf Intel Sys Mol Biol* 3: 127-135 (1995).

Selkov E, Overbeek R, Kogan Y *et al.* Functional analysis of gapped microbial genomes: amino acid metabolism of *Thiobacillus ferrooxidans*. *Proc Natl Acad Sci USA* 97: 3509-3514 (2000).

Index